$$V_{BTPS} = V_{ATPS} \left(\frac{T_B}{T_A} \right) \left(\frac{P_A - P_{H_2O,A}}{} \right)$$

$TV \times C_E = V_A C_A + V_D C_I$

TV: tidal vol.

VA: alveolar vol.

V_D: dead space vol.

C_E: $\qquad$ expired air

C_A: CO_2 conc in alveolar air

C_I: $\qquad$ inspired air

C

$$V_D = TV \left(1 - \frac{C_E}{C_A} \right)$$

$1 \text{ dyne} = 1 g \text{ cm } s^{-2} = 10^{-5} N$

$1 \text{ Poise} = 1 g \text{ cm}^{-1} s^{-1}$

$1 \text{ Pa·s} = 10 P$

$1 \text{ Pa} = kg \, m^{-1} s^{-2}$

Bone lab

P load
L support span
δ midspan deflection

$\sigma_{bending} = \frac{PL r_0}{\pi (r_0^4 - r_i^4)}$

$\epsilon_{bending} = \frac{12 r_0 \delta}{L^2}$

$E = \frac{PL^3}{12\pi (r_0^4 - r_i^4)\delta}$

$toughness = \int_0^{\epsilon_{failure}} \sigma \, d\epsilon$

ADD forces in parallel
disp. in parallel are equal

$\tau = \text{dyne}/cm^2$

$\sigma = \text{dyne}/cm$

$\dot\gamma = s^{-1}$

$\mu = \text{Poise} = \text{dyne·s}/cm^2$

$\Delta P = \rho g h$

Half-space model, micropipette aspiration

$L(t) = \phi \left(\frac{2r}{2\pi} \right) \left(\frac{\Delta P}{E(t)} \right)$

$\phi \approx 2.1$
r = pipette radius
E(t) modulus
L(t) aspiration length

Estimate E of bone by rule of mixtures.

$E_{composite}^{longitudinal} = V_{collagen} E_{collagen} + V_{HA} E_{HA}$

$E_{composite}^{transverse} = \left[\frac{V_{collagen}}{E_{collagen}} + \frac{V_{HA}}{E_{HA}} \right]^{-1}$

Introductory Biomechanics
From Cells to Organisms

Introductory Biomechanics is a new, integrated text written specifically for engineering students. It provides a broad overview of this important branch of the rapidly growing field of bioengineering. A wide selection of topics is presented, ranging from the mechanics of single cells to the dynamics of human movement. No prior biological knowledge is assumed and in each chapter, the relevant anatomy and physiology are first described. The biological system is then analyzed from a mechanical viewpoint by reducing it to its essential elements, using the laws of mechanics, and then linking mechanical insights back to biological function. This integrated approach provides students with a deeper understanding of both the mechanics and the biology than that obtained from qualitative study alone. The text is supported by a wealth of illustrations, tables, and examples, a large selection of suitable problems and many current references, making it an essential textbook for any biomechanics course.

C. Ross Ethier is a Professor of Mechanical and Industrial Engineering, the Canada Research Chair in Computational Mechanics, and the Director of the Institute of Biomaterials and Biomedical Engineering at the University of Toronto, with cross-appointment to the Department of Ophthalmology and Vision Sciences. His research focuses on biomechanical factors in glaucoma and on blood flow and mass transfer in the large arteries. He has taught biomechanics for over 10 years.

Craig A. Simmons is the Canada Research Chair in Mechanobiology and an Assistant Professor in the Department of Mechanical and Industrial Engineering at the University of Toronto, with cross-appointments to the Institute of Biomaterials and Biomedical Engineering and the Faculty of Dentistry. His research interests include cell and tissue biomechanics and cell mechanobiology, particularly as it relates to tissue engineering and heart valve disease.

Cambridge Texts in Biomedical Engineering

Series Editors
Professor Mark Saltzman   *Yale University*
Professor Shu Chien   *University of California, San Diego*
Professor William Hendee   *Medical College of Wisconsin*
Professor Roger Kamm   *Massachusetts Institute of Technology*
Professor Robert Malkin   *Duke University*
Professor Alison Noble   *Oxford University*

Cambridge Texts in Biomedical Engineering provides a forum for high-quality, accessible textbooks targeted at undergraduate and graduate courses in biomedical engineering. It will cover a broad range of biomedical engineering topics from introductory texts to advanced topics including, but not limited to, biomechanics, physiology, biomedical instrumentation, imaging, signals and systems, cell engineering, and bioinformatics. The series will blend theory and practice, aimed primarily at biomedical engineering students but will be suitable for broader courses in engineering, the life sciences and medicine.

Introductory Biomechanics

From Cells to Organisms

C. Ross Ethier and **Craig A. Simmons**

University of Toronto, Canada

CAMBRIDGE
UNIVERSITY PRESS

CAMBRIDGE UNIVERSITY PRESS
Cambridge, New York, Melbourne, Madrid, Cape Town, Singapore, São Paulo

Cambridge University Press
The Edinburgh Building, Cambridge CB2 2RU, UK

Published in the United States of America by Cambridge University Press, New York

www.cambridge.org
Information on this title: www.cambridge.org/9780521841122

First published 2007

Printed in the United States of America

A catalog record for this publication is available from the British Library

ISBN-13   978-0-521-84112-2 hardback

To my family, who make it all worthwhile.

C. ROSS ETHIER

To Deborah,
and to my parents, who inspired my love of learning.

CRAIG A. SIMMONS

Contents

Color plate section between pages 118 and 119

About the cover

The cover contains images that together represent the broad scope of modern biomechanics. The figures are as follows:

- Main image: A fluorescent immunohistochemical image of an endothelial cell isolated from the surface of a pig aortic heart valve and grown in culture. Within the cell, the nucleus is stained blue and vimentin filaments are stained green. Vimentin is an intermediate filament protein of the cellular cytoskeleton that plays an important role in cellular mechanics.
- Left top: An intermediate stage from a simulation of the forced unfolding of repeats 4 and 5 of chain A of the protein filamin. Filamin is an actin cross-linking protein and therefore plays a role in the biomechanics of the cytoskeleton. The simulation was based on the crystal structure of part of filamin [1], and was carried out in NAMD [2] and visualized using the VMD package [3]. (Image courtesy of Mr. Blake Charlebois.)
- Left middle: A sketch by the Swiss anatomist Hermann von Meyer of the orientation of trabecular bone in the proximal human femur. This sketch was accompanied in the original article by a sketch of the principal stress trajectories in a crane having a shape similar to the femur. Together these sketches are believed to have inspired "Wolff's Law" of bone remodelling. From [4].
- Left lower: The distribution of mass transfer rates from flowing blood to cultured vascular endothelial cells. The contoured quantity (the Sherwood number) was computed by first measuring the topography of the endothelial cells using atomic force microscopy and then solving the convection-diffusion equation in the blood flowing over the cells. Mass transfer from blood to endothelial cells is important in cell-cell signalling. (Image courtesy of Mr. Ji Zhang.)

References

1. G. M. Popowicz, R. Muller, A. A. Noegel, M. Schleicher, R. Huber and T. A. Holak. Molecular structure of the rod domain of dictyostelium filamin. *Journal of Molecular Biology*, **342** (2004), 1637–1646.

2. L. Kale, R. Skeel, M. Bhandarkar, R. Brunner, A. Gursoy, N. Krawetz *et al.* NAMD2: Greater scalability for parallel molecular dynamics. *Journal of Computational Physics*, **151** (1999), 283–312.

3. W. Humphrey, A. Dalke and K. Schulten. VMD: Visual Molecular Dynamics. *Journal of Molecular Graphics*, **14** (1996), 33–38.

4. J. Wolff. Über die innere Architectur der Knochen und ihre Bedeutung für die Frage vom Knochenwachsthum. *Archiv für Pathologische Anatomie und Physiologie und für Klinische Medicin*, **50** (1870), 389–450.

Preface

For some years, we have taught an introductory course in biomechanics within the Department of Mechanical and Industrial Engineering at the University of Toronto. We have been unable to find a textbook suitable for the purpose of introducing engineers and others having a "hard science" background to the field of biomechanics. That is not to say that excellent books on biomechanics do not exist; in fact, there are many. However, they are typically at a level that is too advanced for an introductory course, or they cover too limited a subset of topics for purposes of an introductory course.

This book represents an attempt to fill this void. It is not meant to be an extensive treatise on any particular branch of biomechanics, but rather to be an introduction to a wide selection of biomechanics-related topics. Our hope is that it will aid the student in his or her introduction to the fascinating world of bioengineering, and will lead some to pursue the topic in greater detail.

In writing this book, we have assumed that the reader has a background in engineering and mathematics, which includes introductory courses in dynamics, statics, fluid mechanics, thermodynamics, and solid mechanics. No prior knowledge of biology, anatomy, or physiology is assumed, and in fact every section begins with a review of the relevant biological background. Each chapter then emphasizes identification and description of the essential aspects of the related biomechanics problems. Because of the introductory nature of this book, this has led in some cases to a great deal of simplification, but in all instances, we have tried to maintain a firm link to "biological reality."

We wish to thank Professor David F. James, of the Department of Mechanical and Industrial Engineering, University of Toronto. He first developed the introductory course in biomechanical engineering at the University and his course notes provided the inspiration for parts of this book. Professors James E. Moore Jr. and Takami Yamaguchi provided important material for Ch. 1. We have benefited greatly from interactions with our students, who sometimes are the best teachers, and our colleagues and mentors.

We shall be most grateful to students who, upon discovering errors in the text, bring them to our attention.

1 Introduction

Biomechanics is a branch of the field of bioengineering, which we define as the application of engineering principles to biological systems. Most bioengineering is applied to humans, and in this book the primary emphasis will be on *Homo sapiens*. The bioengineer seeks to understand basic physiological processes, to improve human health via applied problem solving, or both. This is a difficult task, since the workings of the body are formidably complex. Despite this difficulty, the bioengineer's contribution can be substantial, and the rewards for success far outweigh the difficulties of the task.

Biomechanics is the study of how physical forces interact with living systems. If you are not familiar with biomechanics, this might strike you as a somewhat esoteric topic, and you may even ask yourself the question: Why does biomechanics matter? It turns out that biomechanics is far from esoteric and plays an important role in diverse areas of growth, development, tissue remodeling and homeostasis. Further, biomechanics plays a central role in the pathogenesis of some diseases, and in the treatment of these diseases. Let us give a few specific examples:

- How do your bones "know" how big and strong to be so that they can support your weight and deal with the loads imposed on them? Evidence shows that the growth of bone is driven by mechanical stimuli [1]. More specifically, mechanical stresses and strains induce bone cells (*osteoblasts* and *osteoclasts*) to add or remove bone just where it is needed. Because of the obvious mechanical function played by bone, it makes good sense to use mechanical stress as the feedback signal for bone growth and remodeling. But biomechanics also plays a "hidden" regulatory role in other growth processes, as the next example will show.

- How do our arteries "know" how big to be so that they can deliver just the right amount of blood to their distal capillary beds? There is good evidence that this is determined in large part by the mechanical stress exerted on the artery wall by flowing blood. Endothelial cells lining the inner arterial surface sense this shear stress and send signals to cells deeper in the artery wall to direct the remodeling of the artery so as to enlarge or reduce its caliber [2].

- What about biomechanics in everyday life? Probably the most obvious application of biomechanics is in locomotion (walking, running, jumping), where our muscles generate forces that are transferred to the ground by bones and soft connective tissue. This is so commonplace that we rarely think about it, yet the biomechanics of locomotion is remarkably complex (watch a baby learning to walk!) and still incompletely understood.

- Locomotion happens on many scales, from whole organisms all the way down to individual cells. Unicellular organisms must be able to move so as to gather nutrients, and they have evolved a variety of clever strategies to accomplish this task [3]. In multicellular organisms, the ability of single cells to move is essential in processes such as repair of wounds, capture of foreign pathogens, and tissue differentiation. Force generation at the cellular level is a fascinating topic that is the subject of much active research.

- Cells can generate forces, but just as importantly, they can sense and respond to forces. We alluded to this above in the examples of bone remodeling and arterial caliber adjustment, but it is not only endothelial and bone cells that can sense forces. In fact, the ability of mechanical stress to elicit a biological response in cells seems to be the rule rather than the exception, and some cells are exquisitely specialized for just this task. One remarkable example is the hair cells in the ear. These cells have bundles of thin fibers (the *stereocilia*) that protrude from the apical cell surface and act as sensitive accelerometers; as a result, the hair cells are excited by sound-induced vibrations in the inner ear. This excitation produces electrochemical signals that are conducted by the auditory nerve to the auditory centers in the brain in a process that we call hearing [4,5].[1]

- The examples above show that biomechanics is important in homeostasis and normal function. Unfortunately, biomechanics also plays a role in some diseases. One example is glaucoma, an ocular disease that affects about 65 million people worldwide [6]. Normally the human eye is internally pressurized, a fact that you can verify by gently touching your eye through the closed eyelid. In most forms of glaucoma, the pressure in the eye becomes elevated to pathological levels, and the resulting extra biomechanical load somehow damages the optic nerve, eventually leading to blindness [7]. A second example is atherosclerosis, a common arterial disease in which non-physiological stress distributions on endothelial cells promote the disease process [8].

- What about biomechanics in the treatment of disease and dysfunction? There are obvious roles in the design of implants that have a mechanical function,

---

[1] Actually, the function of the hair cells is even more amazing than it first appears. The outer hair cells are active amplifiers, changing their shape in response to mechanical stimulation and thus generating sounds. The net effect is to apply a frequency-selective boost to incoming sounds and hence improve the sensitivity of the ear.

such as total artificial hips [9], dental implants [10], and mechanical heart valves [11]. In the longer term, we expect to treat many diseases by implanting engineered replacement tissue into patients. For tissues that have a mechanical function (e.g., heart valves, cartilage), there is now convincing evidence that application of mechanical load to the tissue while it is being grown is essential for proper function after implantation. For example, heart valves grown in a bioreactor incorporating flow through the valve showed good mechanical properties and function when implanted [12]. Cartilage subjected to cyclic shearing during growth was stiffer and could bear more load than cartilage grown without mechanical stimulation [13]. We expect that biomechanics will become increasingly important in tissue engineering, along the way leading to better fundamental understanding of how cells respond to stresses.

The above examples should give a flavor of the important role that biomechanics plays in health and disease. One of the central characteristics of the field is that it is highly interdisciplinary: to be called biomechanics, there must be elements of both mechanics and biology (or medicine). Advances in the field occur when people can work at the frontier of these two areas, and accordingly we will try to give both the "bio" and the "mechanics" due consideration in this book.

Another characteristic feature of biomechanics is that the topic is fairly broad. We can get a sense of just how broad it is by looking at some of the professional societies that fall under the heading of biomechanics. For example, in Japan alone, at least six different professional societies cover the field of biomechanics.[2] Obviously we cannot, in a single book, go into detail in every topic area within such a broad field. Therefore, we have given an introduction to a variety of topics, with the hope of whetting readers' appetites.

1.1 A brief history of biomechanics

We can learn more about the field of biomechanics by looking at its history. In one sense, biomechanics is a fairly young discipline, having been recognized as an independent subject of enquiry with its own body of knowledge, societies, journals, and conferences for only around 30–40 years. For example, the "Biomechanics and Human Factors Division" (later to become the "Bioengineering Division") of the American Society of Mechanical Engineering was established in late 1966. The International Society of Biomechanics was founded August 30, 1973; the European

---

[2] These are the Japanese Society of Biomechanics, the Bioengineering Division of the Japan Society of Mechanical Engineers, the Japan Society of Medical Electronics and Biological Engineering, the Association of Oromaxillofacial Biomechanics, the Japanese Society for Clinical Biomechanics and the Japanese Society of Biorheology.

Society of Biomechanics was established May 21, 1976, and the Japanese Society of Biomechanics was founded December 1, 1984. On the other hand, people have been interested in biomechanics for hundreds of years, although it may not have been called "biomechanics" when they were doing it. Here we take a quick look back through history and identify some of the real pioneers in the field. Note that the summary below is far from exhaustive but serves to give an overview of the history of the field; the interested reader may also refer to Chapter 1 of Fung [14] or Chapter 1 of Mow and Huiskes [15].

Galileo Galilei (1564–1642) was a Pisan who began his university training in medicine but quickly became attracted to mathematics and physics. Galileo was a giant in science, who, among other accomplishments, was the first to use a tele-scope to observe the night sky (thus making important contributions in astronomy) and whose synthesis of observation, mathematics, and deductive reasoning firmly established the science that we now call mechanics.[3] Galileo, as part of his studies on the mechanics of cantilevered beams, deduced some basic principles of how bone dimensions must scale with the size of the animal. For example, he realized that the cross-sectional dimensions of the long bones would have to increase more quickly than the length of the bone to support the weight of a larger animal [17]. He also looked into the biomechanics of jumping, and the way in which loads are distributed in large aquatic animals, such as whales. However, Galileo was really only a "dabbler" in biomechanics; to meet someone who tackled the topic more directly, we must head north and cross the English Channel.

William Harvey (1578–1657) was an English physician who made fundamen-tal contributions to our understanding of the physiology of the cardiovascular system, and who can be rightly thought of as one of the first biomechanicians (Fig. 1.1). Before Harvey, the state of knowledge about the cardiovascular system was primitive at best, being based primarily on the texts of the Roman physician Galen (129–199?). Galen believed that the veins distributed blood to the body, while arteries contained pneuma, a mixture of "vital spirits," air, and a small amount of blood. It was thought that the venous and arterial systems were not in communi-cation except through tiny perforations in the interventricular septum separating the two halves of the heart, so the circulatory system did not form a closed loop. Venous blood was thought to be produced by the liver from food, after which it flowed outward to the tissues and was then consumed as fuel by the body.[4]

Harvey was dissatisfied with Galen's theories, and by a clever combination of arguments and experimentation proved that blood must travel in a closed circuit

---

[3] Charles Murray, in his remarkable survey of human accomplishment through the ages [16], ranks Galileo as the second-most accomplished scientist of all time, behind (who else?) Newton.

[4] It is easy to look back and ask: How could Galen have been so wrong? The answer is that he was influenced by his predecessors; prior to Galen it was thought that arteries were filled with air and that the veins originated in the brain, for example. The lesson to be learned: question dogma!

Figure 1.1

Portraits of Drs. William Harvey (left) and Stephen Hales (right). Both were early biomechanicians; Harvey was a noted English physician, while Hales was a Reverend and "amateur" scientist. Both portraits, courtesy of the Clendening History of Medicine Library and Museum, University of Kansas Medical Center [18].

in the cardiovascular system. For example, he carried out careful dissections and correctly noted that all the valves in veins acted to prevent flow away from the heart, strongly suggesting that the function of the veins was to return blood to the heart. For our purposes, his most intriguing argument was based on a simple mass balance: Harvey reasoned that the volumetric flow of blood was far too large to be supplied by ingestion of food. How did he do this? Using a sheep's heart, he first estimated the volume of blood pumped per heart beat (the stroke volume) as two ounces of blood. Knowing the heart rate, he then computed that the heart must be pumping more than 8600 ounces of blood per hour, which far exceeds the mass of food any sheep would be expected to eat! In his words (italics added) [19]:

Since all things, both argument and ocular demonstration, show that the blood passes through the lungs and heart by the force of the ventricles, and is sent for distribution to all parts of the body, where it makes its way into the veins and porosities of the flesh, and then flows by the veins from the circumference on every side to the center, from the lesser to the greater veins, and is by them finally discharged into the vena cava and right auricle of the heart, *and this in such a quantity or in such a flux and reflux thither by the arteries, hither by the veins, as cannot possibly be supplied by the ingesta, and is much greater than can be required for mere purposes of nutrition*; it is absolutely necessary to conclude that the blood in the animal body is impelled in a circle, and is in a state of ceaseless motion.

By these and additional arguments [20], Harvey deduced the closed nature of the cardiovascular system (although he was unable to visualize the capillaries). For our purposes, Harvey is notable because he was one of the first physicians to use a combination of *quantification*, deductive reasoning, and experimentation to understand a clinically important medical topic. Such approaches are commonplace today but were revolutionary in Harvey's time and even caused him to be strongly criticized by many prominent physicians.

Giovanni Alfonso Borelli (1608–1679) is variously described as a mathematician, physicist, and physiologist, which is surely a testament to the breadth of his interests. He worked at various universities throughout Italy, coming in contact with Galileo. Notably, he spent 10 years in Pisa, where he worked with the famous anatomist Malpighi (responsible for the discovery of the capillaries). Later in his career, Borelli became interested in the mechanics of animal motion, and is best known for his two-volume work on this topic, *On the Movement of Animals (De Motu Animalium)*, published posthumously in 1680 (Volume I) and 1681 (Volume II). In addition to the novelty of the material in these books, they are notable for their wonderfully detailed figures illustrating biomechanical concepts such as locomotion, lifting, and joint equilibrium (Fig. 1.2). Borelli used the principles of levers and other concepts from mechanics to analyze muscle action. He also determined the location of the center of gravity of the human body and formulated the theory that forward motion involved the displacement of the center of gravity beyond the area of support and that the swinging of the limbs saved the body from losing balance [21]. Further, he considered the motor force involved in walking and the location of body support during walking. Borelli was also interested in respiratory mechanics: he calculated and measured inspired and expired air volumes. He was able to show that inspiration is driven by muscles, while expiration is a passive process resulting from tissue elasticity. In honor of his seminal contributions in the field of biomechanics, the career accomplishment award of the American Society of Biomechanics is known as the Borelli Award.

Another early biomechanician was the Reverend Stephen Hales (1677–1761), who made contributions to both plant and animal physiology (Fig. 1.1). He is best known for being the first to measure arterial blood pressure, now a staple of all clinical examinations. He did this by direct arterial cannulation of his horse (in his back yard, no less)! In his words [22,23]:

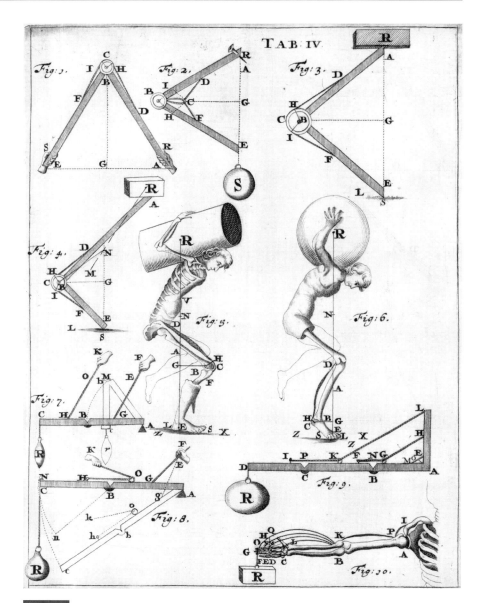

Figure 1.2

Figure from Borelli's classic work, *De Motu Animalium* (*On the Movement of Animals*). Panels 1–4 show how elastic bands (representing muscles) can interact with two pivoting levers (representing bones) in a variety of geometric configurations. Panels 5 and 6 demonstrate how the muscle and bone configurations act in humans carrying loads. Panels 7 and 8 show various pulley arrangements, while panels 9 and 10 show how muscle action in the human arm supports a weight R. (We will revisit this subject in Ch. 8.) The concepts may not seem advanced to modern students, but to put things into context, it should be remembered that the first volume of *De Motu Animalium* was published seven years before the appearance of Newton's *Principia*.

I caused a mare to be tied down alive on her back . . . having laid open the left crural [femoral] artery about 3 inches from her belly, I inserted into it a brass pipe whose bore was 1/6 of an inch in diameter; and to that, by means of another brass pipe . . . I fixed a glass tube of nearly the same diameter, which was 9 feet in length; then untying the ligature on the artery, the blood rose in the tube 8 feet 3 inches perpendicular above the level of the left ventricle of the heart . . . when it was at its full height, it would rise and fall at and after each pulse 2, 3, or 4 inches.

Hales also improved Harvey's estimate of cardiac stroke volume by pouring wax at controlled pressure into the heart's main pumping chamber (the left ventricle) to make a casting. He then measured the volume of the wax cast by immersing it in water, and measured its surface area by carefully covering it with small pieces of paper covered with a measuring grid. Together with his measurements of blood pressure, Hales then used these results to provide the first estimate of left ventricular systolic (pumping) pressure, and a remarkably accurate estimate of the blood velocity in the aorta (0.5 m/s).

Jean Léonard Marie Poiseuille (1797–1869; Fig. 1.3) was a French engineer and physiologist who was also interested in blood flow [25]. In his thesis [27], he described how he simplified and improved the measurement of blood pressure. His contributions were two-fold: first, he developed the U-tube mercury manometer, which did away with the need for an unwieldy 9 foot-long tube. Second, he used potassium carbonate as an anticoagulant [28]. He was surprised to discover that the pressure drop from the aorta to arteries with diameters as small as 2 mm was negligible [29]; we know now that most of the pressure drop in the circulatory system occurs in vessels with diameter smaller than 2 mm. Poiseuille then became interested in laminar flow in small tubes and carried out experiments on the flow of water through artificial glass capillaries with diameters as small as 30 μm. His results allowed him to deduce the relationship between flow, tube dimensions, and pressure drop, which we know today as the Hagen–Poiseuille law [30]. We will explore the implications of this law for blood flow in Ch. 3.

Thomas Young (1773–1829) was an English physician and physicist (Fig. 1.3). He was remarkably prodigious as a child, having learned to read "with considerable fluency" at the age of two, and demonstrating a knack for languages, such that he had knowledge of English, Greek, Latin, French, Italian, and Hebrew by the age of 13 [26]. He studied medicine and practiced in London while developing and maintaining expertise in a staggering range of areas. For example, he demonstrated the wave theory of light, deciphered some of the first Egyptian hieroglyphics by analysis of the Rosetta stone, helped to establish actuarial science, and lectured on the theory of tides, surface tension, etc. In the biomedical area, he established, with von Helmholtz, the theory of color, discovered and measured astigmatism in the

Figure 1.3

Portraits of Drs. Jean Poiseuille (left) and Thomas Young (right). Both men did important work in physiology and medicine, yet are familiar to engineering students: Poiseuille for his work on steady laminar incompressible flow in a tube of uniform circular cross-section (Hagen–Poiseuille flow) and Young from his work on the elasticity of bodies (Young's modulus of elasticity). Poiseuille portrait reproduced with permission from [24] as modified by Sutera [25]; Young portrait by Sir Thomas Lawrence, engraved by G. R. Ward, as shown in Wood [26].

eye, and deduced that the focussing power of the eye resulted from changes in the shape of the lens. He devised a device for measuring the size of a red blood cell, with his measurements showing a size of 7.2 μm [26], a value that is remarkably accurate (see Ch. 3). He also studied fluid flow in pipes and bends, and the propagation of impulses in elastic vessels, and then applied this to analysis of blood flow in the arteries. He correctly deduced that peristaltic motion of the artery wall did not contribute to the circulation of blood, and instead that the motive power must come from the heart [31]. He is most familiar to engineering students for defining the modulus of elasticity, now known as Young's modulus in his honor.

Julius Wolff (1836–1902) and Wilhelm Roux (1850–1924) were German physicians (Fig. 1.4). Of the two, Wolff is better known to biomedical engineers because

Figure 1.4

Portraits of Drs. Julius Wolff (left) and Wilhelm Roux (right). Both were German physicians who were interested in how mechanical forces could influence the structure and development of bone. Both portraits, courtesy of the Clendening History of Medicine Library and Museum, University of Kansas Medical Center [18].

of his formulation of "Wolff's law" of bone remodeling. Legend has it [32] that the structural engineer Karl Culmann saw a presentation by the anatomist Hermann von Meyer, in which von Meyer described the internal architecture of the bone in the head of the femur. Culmann was struck by the similarity between the pattern of solid elements in the cancellous ("spongy") bone of the femur and the stress trajectories[5] in a similarly shaped crane that he was designing (Fig. 1.5).[6] Based on von Meyer's paper describing this similarity [33], as well as other data available at the time, Wolff hypothesized that bone was optimized to provide maximum strength for a minimum mass. He then went on to formulate his "law" of bone

---

[5] A stress trajectory is an imaginary line drawn on a surface that is everywhere tangent to the principal stress directions on the surface. Stress trajectories help to visualize how the stress is carried by an object, and they can be used as the basis of a graphical procedure for determining stress distributions in bodies. This graphical solution method is now obsolete, having been replaced by computational methods.

[6] This certainly emphasizes the importance of interdisciplinary interaction in biomedical engineering!

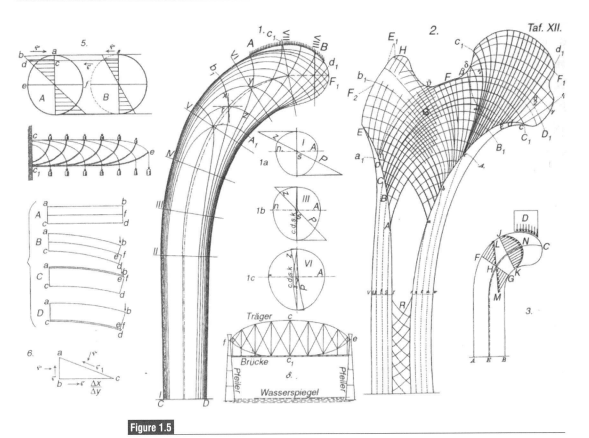

Figure 1.5

Comparison of internal architecture of cancellous bone in the head of a femur (large drawing at right) and the stress trajectories in the head of a crane (large drawing at left). The smaller drawings provide details of the mechanics of the crane and the stress distributions in various structures. This picture originally appeared in the article by von Meyer [33], as reproduced in [32].

remodeling [34,35]:[7] "Every change in the form and the function of a bone or of their function alone is followed by certain definite changes in their internal architecture, and equally definite secondary alterations in their external confirmation, in accordance with mathematical laws." In simpler terms, Wolff stated that bone will adapt its internal architecture in response to external constraints and loads. Wolff went on to claim a rigorous similarity between cancellous bone architecture and stress trajectories. Cowin has shown that this is based on a false comparison of apples and oranges (i.e., a continuous material vs. a porous one) [32] and argued persuasively that Wolff gets rather too much credit for his remodeling law, possibly because he was a more prolific writer on this topic. Cowin [32] suggested that the anatomist Roux should get at least as much credit as Wolff for this "law" of

[7] Wolff's original book was in German [34], but an English translation exists [35].

remodeling, and we are inclined to agree. Roux was very interested in developmental biology and physiology, carrying out his doctoral work on the factors governing the bifurcation of blood vessels [36]. He was convinced that mechanical and physical principles played important roles in development (e.g., [37]) and carried out seminal experimental studies in this area. In 1894, he also founded a journal entitled *Archiv für Entwicklungsmechanik* (*Archives of Developmental Mechanics*) and served as its editor for many years, stirring up his fair share of controversy along the way [38]. He may be rightly thought of as the first developmental biomechanician.

In the years since Wolff and Roux, there have been many advances in biomechanics. This is not the place to try to provide a complete history of biomechanics, and we will not list all of the outstanding investigators who have worked (or are working) in this fast-growing field. However, it is worth mentioning one other investigator. An important event in the maturation of the field of biomechanics was the publication, in 1981, of the book *Biomechanics: Mechanical Properties of Living Tissues*, by Yuan-Cheng Fung (Fig. 1.6). Fung was born in 1919 and was trained as an aeronautical engineer, a field in which he made many technical contributions in the early years of his career. However, in the late 1950s, he became interested in biomechanics and consequently changed his research focus away from aeronautics. In addition to his 1981 book, Fung has also written several other books in biomechanics [40,41]. Y.-C. Fung is generally regarded as the "father" of modern biomechanics and has the rare distinction of being a member of the US National Academy of Engineering, the US Institute of Medicine, and the US National Academy of Sciences. Membership in all three of these learned societies is surely a testament to the abilities of Dr. Fung, but it is also a reflection of the highly interdisciplinary nature of biomechanics, which (when done properly) should tightly integrate engineering, medicine, and biology.

1.2 An outline of this book

Western science is traditionally reductionist: we tend to conceptually break down complex systems into smaller functional units, and try to analyze those units and their interconnectivity (although see the aside on systems biology below). We will follow this approach in this book because we believe that this is the best way for the student to be introduced to a topic.

The basic unit of life is the cell, and an understanding of cellular behavior is a cornerstone of modern bioengineering. In Chapter 2, we describe the basic components of the cell, with special emphasis on molecules that play a role in the biomechanical behavior of the cell. We then attempt to synthesize our understanding of these molecular components to answer basic questions about

Figure 1.6

Portrait of Y.-C. Fung, who has played an important role in the establishment of biomechanics as a modern, rigorous discipline, primarily through the publication of an influential series of books on biomechanics. Reproduced with permission from [39].

cellular biomechanics. How stiff is a cell, and how do we measure this parameter? How does the cell anchor itself to substrates and to neighbors? How does the cell respond to external forces, and what implications does this have for tissue organization?

At a higher level, physiologists subdivide the body into *organs*, which are tissues specialized for a specific purpose, and *systems*, which are collections of organs working in concert. In this book, we will also consider the biomechanics of some of the body's systems. There are many such systems; here is a partial list and description of their functions, with emphasis on the biomechanical aspects.

Circulatory system. This system delivers nutrients and picks up waste products from the cells, as well as delivering signaling molecules, such as hormones, between different organs. The flow of blood will be studied in Ch. 3; other biomechanical aspects of the circulatory system are discussed in Ch. 4.

Lymphatic system.   Excess fluid is passively collected from tissues and returned to the heart via a network of ducts and channels that make up the lymphatic system. The lymphatic system is also an important locus for immune function. We will not examine the lymphatic system in detail, although it is briefly touched upon in Ch. 5.

Nervous and sensory systems.   The nervous system consists of the nerves and brain and is responsible for signaling and control within the body. Its operation is highly complex and will not be considered in this book. The sensory organs provide input to the nervous system; we will briefly consider ocular biomechanics in Ch. 6.

Respiratory system.   In order to oxidize foodstuffs, O_2 must be delivered to the blood and CO_2 must be removed from it. This is accomplished by exposing the blood to the air through a very thin membrane of enormous surface area. This membrane is convoluted and folded to form a large number of small sacs within the lung. The respiratory system consists of the lungs, plus structures that assist air passage in and out of the lungs. It will be considered in Ch. 7.

Urinary system.   The urinary system consists of the kidneys, ureters, urinary bladder, and urethra. The kidneys are responsible for removing waste products from blood and for the production of urine, which is then stored in the bladder and excreted through the urethra. The biomechanics of the urinary system is fascinating and complex [42–45]. Unfortunately, consideration of the urinary system is beyond the scope of this book.

Muscular system.   Muscles are specialized tissues that generate force upon appropriate stimulation. Muscular action is required for locomotion (movement of the body), motion of individual body parts, and bulk transport of materials within the body (e.g., pumping of blood by the heart). Muscles, and how they effect motion, will be discussed in Chs. 8 and 10.

Skeletal system.   This framework of bones and soft connective tissues (cartilage, ligaments, and tendons) provides a rigid, supportive, and protective structure for the body. The bony skeleton also provides attachments for muscles, serves as a system of levers for movement and locomotion, and has important metabolic functions. Bones, cartilage, ligaments, and tendons will be discussed in greater detail in Ch. 9.

Digestive system.   The digestive system comprises the gastrointestinal tract (mouth to anus) plus the liver, gall bladder, and pancreas. The digestive system is responsible for ingestion and breakdown of food, delivery of foodstuffs to the blood, and waste excretion. We will not consider the digestive system in this book.

Immune system.   This system consists of specialized cells and molecules distributed throughout the body (in organs such as the spleen and as cells in the bloodstream and interstitial fluid). It is responsible for identification and destruction of foreign entities, including viruses and bacteria. Consideration of the immune system is beyond the scope of this book.

Box 1.1 Systems biology and the integration of information

The reductionist trend in biology has begun to change recently, with the emergence of several high-profile initiatives in systems biology. The goal here is to integrate information at the genomic, protein, and higher levels to understand how biological systems work as functional units. A wonderful example of a systems biology approach is the work of Davidson and coworkers [46], who are mapping the genetic regulatory network in the sea urchin (*Strongylocentrotus purpuratus*) embryo. By using a variety of techniques, including perturbing the function of regulatory genes and observing their effects on embryo development, they have produced maps of genetic regulatory networks that convey a taste of the complexity of life (Fig. 1.7 [color plate]; note this represents the development of only part of the sea urchin embryo, for a small fraction of the developmental period). Readers will note that there are no biomechanical stimuli listed in this diagram; these have yet to be elucidated.

References

1. M. Mullender, A. J. El Haj, Y. Yang, M. A. van Duin, E. H. Burger *et al.* Mechanotransduction of bone cells in vitro: mechanobiology of bone tissue. *Medical and Biological Engineering and Computing*, **42** (2004), 14–21.

2. P. F. Davies. Flow-mediated endothelial mechanotransduction. *Physiological Reviews*, **75** (1995), 519–560.

3. J. O. Kessler. The dynamics of unicellular swimming organisms. *ASGSB Bulletin*, **4** (1991), 97–105.

4. W. E. Brownell, A. A. Spector, R. M. Raphael and A. S. Popel. Micro- and nanomechanics of the cochlear outer hair cell. *Annual Review of Biomedical Engineering*, **3** (2001), 169–194.

5. R. A. Eatock. Adaptation in hair cells. *Annual Review of Neuroscience*, **23** (2000), 285–314.

6. H. A. Quigley. Number of people with glaucoma worldwide. *British Journal of Ophthalmology*, **80** (1996), 389–393.

7. C. R. Ethier, M. Johnson and J. Ruberti. Ocular biomechanics and biotransport. *Annual Review of Biomedical Engineering*, **6** (2004), 249–273.

8. R. Ross. The pathogenesis of atherosclerosis: a perspective for the 1990s. *Nature*, **362** (1993), 801–809.

9. J. P. Paul. Strength requirements for internal and external prostheses. *Journal of Biomechanics*, **32** (1999), 381–393.

10. J. T. Steigenga, K. F. al Shammari, F. H. Nociti, C. E. Misch and H. L. Wang. Dental implant design and its relationship to long-term implant success. *Implant Dentistry*, **12** (2003), 306–317.

11. Q. Yuan, L. Xu, B. K. Ngoi, T. J. Yeo and N. H. Hwang. Dynamic impact stress analysis of a bileaflet mechanical heart valve. *Journal of Heart Valve Disease*, **12** (2003), 102–109.

12. S. P. Hoerstrup, R. Sodian, S. Daebritz, J. Wang, E. A. Bacha, D. P. Martin *et al.* Functional living trileaflet heart valves grown in vitro. *Circulation*, **102** (2000), III44–III49.

13. S. D. Waldman, C. G. Spiteri, M. D. Grynpas, R. M. Pilliar and R. A. Kandel. Long-term intermittent shear deformation improves the quality of cartilaginous tissue formed in vitro. *Journal of Orthopaedic Research*, **21** (2003), 590–596.

14. Y. C. Fung. *Biomechanics: Mechanical Properties of Living Tissues*, 2nd edn (New York: Springer Verlag, 1993).

15. V. C. Mow and R. Huiskes (ed.) *Basic Orthopaedic Biomechanics and Mechano-Biology*, 3rd edn (Philadelphia, PA: Lippincott Williams & Wilkins, 2005).

16. C. Murray. *Human Accomplishment: The Pursuit of Excellence in the Arts and Sciences, 800 BC to 1950* (New York: HarperCollins, 2003).

17. A. Ascenzi. Biomechanics and Galileo Galilei. *Journal of Biomechanics*, **26** (1993), 95–100.

18. Clendening History of Medicine Library and Museum. *Clendening Library Portrait Collection*. Available at http://clendening.kumc.edu/dc/pc/ (2005).

19. W. Harvey. *Exercitatio anatomica de motu cordis et sanguinis in animalibus.* [*An Anatomical Study of the Motion of the Heart and of the Blood in Animals.*] (1628).

20. W. C. Harrison. *Dr. William Harvey and the Discovery of Circulation* (New York: MacMillan, 1967).

21. A. J. Thurston. Giovanni Borelli and the study of human movement: an historical review. *ANZ Journal of Surgery*, **69** (1999), 276–288.

22. S. Hales. *Statical Essays, Containing Haemastaticks.* [With an Introduction by Andre Cournand.] (New York: Hafner, 1964).

23. S. Hales. Foundations of anesthesiology. *Journal of Clinical Monitoring and Computing*, **16** (2000), 45–47.

24. M. Brillouin. Jean Leonard Marie Poiseuille. *Journal of Rheology*, **1** (1930), 345–348.

25. S. P. Sutera and R. Skalak. The history of Poiseuille law. *Annual Review of Fluid Mechanics*, **25** (1993), 1–19.

26. A. Wood. *Thomas Young, Natural Philosopher 1773–1829* (New York: Cambridge University Press, 1954).

27. J. L. M. Poiseuille. Recherches sur la force du coeur aortique. *Archives Générales de Médecine*, **8** (1828), 550–554.

28. K. M. Pederson. Poiseuille. In *XI: Dictionary of Scientific Biography*, ed. C. C. Gillispie (New York: Charles Scribner's Sons, 1981), pp. 62–64.

29. H. J. Granger. Cardiovascular physiology in the twentieth century: great strides and missed opportunities. *American Journal of Physiology*, **275** (1998), H1925–H1936.

30. J. L. M. Poiseuille. Recherches expérimentales sur le mouvement des liquides dans les tubes de très-petits diamètres. *Comptes Rendus Hebdomadaires des Séances de l'Académie des Sciences*, **11** (1840), 961–967, 1041–1048.

31. T. Young. The Croonian Lecture. On the functions of the heart and the arteries. *Philosophical Transactions of the Royal Society*, **99** (1809), 1–31.

32. S. C. Cowin. The false premise in Wolff's law. In *Bone Mechanics Handbook*, 2nd edn, ed. S. C. Cowin. (Boca Raton, FL: CRC Press, 2001), pp. 30.1–30.15.

33. G. H. von Meyer. Die Architektur Der Spongiosa. *Archiv für Anatomie, Physiologie und Wissenschaftliche Medizin*, **34** (1867), 615–628.

34. J. Wolff. *Das Gesetz Der Transformation Der Knochen* (Berlin: A Hirschwald, 1891).

35. J. Wolff. *The Law of Bone Remodeling* (New York: Springer Verlag, 1986).

36. W. Roux. Über die verzweigungen der blutgefässe. [On the bifurcations of blood vessels.] Doctoral thesis, University of Jena (1878).

37. H. Kurz, K. Sandau and B. Christ. On the bifurcation of blood vessels: Wilhelm Roux's doctoral thesis (Jena 1878) – a seminal work for biophysical modeling in developmental biology. *Anatomischer Anzeiger*, **179** (1997), 33–36.

38. V. Hamburger. Wilhelm Roux: visionary with a blind spot. *Journal of Historical Biology*, **30** (1997), 229–238.

39. National Academy of Engineering. *Awards*. Available at http://www.nae.edu/nae/awardscom.nsf/weblinks/NAEW-4NHMBK? OpenDocument (2005).

40. Y. C. Fung. *Biomechanics: Motion, Flow, Stress and Growth* (New York: Springer Verlag, 1990).

41. Y. C. Fung. *Biomechanics: Circulation*, 2nd edn (New York: Springer Verlag, 1997).

42. K. E. Andersson and A. Arner. Urinary bladder contraction and relaxation: physiology and pathophysiology. *Physiological Reviews*, **84** (2004), 935–986.

43. M. S. Damaser. Whole bladder mechanics during filling. *Scandinavian Journal of Urology and Nephrology Supplement*, **201** (1999), 51–58.

44. E. J. Macarak, D. Ewalt, L. Baskin, D. Coplen, H. Koo, *et al.* The collagens and their urologic implications. *Advances in Experimental Medicine and Biology*, **385** (1995), 173–177.

45. van Mastrigt R. Mechanical properties of (urinary bladder) smooth muscle. *Journal of Muscle Research and Cell Motility*, **23** (2002), 53–57.

46. E. H. Davidson, J. P. Rast, P. Oliveri *et al.* A genomic regulatory network for development. *Science*, **295** (2002), 1669–1678.

2 Cellular biomechanics

The cell is the building block of higher organisms. Individual cells themselves are highly complex living entities.[1] There are two general cell types: *eukaryotic* cells, found in higher organisms such as mammals, and *prokaryotic* cells, such as bacteria. In this chapter, we will examine the biomechanics of eukaryotic cells only. We will begin by briefly reviewing some of the key components of a eukaryotic cell. Readers unfamiliar with this material may wish to do some background reading (e.g., from Alberts *et al.* [1] or Lodish *et al.* [2]).

2.1 Introduction to eukaryotic cellular architecture

Eukaryotic cells contain a number of specialized subsystems, or *organelles*, that cooperate to allow the cell to function. Here is a partial list of these subsystems.

Walls (*the membranes*). These barriers are primarily made up of lipids in a bilayer arrangement, augmented by specialized proteins. They serve to enclose the cell, the nucleus, and individual organelles (with the exception of the cytoskeleton, which is distributed throughout the cell). The function of membranes is to create compartments whose internal materials can be segregated from their surroundings. For example, the cell membrane allows the cell's interior to remain at optimum levels of pH, ionic conditions, etc., despite variations in the environment outside the cell. The importance of the cell membrane is shown by the fact that cell death almost invariably ensues if the cell membrane is ruptured to allow extracellular materials into the cell.

A framework (*the cytoskeleton*). This organelle consists of long rod-shaped molecules attached to one another and to other organelles by connecting molecules.

---

[1] What defines a living system? A living system must satisfy the following five characteristics:
- it must show complex organization (specialization)
- it must be able to metabolize (assimilate "food," transform it, and excrete it)
- it must show responsiveness, including the ability to adapt to differing conditions
- it must be able to reproduce
- it must have evolutionary capability.

Individual cells exhibit each of these characteristics.

The cytoskeleton gives the cell form, allows it to move, helps to anchor the cell to its substrate and neighbors, and speeds the transport of materials within certain types of cells. We will consider the cytoskeleton in greater detail later in this chapter.

Engines (*the mitochondria*). These organelles produce most of the basic energy-containing molecules from certain substrates such as glucose. Then these energy-containing molecules are used by other subsystems within the cell.

A command center (*the nucleus*). The genetic material which codes for molecules synthesized by the cell is mostly contained in the nucleus, although there is a small amount of mitochondrial DNA. The synthesis of proteins and other important biomolecules is initiated by transcription of genetic information coded in DNA into messenger RNA (mRNA). These mRNA coding sequences leave the nucleus where they are eventually turned into proteins by . . .

Factories (*the endoplasmic reticulum*). These production centers synthesize biomolecules needed by the cell. They take their "orders" from the nucleus in the form of mRNA.

Packaging plants (*the Golgi apparatus*). Proteins produced by the endoplasmic reticulum are not "ready for prime time." They typically must undergo a series of folding steps and post-translational modifications before they have biological activity. This very important task is handled by the Golgi apparatus, which takes the protein output of the endoplasmic reticulum and trims, modifies, and packages it in membrane-delimeted structures (the *vesicles*) that are sent to various locations within or outside the cell. Misfolded and otherwise defective proteins must be disposed of immediately since they are potentially harmful to the cell.

A disposal system (*the lysozomes*). This system of vesicles contains *enzymes* (catalytic proteins) which act to break down metabolic by-products, misfolded proteins, ingested extracellular material, and other unwanted substances.

Clearly a single cell is a remarkable assortment of complex subsystems (Figs. 2.1 and 2.2). It is also a miracle of miniaturization: all of the above systems fit into a neat package having a typical mass of 2×10^{-8} g, and a typical diameter of order 15 μm!

Now that we have a basic overview of the components of a eukaryotic cell, let us look in more detail at cellular biomechanics and mechanobiology. We will start (Section 2.2) with some basic ideas about how the cell uses energy, which we will see resembles energy flow in a thermal energy-generating station in some ways. Then we will delve into more detail about the cytoskeleton of the cell, focussing on its mechanical properties and how it helps to anchor the cell to its surroundings (Sections 2.3 and 2.4). The main focus here will be to give the reader enough biological background to understand Sections 2.5 and onwards.

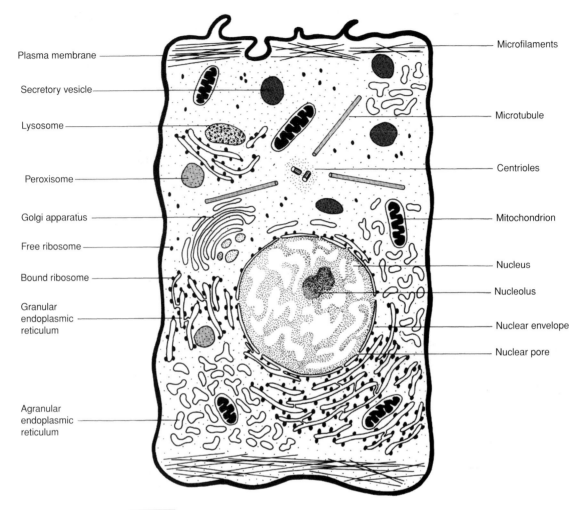

Plasma membrane

Secretory vesicle

Lysosome

Peroxisome

Golgi apparatus

Free ribosome

Bound ribosome

Granular
endoplasmic
reticulum

Agranular
endoplasmic
reticulum

Microfilaments

Microtubule

Centrioles

Mitochondrion

Nucleus

Nucleolus

Nuclear envelope

Nuclear pore

Figure 2.1

Structures and organelles found in most human cells. This diagram is highly schematized but serves to indicate the major features of the cellular organelles. From Vander *et al.* [3]. Reproduced with kind permission of the McGraw-Hill Companies.

When engineers talk about the mechanics of conventional engineering materials, they can refer to handbooks that tabulate the properties of, for example, different types of stainless steel, and describe the internal structure of these steels. Can we do that for cells? Not quite, but a body of data is slowly being accumulated about the "mechanical properties of cells." In Section 2.5, we will tackle the tricky question of how one measures the mechanical properties of a single cell, while in Section 2.6 we will introduce some engineering models that, in combination with experimental data, teach us something about the cell's internal mechanics.

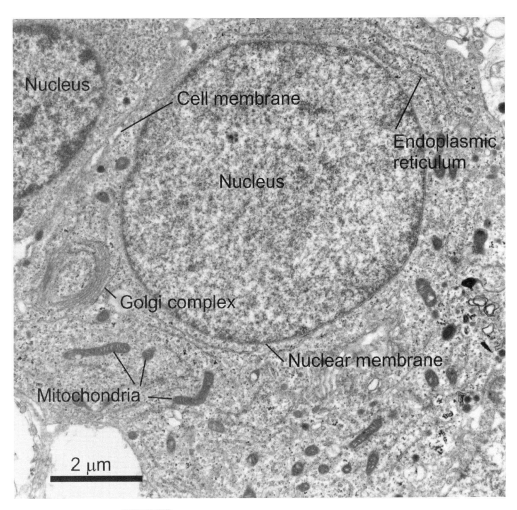

Figure 2.2

Transmission electron micrograph of an insulin-producing pancreatic cell, showing several of the structures depicted schematically in Fig. 2.1. A prominent nucleus delimited by the nuclear envelope (membrane) is present, as are several organelles in the cytoplasm: mitochondria, Golgi complex, and endoplasmic reticulum with associated ribosomes. A second cell is visible at the top left of the image. Sample stained with uranyl acetate and lead citrate. Micrograph courtesy of Mr. Steven Doyle, University of Toronto.

One of the remarkable things about most cells is how *good* they are at sensing relatively small levels of mechanical stimulation, while living in a constantly changing biomechanical environment. How do they do this? Many details of this process, known as *mechanotransduction*, are unknown, but in Section 2.7 we will discuss, in general terms, current thinking on how adherent cells are able to sense and respond to mechanical stimulation.

Finally, we consider the consequences of the response of cells to mechanical stimulation. In Section 2.8, we describe some of the experimental tools that are used to apply mechanical stimuli to tissues or small groups of cells in culture. Using these devices, the effects of mechanical stimulation on several cell types have been determined. In Section 2.9, we present some of the effects on cells from three specific tissues (vascular tissue, cartilage, and bone) and we consider the implications for the whole tissue. Let's get started!

2.2   The cell's energy system

Life requires energy. At the cellular level, energy-consuming tasks include:

- motion, including both cellular shape changes and locomotion of the cell on its substrate
- synthesis of compounds
- transport of ions and other molecules, both within the cell and between the cell and its surroundings

How does the cell utilize food energy? When we eat a meal, the constituent food-stuffs are acted upon by the digestive enzymes and broken down into simpler compounds, transferred into the bloodstream across the intestinal walls, and then transported throughout the body.

Individual cells are therefore presented with a complex mixture of compounds from which they must obtain energy. The cell solves this problem by having specialized "energy plants" (mitochondria), which are able to use compounds such as glucose and fatty acids to produce a common energy-containing molecule that all cellular organelles can use. This common molecule is *adenosine triphosphate* (ATP), formed from *adenosine diphosphate* (ADP) and phosphate (PO_3^{2-}) in the following reaction:

$$ADP + PO_3^{2-} + energy + 2H^+ \rightarrow ATP$$

Energy is stored in the chemical bond between ADP and PO_3^{2-} (Fig. 2.3). A mechanical analogue is a spring, which starts in an uncompressed state (ADP) and is then compressed and held in place by a catch (PO_3^{2-}). The organelles can "release the catch" to produce energy, with by-products ADP and PO_3^{2-}. We say, therefore, that *ATP is the common currency of energy within the cell.*

It is important to note that ADP and phosphate are recycled, to be once more combined in the mitochondria to yield ATP. This is similar to the movement of primary loop cooling water in a thermal generating station, and it emphasizes that ATP is merely a transient carrier of energy within the cell (Fig. 2.4).

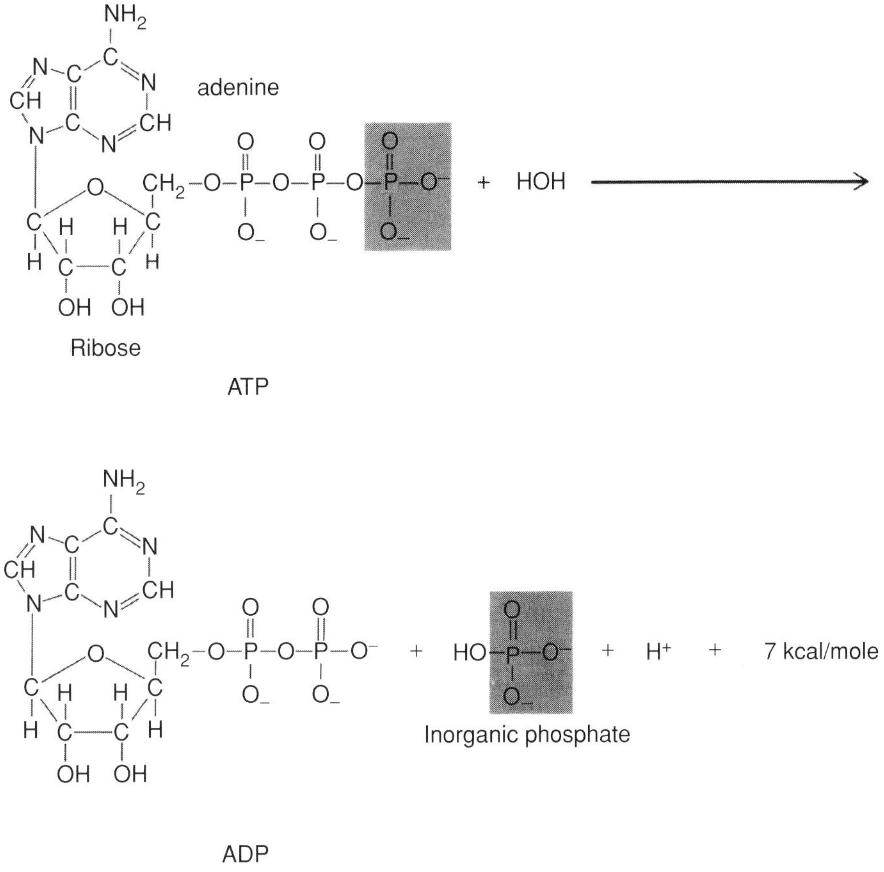

Figure 2.3

Structure of ADP and ATP. The breakdown of ATP to ADP and inorganic phosphate yields 7 kcal/mole energy. From Vander *et al.* [3]. Reproduced with kind permission of the McGraw-Hill Companies.

2.3   Overview of the cytoskeleton

Just as an understanding of a cell's architecture and its energy system is critical to understanding cellular biology, the characteristics of the cytoskeleton are central to understanding a cell's biomechanical behavior. Here we will only give an overview of this fascinating topic. Students are encouraged to consult the references if they wish to learn more about the cytoskeleton.

The cytoskeleton is an elaborate network of fibrous proteins that can adopt a remarkable range of configurations (Figs. 2.5 and 2.6). It:

- establishes and maintains the shape of the cell
- allows the cell to move (the process of *locomotion*)

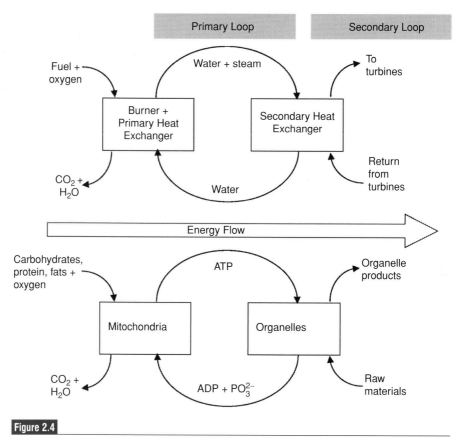

Figure 2.4

Energy flow in a thermal generating station (top) and a cell (bottom). Note that primary loop cooling water is analogous to ATP in that it is a transient vehicle for energy storage which is recycled.

- provides mechanical strength and integrity to the cell
- is central to the intracellular transport of organelles, especially in large cells such as axons
- is essential during cell division, where it plays a key role in many processes, including chromosome separation in mitosis and meiosis.

The cytoskeleton consists of three types of filament, each with a specialized protein composition: *actin filaments* (7–9 nm in diameter), *intermediate filaments* (10 nm in diameter), and *microtubules* (approximately 24 nm in diameter). Actin filaments are also called *microfilaments* or – in skeletal muscle cells – thin filaments. The interaction between all three filament types helps to determine the cell's mechanical behavior. We will briefly review the function of each of these filament types.

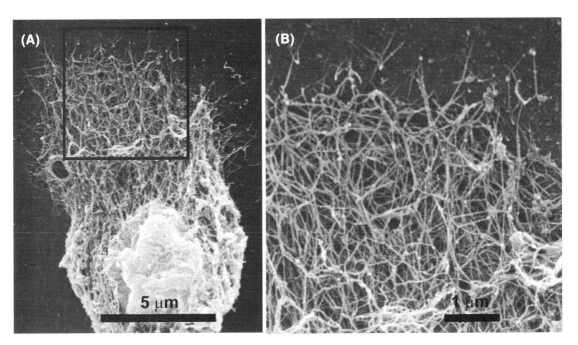

Figure 2.5

Scanning electron micrograph of the actin component of the cytoskeleton within a rat fibroblast adhering to an N-cadherin-coated glass cover slip (A). A high-magnification inset of the boxed region is shown in (B). The rich, highly interconnected actin network is clearly visible. Cells were extracted with a detergent solution, fixed in glutaraldehyde, post-fixed, and gold sputter coated before visualization. Images courtesy of Dr. Tarek El Sayegh, Faculty of Dentistry, University of Toronto.

2.3.1   Actin filaments

Actin exists within the cell in two forms, as a globular protein (G-actin) and as a filamentous protein (F-actin). G-actin has a molecular weight of approximately 43 kDa, and consists of a single polypeptide chain. Monomeric G-actin binds one Ca^{2+} and one molecule of ATP. F-actin is formed by the polymerization of G-actin, which causes the bound ATP to be hydrolyzed to ADP and a phosphate ion (Fig. 2.7, color plate). The ADP remains bound to the actin subunit within the F-actin chain.

F-actin chains are dynamic structures that grow and break down according to their position within the cell and the activities of the cell at any given instant. F-actin filaments are polarized, having an end where G-actin monomers are preferentially added (the fast-growing "barbed" or "+" end) and an end where the filament is either slowly growing or disassembled (the "pointed" or "−" end). Thus, individual actin monomers move along filaments, tending to be added at the + end and

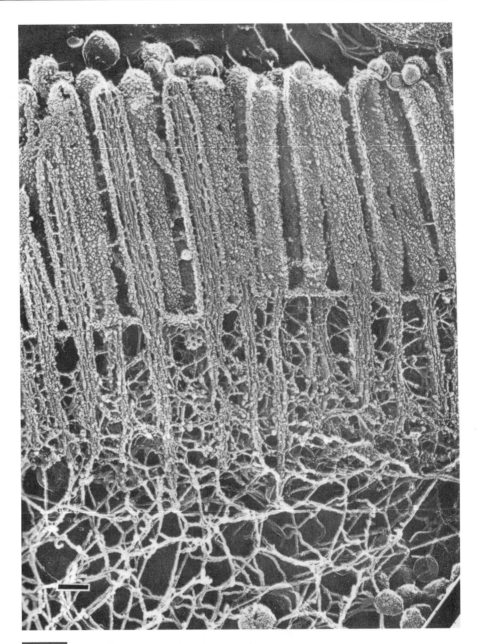

Figure 2.6

Cytoskeleton of the brush border of intestinal epithelial cells. Tight actin filaments are evident in the microvilli, the finger-like structures in the top half of the image. The actin extends from microvilli into the cytoplasm of the cell, where it connects with a network of actin, intermediate filaments, myosin, and other cytoskeletal proteins. The scale bar in the lower left corner is 0.1 μm. Reproduced with permission from Bershadsky and Vasiliev [4] and from the *Journal of Cell Biology*, 1982, 94, 425–443 by copyright permission of the Rockefeller University Press.

Table 2.1. Summary of Young's modulus values for F-actin measured
by various methods. Modified with permission from Janmey *et al.* [13].

Authors	Method	Estimated Young's modulus (N/m²)
Kojima *et al.* [9]	Micro-needle/single filament	1.8×10^9
Huxley *et al.* [10]	X-ray diffraction frog/muscle	2.5×10^9
Higuchi *et al.* [11]	Optical diffraction/rabbit skeletal muscle fiber	Not in paper, by comparison: $\sim 2 \times 10^9$
Wakabayashi *et al.* [12]	X-ray diffraction/frog muscle	Not in paper, by comparison: $\sim 2 \times 10^9$

moving to the − end, in a process known as *treadmilling* (Fig. 2.8, color plate).
The polymerization and breakdown of F-actin are regulated by several proteins,
including actin depolymerizing factor/cofilin, members of the gelsolin/villin protein family, and CapZ. The lifetime of actin filaments, the length of the filaments,
the percentage of actin in polymeric form, and the number of barbed ends change
as a function of the cell's activity. For example, in confluent bovine aortic endothelial cells, the mean filament lifetime is approximately 40 min and about 70% of
the cell's total actin is present as F-actin [8]. Confluent cells tend to be relatively
quiescent, exhibiting only modest amounts of cellular movement. In contrast, in
subconfluent endothelial cells, which are more active, the mean filament lifetime
is only approximately 8 min and only approximately 40% of the cell's actin is
present in polymerized form [8].

F-actin exists within the cell in several different forms. In all cells, it is present
in a thin layer adjacent to the cell membrane, in the so-called *cortical actin* layer.
This layer helps to anchor transmembrane proteins to cytoplasmic proteins (see
Section 2.4) and generally provides mechanical strength to the cell. It also anchors
the centrosomes at opposite ends of the cell during mitosis, and helps the cell to split
into two parts during cytokinesis. In many cells, actin is also present in long bundles
that criss-cross the cell, known as *stress fibers*. Stress fibers presumably reinforce
the cortical actin layer and are also important for the transport of organelles and
cellular locomotion. Finally, in skeletal muscle cells, actin is highly abundant
and interacts with myosin filaments (thick filaments) to create the actin–myosin
complex, responsible for muscle contraction.

Because of its biomechanical importance, a number of authors have experimentally estimated the Young's modulus of F-actin filaments. Some of these measurements are summarized in Table 2.1, which shows a fairly tight range of values of
order 2 GPa, approximately 100 times less than the modulus for steel.

2.3.2 Intermediate filaments

There are currently six known classes of intermediate filament protein. Each class contains multiple members, so that the intermediate filaments together make up a very diverse group of molecules whose expression is cell-type specific [2]. For example, the *keratins* are expressed in epithelial cells, while glial cells express *glial fibrillary acid protein* (GFAP), neurons express various *neurofilaments*, and endothelial cells express *vimentin*.

The biology and biomechanics of intermediate filaments are not as well understood as those of microfilaments and microtubules [14,15]. We know that intermediate filaments are more stable than other components of the cytoskeleton, although they can and do alter their configuration under the right circumstances (e.g., during cell division). Because of this stability, they provide a supporting network for the cell; for example, the nucleus is stabilized by a network of intermediate filaments made up of *lamin*. Intermediate filaments interact with other cytoskeletal elements, and dysfunction of intermediate filaments leads to the inability of the cell to withstand mechanical forces [16].

2.3.3 Microtubules

Microtubules are hollow cylindrical structures assembled from dimers of the proteins α-tubulin and β-tubulin (Fig. 2.9, color plate). They are dynamic structures, with the potential to grow or shrink at each end by the addition/removal of dimers. Like F-actin, microtubules are polarized: the + end is always more active than the other end. In vitro under appropriate conditions, tubulin monomers spontaneously polymerize to form a network of filaments (Fig. 2.10). Similar networks are present in cells.

Microtubules play a key role in the structure of cilia and flagella, where they provide structural rigidity and assist in generating motion. Microtubules are often associated with molecular motors, which are proteins designed to travel along the microtubule, usually to help to transport something. These molecular motors include:

- the *kinesins*, which move towards the + end of the tubule
- the *dyneins*, which move towards the − end of the tubule.

Molecular motors can be used for more than just transporting materials along the microtubules. Figure 2.11 shows a micrograph of a cilium, demonstrating the characteristic distribution of nine bundles of two microtubules around the periphery of

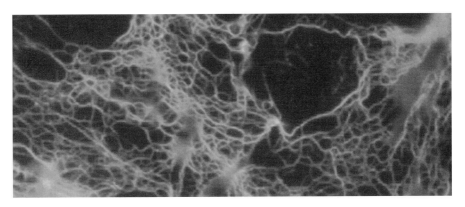

Figure 2.10

Micrograph of fluorescently labeled tubulin polymerized in vitro to show network formation. Tetramethylrhodamine-labeled tubulin was copolymerized with unlabeled tubulin, fixed, and photographed. It is clear that under the right conditions, microtubule formation can happen spontaneously. Image copied with permission from the Molecular Probes image gallery, http://probes.invitrogen.com/. Copyright Molecular Probes, Inc.

the cilium and two central microtubules. This micrograph also shows interconnecting filaments between the central and peripheral microtubules. These connecting filaments are motor proteins. Forces generated by these motor proteins, in conjunction with the structural integrity of the microtubules, produce the characteristic back-and-forth motion of the cilium.

Microtubules also play a major role during cell division, where they form a basket-like network called the *mitotic apparatus* (Fig. 2.12). To accomplish this task, the microtubules organize themselves around a pair of organelles called the *centrosomes* (also known as the microtubule organizing centers). Each centrosome contains two microtubule aggregates oriented at right angles to one another, called the *centrioles* (Fig. 2.1). In a remarkable process, the microtubules in this apparatus attach to and orient the chromosomes containing the previously duplicated genetic material of the cell, and then pull the two halves of each chromosome apart to form two complete daughter sets of genetic material. It is hard to think of a process in which mechanics is more central to life.

2.4 Cell–matrix interactions

To function properly, cells must stay attached to their substrates and to their neighbors. (The exceptions are circulating cells, such as formed elements in the blood.) This is not as straightforward as it sounds. Think about a vascular endothelial cell in a large artery: it lives on a continuously deforming substrate to which it must

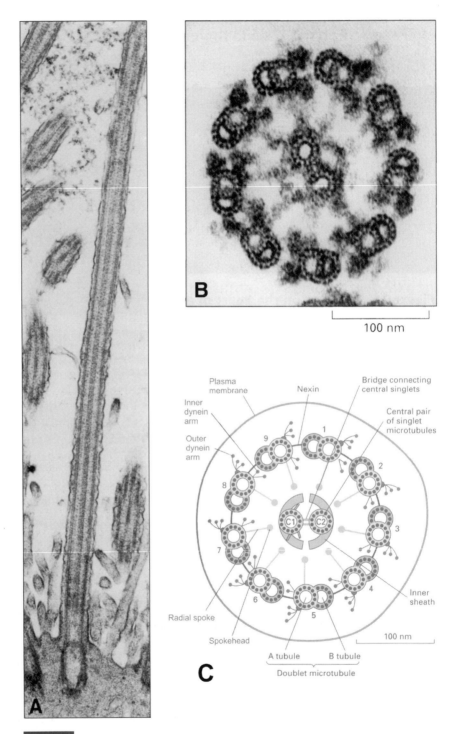

Figure 2.11

(A) Electron micrograph of the longitudinal cross-section of a cilium from a human bronchus epithelial lining cell. (B) Electron micrograph of the transverse cross-section of a flagellum of a green alga cell showing the characteristic "9 + 2" arrangement of microtubules. This microtubule arrangement is found in almost all flagella and cilia from single cell organisms to humans. (C) Diagram of the cross-section of a flagellum, showing the complex internal structure, which is very similar to that of a cilium. (A) reproduced with permission from Cormack [18]; (B) from Alberts *et al.* [1]; (C) from Lodish *et al.* [2], with permission of W. H. Freeman.

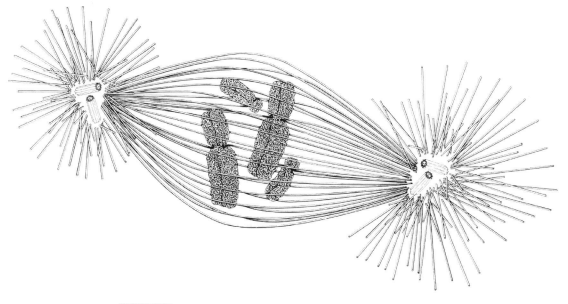

Figure 2.12

Schematic drawing of a mitotic apparatus. Microtubules emanate from two centrosomes, forming a basket-like network around cellular DNA within the duplicated chromosomes (dark objects in center of picture). Other microtubules will attach to each duplicated chromosome; these microtubules pull apart the chromosomes and produce two daughter sets of identical genetic material. From DuPraw [19], with permission from Elsevier.

adhere, while maintaining close contact with its neighboring endothelial cells. Furthermore, it must be able to detach locally from its substrate, for example to crawl to cover a denuded region of the artery wall following injury or to allow a circulating macrophage to enter the artery wall. In addition, there is now much evidence that the cell senses some aspects of its biomechanical environment through deformations of its substrate, and that interaction with its substrate influences cytoskeletal structure, phagocytosis (ingestion of foreign material), and signal transduction.

There are two kinds of attachment that cells make: to neighboring cells and to their substrates. Here we will concentrate on the latter. To understand such interactions better we must first learn a little about the extracellular materials that form the cell's substrate. This extracellular material is a complex mixture of biopolymers that is collectively known as the extracellular matrix (ECM). Immediately adjacent to some cells (e.g., endothelial cells, epithelial cells, muscle cells, fat cells, Schwann cells of the nervous system), there is a specialized layer known as the *basal lamina*, or *basement membrane*, which is a specialized region of the ECM.

In general, the ECM is very complex and consists of a large number of highly specialized macromolecules. Some of the important components are:

Collagens. Collagen is a fibrous protein that imparts structure and rigidity to tissue. It occurs in many different types[2], and is the most abundant protein in higher vertebrates; for example, Lehninger [21] states that collagen makes up one third or more of total body protein. We will consider the structure of collagen in detail in Section 9.8; here we provide a brief overview of several of the more important types of collagen:

- *collagen type I*: forms large structural bundles that are present in tendons, ligaments, and other tissues subject to significant mechanical loading; has high tensile strength (tangent Young's modulus $\sim 1 \times 10^9$ Pa)
- *collagen type IV*: forms X-shaped complexes that associate together to create a highly interconnected fibrous network; is abundant in basement membrane
- *collagen type VI*: widely distributed throughout the ECM, this collagen type seems to play an intermediate role between the interstitial collagens (types I–III) and cells, helping cells to form attachments to the surrounding matrix [22].

Elastin. This fibrous elastic protein acts to impart elasticity and resilience to tissue. It is much more compliant than collagen (Young's modulus $\sim 3 \times 10^4$ Pa) and is present in significant amounts in the walls of large arteries, lungs and skin.

Proteoglycans. Proteoglycan is a generic term denoting a protein that has one or more glycosaminoglycan (GAG) side chains. The physicochemical properties of the proteoglycans are in large part determined by the GAGs, which are themselves large biopolymers made up of negatively charged carbohydrate repeat units. The GAGs are avidly hydrophilic and will imbibe and retain very large volumes of water. Common proteoglycans are aggrecan (found in large amounts in cartilage), heparin sulfate (found in basal lamina), and chondroitin sulfate.

Hyaluronan. Hyaluronan (also called hyaluronate and hyaluronic acid) is an extracellular GAG that is unusual in not being covalently linked to a protein, although it associates non-covalently with proteins in the extracellular matrix. Its large size and strong negative charge means that it is very effective at taking up water and keeping tissue hydrated.

Adhesion proteins. Laminins and fibronectin are adhesion proteins:

- *laminin*: there are 11 known isoforms of laminin [23], all of which are cross-shaped proteins that are abundant in basal lamina and bind collagen type IV and other matrix molecules; they are key components of the basement membrane
- *fibronectin*: this large protein has binding domains for collagen, cell-surface binding molecules (integrins; see below) and heparin sulfate; it is the "all-purpose glue" of the ECM.

When these components are put together, a composite structure results (Fig. 2.13). In this composite material, collagen and elastin act together to provide mechanical

---

[2] In October 2002 the 26th type of collagen was reported [20].

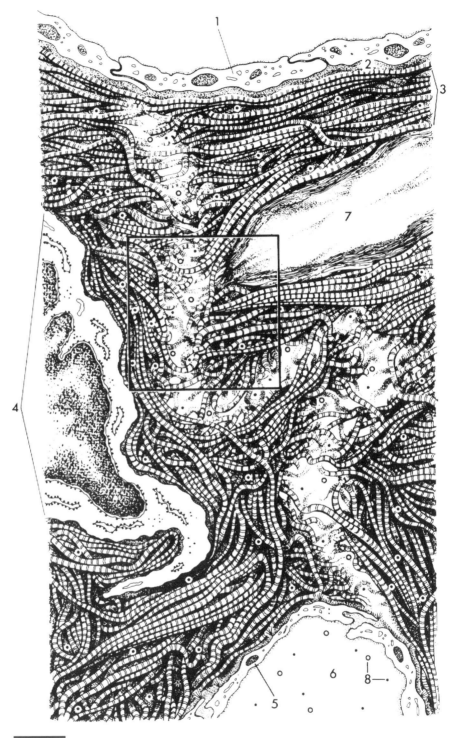

Figure 2.13

A highly schematic diagram showing some of the major components of the extracellular matrix, with a flow pathway from a vascular capillary (upper portion of figure) to a lymphatic capillary (lower portion of figure). 1, an endothelial cell in a capillary wall; 2, capillary endothelial basement membrane; 3, interstitial matrix; 4, interstitial cell; 5, lymphatic capillary endothelial cell; 6, lymphatic fluid; 7, elastic fiber; 8, plasma proteins in lymphatic fluid. Reproduced with permission from Bert and Pearce [24].

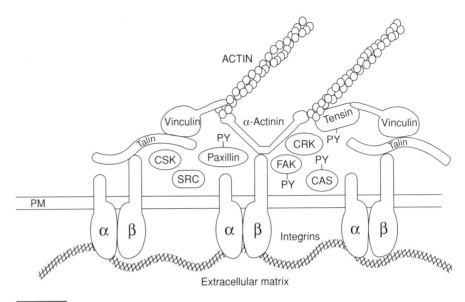

ACTIN

Vinculin   α-Actinin   Tensin   Vinculin

Talin   PY   CRK PY   Talin

CSK   Paxillin   FAK PY

SRC   PY CAS

PM

α β   α β   Integrins   α β

Extracellular matrix

Figure 2.14

Schematic description of how actin interacts with transmembrane proteins (integrins). Integrins are responsible for binding to extracellular molecules, primarily extracellular matrix components. To ensure the mechanical integrity of this attachment, it is necessary that the integrins be somehow connected to the cytoskeleton of the cell. As can be seen, this actin–integrin binding is effected through a complex set of proteins, including vinculin, talin, tensin, and α-actinin. CRK, Crk protein; CSK, Csk protein; CAS, Crk-associated substrate; FAK, Focal adhesion kinase; PY, phosphotyrosine; PM, plasma membrane. See Fig. 2.15 for a more complete summary of the proteins participating in cell–matrix adhesion. Reproduced with permission from Vuori [26].

integrity and tissue structure, and to provide mechanical "scaffolding" for the resident cells, while hyaluronan and the proteoglycans fill the empty spaces and hold water in the tissue. The role of the bound water in the tissue is crucial, since this provides the pathway whereby nutrients and other species are transported to and from the cells. The entire structure can be thought of as a biological "sponge" loaded with water [25].

Cells express transmembrane proteins called *integrins* that are responsible for binding to specific components in the matrix and attaching the cell to its substrate. The integrins are heterodimers (formed by two dissimilar subunits, an α chain and a β chain) that have an extracellular binding domain, a transmembrane domain, and a cytosolic domain. There are at least 22 different integrins in mammals, each of which specifically recognizes binding domains on collagen, laminin, fibronectin, and/or other matrix proteins [17,26]. The cytosolic domain of the integrin associates with a large protein grouping called a focal adhesion complex, which in turn attaches to F-actin filaments (Figs. 2.14 and 2.15 [color plate]). In this way, the actin filaments of the cytoskeleton are ultimately connected to the

Table 2.2. Typical magnitudes of quantities measured on the cellular scale
(last column)

Quantity	SI units	"Micro SI" units (suitable for cells)
Distance	m	μm
Force	N	pN ($= 10^{-12}$ N) to nN ($= 10^{-9}$ N)[a]
Pressure, stress	Pa ($= $ N/m^2)	pN/μm^2 ($= 1$ Pa) to nN/μm^2 ($= 1$ kPa)

[a] Forces of molecular bonds and those exerted by "soft" cells are in the picoNewton range. "Stiff" cells can exert forces in the nanoNewton range.

fibrous matrix outside the cell. We will consider the important role integrins play in cellular biomechanics and the sensing of mechanical signals later in this chapter.

2.5 Methods to measure the mechanical properties of cells and biomolecules

To help in understanding cellular biomechanics we need experimental data on the mechanical properties of individual cells, such as Young's modulus, shear modulus, etc. It is also useful to have measurements of the mechanical strength of individual molecules (e.g., actin filaments) and of bonds between molecules, e.g. between a receptor and its ligand. How does one *measure* Young's modulus for a cell or a molecule? As you may imagine, it is not trivial. Consider, for instance, the magnitudes we are dealing with at the cellular level. These are discussed very nicely by Hochmuth in a review article [28] and are summarized in Table 2.2. It can be seen that the forces are pretty small! Nonetheless, there are a number of techniques for measuring the mechanical properties of single cells and molecules. Some of these techniques are summarized schematically in Fig. 2.16. Each technique is suited to a different force range and length scale. Below we review some of them.

2.5.1 Atomic force microscopy

The atomic force microscope (AFM) is a powerful tool for measurement of forces and displacements on both molecular and cellular scales [30,31]. The key element of the microscope is a tapered probe, typically made from silicon or silicon nitride, which is attached to a cantilever arm. When the probe tip interacts with a sample, the arm is deflected. This deflection can be measured by sensing the position of a laser beam that reflects off the cantilever arm (Fig. 2.17). By suitable placement of the laser and detector to take advantage of the "optical level arm" effect, displacements of less than 1 nm can be measured. The arm is attached to a

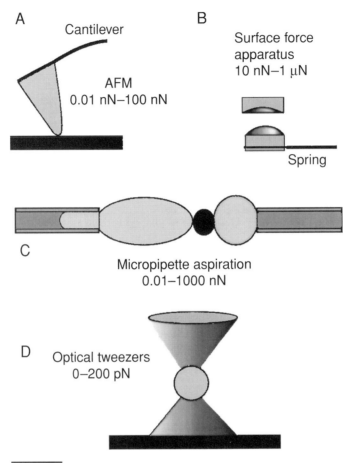

Figure 2.16

Schematic diagrams of force probes used to measure the mechanical properties of single cells and protein interaction forces. (A) An atomic force microscope (AFM) shows the probe tip, attached to the cantilever force transducer, coming in contact with a substrate. (B) A surface force apparatus, showing the crossed cylinders of the apparatus and the force-transducing spring. (C) The bioforce probe, consisting of a cell partially aspirated into a micropipette. A bead (center black sphere) is attached to one cell (left), and a force between the bead and a second cell or glass bead (right) is exerted by aspirating the (left) cell into the pipette. (D) Optical tweezers. A bead is held in the optical trap, such that radiation pressure exerted on the bead applies force to adhesive contacts between materials on the bead surface and the substrate. Modified with permission from Leckband [29].

piezoelectric element that can move in the vertical and transverse (lateral) directions. In biological applications, the AFM is typically used in conjunction with an optical microscope that can be used for real-time visualization, for example, to help to localize the AFM probe over a target cell. One of the great advantages of the AFM is that imaging can be carried out on living cells or intact molecules in an aqueous environment. This is in contrast to other high-resolution microscopy techniques (e.g., scanning electron microscopy) that require that the samples be fixed and imaged in a near-vacuum.

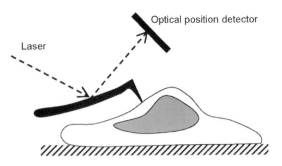

Figure 2.17

Schematic description of the tip of the atomic force microscope contacting a cell, showing the path of the laser beam. Deflection of the laser beam is sensed and related to tip deflection by the known geometry of the optical pathway. Tip deflection is then related to applied force by knowledge of the mechanical characteristics of the cantilever arm supporting the tip.

To understand the operation of the AFM it is important to delve into the operation of the cantilever arm in more detail. If the arm undergoes only small deflection in the vertical direction, then it can be treated as a linear spring with stiffness k_c. In this case, the force F needed to create a vertical deflection z can be written as

$$F = k_c z. \tag{2.1}$$

Typical values for k_c are about 0.02 to 5 Newton per meter (N/m), which is pretty soft! This has interesting consequences for the minimum force that can be resolved by the AFM [32]. Remember that the probe tip will typically be in an aqueous environment and will, therefore, be subject to random collisions from water molecules undergoing thermal motion. The average energy of a molecule at absolute temperature T is given by $\frac{1}{2}k_B T$, where k_B is Boltzmann's constant, 1.3807×10^{-23} J/K. We know that the energy stored in a spring of stiffness k_c as it deflects by an amount z is $\frac{1}{2}k_c z^2$. Equating these two energies allows us to solve for the fluctuations, z, that will occur as a result of random collisions with water molecules as:

$$z = \sqrt{\frac{k_B T}{k_c}}. \tag{2.2}$$

By Equation (2.1) this corresponds to a force

$$F = \sqrt{k_c k_B T} , \tag{2.3}$$

which represents the thermal "noise" that will be continually measured by the AFM tip. Forces appreciably smaller than this are not resolvable. For a sample at $37\,^\circ$C (310 K), and an arm stiffness of 0.05 N/m, this corresponds to approximately 15 pN.

How is the AFM used to obtain information about biological specimens? Although the AFM can be used to image many types of specimen, here we will focus on measurements made on adherent cells. Consider first the "simple" problem of mapping out the topography of the cell. There are a number of ways, or modes, in which the AFM can be used in such an application. The simplest of these modes is the so-called contact mode, in which the probe tip is scanned over the surface of the cell in a raster-like pattern. At each raster location, the deflection of the cantilever arm is measured (based on the reflected laser beam's position) and the arm's vertical position is adjusted by the piezoelectric element in a feedback loop so that the cantilever arm experiences a constant deflection. From Equation (2.1), this implies that a constant force is being applied to the cell's surface by the probe tip. Since the vertical position applied by the piezoelectric element is known, this produces a "map" of cell topography, or more specifically, a set of height values at which the cell exerts a constant reaction force on the probe tip (Fig. 2.18, color plate). Unfortunately, this method of measuring cellular topography has a significant drawback: the probe tip often applies large lateral forces to the cell. These lateral forces can deform or even damage unfixed cells or other "soft" specimens. This has led to the development of more sophisticated scanning modes. For example, in "tapping mode," the cantilever arm is acoustically or magnetically driven so that it vibrates at or near its resonant frequency, and it is then moved vertically so that it approaches the cell surface. Far from the cell, the arm undergoes oscillations whose amplitude is determined by the magnitude of the driving signal. However, as the probe tip approaches the cell it begins to interact with the cell, which changes the magnitude of the arm's vibrations. The position where this change in magnitude occurs is a measure of the topography of the cell at that lateral location. Because the probe tip just taps the specimen, rather than being dragged along it, this produces smaller lateral forces on the cell and less potential for damage.

Other interesting measurements can be made with the AFM on cells. For example, the probe tip can be vertically traversed towards the sample while the cantilever arm deflection is measured.[3] This is known as "force mapping mode" and produces a curve of applied force versus surface deflection (a "force curve;" Fig. 2.19). To understand this curve, we must remember that the probe tip is much harder than the relatively soft biological specimens, so that we effectively have a rigid cone penetrating into the cell, which we will treat as being locally planar and linearly elastic. This is a classical problem in contact mechanics, which was treated by Hertz in 1882. The key result is that the deflection of the cell at the center of the probe tip, δ, due to an applied force F is

$$\delta^2 = \frac{\pi}{2} \frac{F(1 - \nu^2)}{E \tan \alpha}, \tag{2.4}$$

---

[3] An early version of an apparatus that functioned in a similar fashion was known as a "cell poker."

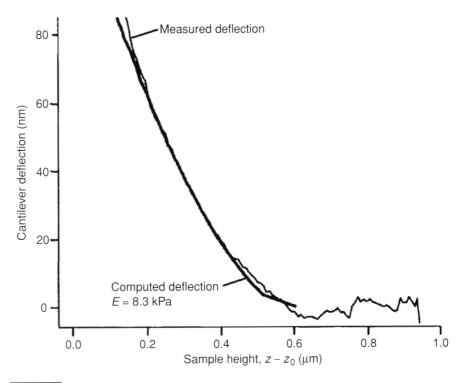

Figure 2.19

Example force curve for measurement of cellular stiffness of an activated human platelet using the atomic force microscope. The vertical axis is proportional to the force applied to the cell by the probe tip. The horizontal axis is the vertical offset of the base of the cantilever arm applied by the piezoelectric actuator, which we denoted by $z - z_0$ in Equation (2.5). The thin line is the measured deflection, while the thick line is the fit of a modified form of Equation (2.5) to the data. Notice the variations in the measured deflections from thermal noise. From Radmacher [32], reprinted with permission from the IEEE.

where E and ν are the modulus and Poisson ratio for the cell and α is the known cone half-angle.[4] A generalization of this solution that is useful for our purposes is [35]:

$$z - z_0 = \frac{F}{k_c} + \sqrt{\frac{\pi}{2} \frac{F(1 - \nu^2)}{E \tan \alpha}}. \qquad (2.5)$$

Here z_0 is the probe height at which the applied force becomes non-zero, and the first term on the right-hand side accounts for the deflection of cantilever arm. By assuming a value for ν (usually 0.5), the measured force–displacement data can

---

[4] The reader should be aware that the coefficient $\pi/2$ in Equation (2.4) and under the radical sign in Equation (2.5) is often erroneously given as $2/\pi$ in the AFM literature. The form shown here is correct, as can be confirmed by consulting the original solutions of the Hertz contact problem for a conical indenter (e.g., [33]) or any textbook on contact mechanics (e.g., [34]).

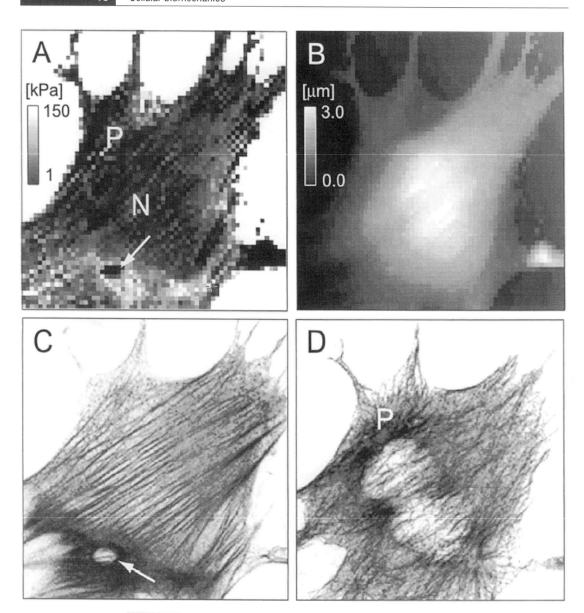

Figure 2.20

Elasticity map (A) and corresponding topography map (B) of a living NIH3T3 fibroblast. The nuclear portion (N) is the softest, with a stiffness of approximately 4 kPa. A small softer "island" was observed in the perinuclear region (marked by arrow). (C,D) Immunofluorescence images of actin filaments (C) and microtubules (D) for the same cell shown in A and B. A bi-lobed nucleus is visible in the microtubule image. The small open area observed in the actin image (arrow in C) corresponds to the soft "island" observed in the elasticity map. Part of the cell with low actin density and high density of microtubules shows a low Young's modulus (marked P in A and D). Each image size is 80 μm square. Reprinted from Haga *et al.* [35], with permission from Elsevier.

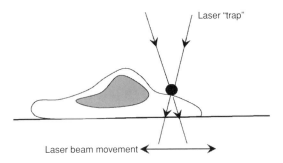

Laser "trap"

Laser beam movement

Figure 2.21

Schematic description of the "optical tweezers" manipulating a bead that is attached to a cultured cell. Bead position is monitored by microscopy and so the relative position of the bead and the optical trap center are known.

be fit by Equation (2.5) to obtain z_0 and E. It will be appreciated that the value of E so obtained reflects a local Young's modulus and can be expected to vary over the surface of the cell. For example, the local stiffness has been related to the local cytoskeletal structure in fibroblasts (Fig. 2.20).

What sort of maximum spatial resolution can we expect with the AFM? A typical probe tip radius is 10–50 nm, but unfortunately this lateral resolution is not achievable when measuring cellular stiffness. That is because the probe sinks into the relatively soft cell and consequently measures over a larger area than just the tip. The magnitude of this effect depends on the applied force and the cell stiffness; typical values for lateral resolution are in the range of tens to several hundreds of nanometers [32]. This is small enough to give reasonable resolution when mapping stiffness over a cell.

2.5.2 Optical trapping ("optical tweezers")

Photons carry momentum; consequently when light shines on a surface there is an effective force exerted on that surface. Usually this effect is very small and can be neglected. However, for intense light shining on a small particle, forces in the range of 1 to 200 pN can be generated. An extension of this concept is to create a specially focussed light beam that creates a potential "well" that traps a bead or small particle, typically 1–2 μm in diameter.

This principle can be used for biological measurements by coating the bead with fibronectin (or some other molecule that will bind to receptors on the cell surface) so that the bead adheres to a given location on the cell. The beam is then moved laterally, and the motion of the bead is observed microscopically (Fig. 2.21). From knowledge about the characteristics of the light trap, the force exerted on the bead by the moving light beam can be determined from the bead position relative to

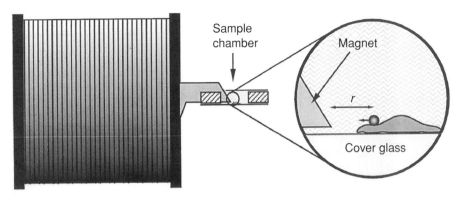

Figure 2.22

Schematic description of the central measuring unit of a magnetic bead microrheometer based on an electromagnet. The electromagnet is attached to a microscope stage and consists of a coil (1200 turns of 0.7 mm copper wire) and a soft iron core. This core extends beyond the windings, forming a pole piece that penetrates the sample chamber. The tip of the pole piece can be positioned at distances of (r) of 10 to 100 μm from a magnetic bead attached to a cell. This produces maximal forces of 10 000 pN on a paramagnetic bead of 4.5 μm diameter. From Bausch *et al.* [42] with kind permission of the authors and the Biophysical Society.

the center of the optical trap. Since the displacement of the bead is simultaneously monitored, this information can be used to determine the local stiffness of the cell. Essentially the same technique can be used to measure the force exerted by individual molecules. Optical trap techniques have been used for a variety of biomechanical measurements; for reviews see [36–38].

2.5.3 Magnetic bead microrheometry

In the magnetic bead microrheometry technique, a paramagnetic bead, typically 4–5 μm in diameter, is coated with fibronectin or some other suitable molecule. The fibronectin binds to integrins on the cell surface, providing a direct link between the bead and the actin cytoskeleton of the cell (Fig. 2.14). The paramagnetic bead is then subjected to a magnetic field that either twists it (*magnetic twisting cytometry* [39–41]) or displaces it (*magnetic bead microrheometry*; Fig. 2.22). Here we will concentrate on magnetic bead microrheometry, where the magnetically induced motion of the bead is visualized by light microscopy. Information about bead displacement and applied force is used to estimate local mechanical properties of the cell. One advantage of this approach is that it can produce forces in the 100–10,000 pN range, which is relatively large. An extension of this approach involves "seeding" the surface of the cell with latex microspheres and watching their displacement in the neighborhood of the magnetic bead. This gives information about the length scale over which displacements are coupled within the cell.

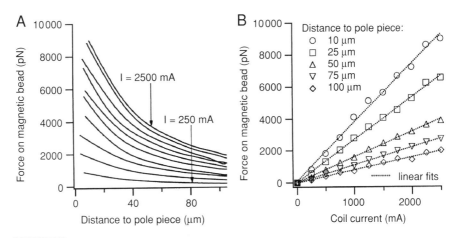

Figure 2.23

An example of force calibration data for the microrheometer shown in Fig. 2.22. (A) Force versus distance for a bead of diameter 4.5 μm, plotted for coil currents between 250 and 2500 mA. (B) The same data replotted as force versus current for distances between 10 and 100 μm. A linear relationship between the coil current and force on the paramagnetic bead is evident. From Bausch *et al.* [42] with kind permission of the authors and the Biophysical Society.

An important issue is how the system will be calibrated. This is accomplished by suspending a bead in a liquid of known viscosity and applying a magnetic field to the bead. The resulting displacement of the bead can be tracked optically as a function of time, and then Stokes' law can be used to estimate the force on the bead.[5] From this, the force profile as a function of applied current and distance from the magnet's pole piece can be determined (Fig. 2.23).

Figure 2.24 shows some of the data gathered using this method. It is clear that when a force is applied to the bead, there is a nearly immediate displacement of the bead (phase I in Fig. 2.24), followed by a gradual creep (phases II and III). This is consistent with the cytoskeleton and cytoplasm demonstrating viscoelastic behavior. In Section 2.6.1 we show how construction of suitable models allows us to determine stiffness and damping coefficients for the elastic and viscous elements in the cell from data such as those shown in Fig. 2.24. An example of how this technique can be used to image the strain field around a single paramagnetic bead bound to a cell is shown in Fig. 2.25.

2.5.4 Micropipette aspiration

The final methodology that we will consider is micropipette aspiration. This is one of the oldest techniques for measuring cellular (and subcellular) biomechanical

---

[5] Stokes' law states that the terminal velocity experienced by a spherical particle of radius a acted upon by a force F as it translates in a fluid of viscosity μ is given by $U = F/(6\pi\mu a)$, so long as the particle Reynolds number, $Re_a = Ua\rho/\mu$, is much less than one.

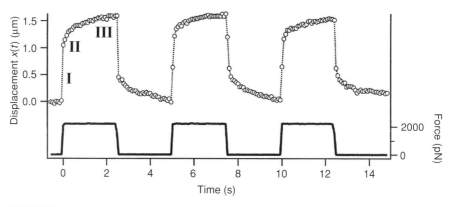

Figure 2.24

Typical creep response and relaxation curves measured for a 4.5 μm diameter bead bound to the membrane of a 3T3 fibroblast. The bead was coated with fibronectin and is therefore presumed to be bound to an integrin receptor. Force pulses of amplitude $F = 2000$ pN and duration $\Delta t = 2.5$ s were applied, as shown in the lower portion of the graph. A rapid (elastic) displacement is evident in phase I, followed by creep in phases II and III. From Bausch *et al.* [42] with kind permission of the authors and the Biophysical Society.

properties, dating back to the middle of the twentieth century [43]. There are different ways in which micropipette aspiration can be used, but all approaches employ a very fine glass micropipette, typically having internal diameter of 1–10 μm and with a tip that can be moved about by a micromanipulator. A small, known pressure can be applied to the pipette through a fluid-filled reservoir attached to the pipette. By controlling the reservoir height, this pressure can be altered during the course of an experiment (Fig. 2.26).

In the simplest form of micropipette aspiration, the pipette tip is brought into contact with a cell and a small suction (aspiration) pressure is generated by the reservoir. Direct microscopic observation reveals the cell's deformation and motion in real-time as it is acted upon by the micropipette. From knowledge of the reservoir height and the pipette tip cross-sectional area, the force applied to the cell can be determined. This apparatus can generate forces ranging from 10 pN to about 10^4 nN, which is sufficient to cause appreciable cellular deformation (Fig. 2.27). The smallest resolvable force in this technique depends on the precision with which the reservoir can be positioned; typical vertical positioning accuracy is several μm, which translates into a force of order 1–10 pN for a 10 μm diameter pipette.

Biomechanical measurements using this technique have been made on several cell types, including neutrophils [28,44,46–50], red cells [28,51–58] and outer hair cells [45,59–61]. Micropipette aspiration is particularly well suited for measurement of mechanical properties of the mammalian red cell, since erythrocytes are one of the mechanically "simplest" cell types. Let us consider red cell

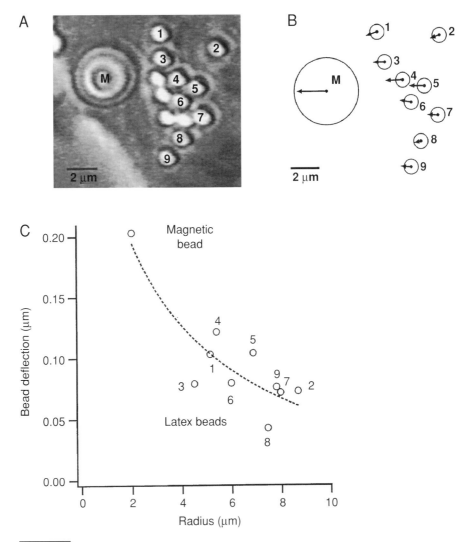

Figure 2.25

Sample data showing how the displacement field induced by a paramagnetic bead is mapped on the plasma membrane of a fibroblast. (A) Microphotograph showing one magnetic bead (M, radius 2.25 μm) and several non-magnetic particles (1–9, radius 0.5 μm) attached to a region of the cell membrane. Notice that there are several non-numbered particles. These are ignored, since their deflection amplitude could not be accurately measured due to technical problems. (B) Scale drawing of the corresponding displacement field after a 1 second duration force pulse. The bead deflections are enlarged by a factor of 10. (C) Plot of the bead deflection ($u_r/\cos\theta$) versus distance between the center of the paramagnetic bead and the latex beads, r. Here u_r is the displacement of a latex bead in the direction of the magnetic bead and θ is the angle between the r-vector and the magnetic field vector. The dotted line is the results of a model of cellular deformation. It can be seen that the deflections $u_r(r)$ show a lot of scatter about the theoretical result. This might be caused by structural cytoskeletal components within the cell altering the local deformation field, by variations in the strength of attachment of the small beads to the integrin receptors, or by differences in the strength of the coupling between the integrin receptors and the cytoskeleton. From Bausch *et al.* [42] with kind permission of the authors and the Biophysical Society.

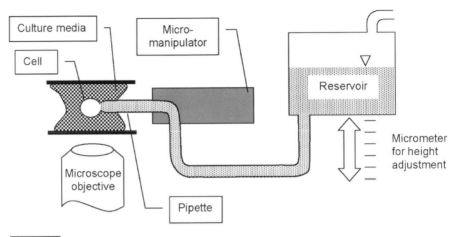

Figure 2.26

Schematic overview of apparatus used for micropipette aspiration of a living cell (not to scale). Adapted from Shao and Hochmuth [44] with kind permission of the authors and the Biophysical Society.

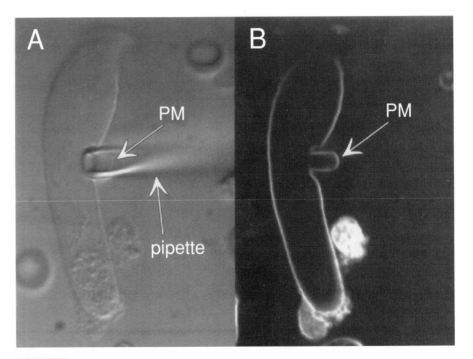

Figure 2.27

Micropipette aspiration studies of guinea pig outer hair cells. The outer hair cell resides within the cochlea and is responsible for several biomechanical functions, most importantly in the amplification of incoming sounds through active remodeling. It is one of the most intriguing cell types from a biomechanical viewpoint. In this set of images, the plasma membrane of the hair cell has been labeled with a fluorescent dye (Di-8-ANEPPS). (A) A conventional transmitted light image of a hair cell undergoing micropipette aspiration at a suction pressure of -11 cmH$_2$0. (B) Corresponding fluorescent image, where the aspiration of the cell's plasma membrane (PM) into the pipette lumen is clearly visible. From Oghalai et al. [45]. Copyright 1998 by the Society of Neuroscience.

biomechanics in more detail, which requires a small detour to describe the structure of the erythrocyte.

Formation of a human erythrocyte begins when stem cells within the bone marrow differentiate to create an erythroblast. This cell expels its nucleus to produce a reticulocyte, which then enters the circulation and loses its organelles to become an erythrocyte. The erythrocyte has only a membrane-associated (cortical) cytoskeleton, with no cytoskeletal network extending into the interior of the cell. The major component of this cortical cytoskeleton is a fibrous protein called *spectrin*, which forms tetramers that then cross-link to other tetramers to form a "hub-and-spoke" network. The cross-linking is provided by a protein complex that contains short actin filaments and a variety of other proteins (Fig. 2.28, color plate). The entire network is attached to integral membrane proteins by a protein called *ankyrin*, so that the membrane is reinforced by, and tightly bound to, the cortical cytoskeleton. This architecture means that the red cell is very flexible (and hence able to fit through small capillaries throughout the circulatory system). Biomechanically, the absence of any significant internal structure means that the cell can be treated as a reinforced "bag of hemoglobin."

When red cells are aspirated by a micropipette, a portion of the red cell is drawn inside the lumen of the pipette and is elongated (Fig. 2.29). If the applied suction pressure is low enough to avoid red cell rupture, an equilibrium is established in which the applied suction pressure is balanced by mechanical stresses within the cortical cytoskeleton of the red cell. One measure of the cell's elastic properties is the distance that this aspirated segment extends into the pipette, L. Experimental measurements of this distance as a function of the applied pressure are shown in Fig. 2.30, where we see that there is an approximately linear relationship between L and applied pressure, at least over the range of pressures plotted.

Another useful property that can be measured in an aspiration experiment is the density of spectrin as a function of position on the cell's surface. The deformed red cell has a non-uniform spectrin density because the strain field over the red cell surface is non-uniform: in some areas the spectrin network is highly extended, while in others it is less extended or even compressed. This can be quantified by fluorescently labeling components of the cortical cytoskeleton and then measuring the fluorescence distribution before and after micropipette aspiration. Discher and colleagues [64] presented such measurements by plotting the mass of spectrin per unit area of membrane, normalized to the value far from the pipette, for different aspirated segment lengths. Figure 2.31 shows that the density of spectrin near the tip of the aspirated segment can fall to approximately 30% of its normal value, while the spectrin density near the entrance to the pipette can nearly double. It is clear that the strain magnitudes within the red cell cytoskeleton are huge compared with "traditional" engineering applications.

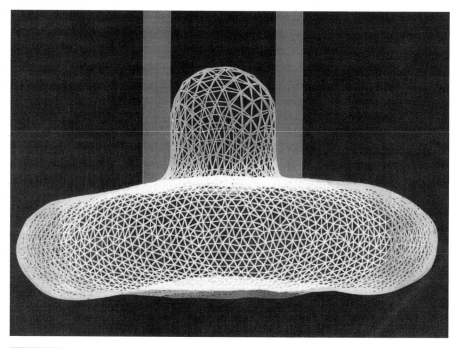

Figure 2.29

Simulation of a flaccid red cell being aspirated by a micropipette. The spectrin network within the cell is modeled as a network of non-Hookean springs, shown in this image as white segments. There are 6110 vertex nodes, each of which is connected to six neighbors by spectrin strands. From Discher *et al.* [62] with kind permission of the authors and the Biophysical Society.

Can we use data of the type shown in Figs. 2.30 and 2.31 to deduce mechanical properties of the cortical cytoskeleton? The answer is yes. However, because of the complexity of the cytoskeletal deformation, this is far from trivial (even for the "simple" erythrocyte), and requires a model of cytoskeletal deformation during aspiration. Interested readers are referred to the excellent summary provided by Boal [65] for further details; here we simply summarize the main results. It is fairly easy to show that bending stresses contribute a relatively small amount to the mechanical behavior of the cell, at least for typical pipette diameters [65]. Therefore, the most important modes of deformation of the cell membrane/cytoskeleton during pipette aspiration are an in-plane shear and an in-plane dilation (area change).[6] Because everything is happening in the local plane of the

---

[6] A shearing deformation is one that preserves the area of a small element of membrane, but changes internal angles. A dilatational deformation changes the area of the membrane element without changing internal angles. See Boal [65] for a more complete discussion of this point.

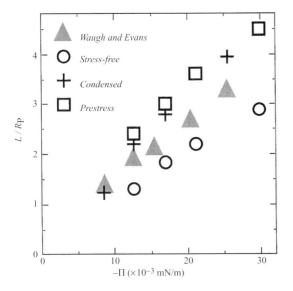

Figure 2.30

Length of the aspirated segment, L, as a function of aspiration pressure for micropipette aspiration of a flaccid human red cell. The aspirated segment length has been non-dimensionalized by inner pipette radius R_p. The quantity Π is defined to be $2PR_p$, where P is the aspiration pressure (taken to be < 0 for suction). This implies that Π values < 0 correspond to tension in the cortical cytoskeleton/membrane. The grey triangles are experimental data from Waugh and Evans [63]; other symbols are simulation results from Discher *et al.* [62] using the model shown in Fig. 2.29. From Discher *et al.* [62] with kind permission of the authors and the Biophysical Society.

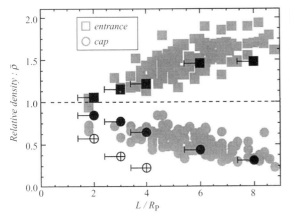

Figure 2.31

Relative density of cortical cytoskeleton as a function of aspirated segment length for the tip region of the aspirated segment ("cap") and near the entrance to the pipette ("entrance"). The relative density is the amount of cytoskeleton per unit membrane area normalized by the value far from the pipette. Grey symbols are experimental measurements based on fluorescence (see text) from Discher *et al.* [64]; other symbols are simulation results from Discher *et al.* [62] using the model shown in Fig. 2.29. From Discher *et al.* [62] with kind permission of the authors and the Biophysical Society.

membrane/cytoskeleton, it makes sense to work with stresses defined as forces per unit length, rather than the more familiar force per unit area. In this approach, we draw a small line segment on the cell membrane and express the in-plane stress as the force acting across that line segment divided by the length of the segment. This implies that elastic moduli will also have units of force per unit length (or energy per unit membrane area), rather than the more familiar force per unit area we use for bulk materials. For a planar isotropic material, there are two such constants: an area deformation modulus, K_A, and a shear modulus, K_S. Experimental measurements indicate that the area modulus is twice the value of the shear modulus and that the shear modulus is 6×10^{-6} to 9×10^{-6} J/m^2 (see summary in Boal [65]). This is approximately three to four orders of magnitude smaller than values measured in some other cell types (outer hair cells, fibroblast), consistent with the idea of the erythrocyte being a highly deformable cell.

Extensions of the micropipette aspiration technique

The micropipette aspiration technique described above is a powerful experimental tool. It can be made more powerful by using a second pipette, which is employed to manipulate a second cell or a microsphere coated with specific molecules. Here we will consider the case in which one of the pipettes manipulates a microsphere, while the other pipette manipulates a cell. By holding the microsphere fixed, allowing the cell and the microsphere to contact one another, and then measuring the force needed to pull the microsphere away from the cell, the strength of specific ligand–receptor bonds can be measured. For example, Shao and Hochmuth [44] used microspheres coated with antibodies directed against cell-surface molecules present on neutrophils, specifically CD18 (β2-integrins), CD62L (also known as L-selectin), or CD45. They then brought one of these coated microspheres into contact with a human neutrophil, immobilized the microsphere with one pipette, and aspirated the neutrophil into a second pipette with a diameter that was very slightly greater than the neutrophil diameter. They observed that if the aspiration force was less than approximately 45 pN, the cell and bead remained attached and essentially motionless for several seconds before moving apart. However, if the applied force was greater than this, the neutrophil and the bead began to move apart from each other, but at a slower rate than would be expected if they were uncoupled. How is this possible? The answer is that the neutrophil extends a thin process, called a membrane tether, which acts like an elastic string connecting the bead and the cell. This allows the neutrophil to be aspirated into one of the pipettes, but at a speed less than would occur if the neutrophil were moving freely. Eventually the tether detaches from the bead, and then the neutrophil moves freely once again.

In order to characterize the aspiration force acting on the aspirated neutrophil in this case, it is necessary to consider the case of a non-stationary cell within the lumen of the pipette. Suppose that U_t is the observed velocity of the cell moving within the pipette and U_∞ is the velocity that the cell would have if there was no tethering force applied to the cell, for the same suction pressure. Then the force exerted on the cell in a micropipette of internal radius R_p by an imposed suction pressure Δp is given by

$$F = \Delta p \, \pi \, R_p^2 \left(1 - \frac{U_t}{U_\infty} \right) \tag{2.6}$$

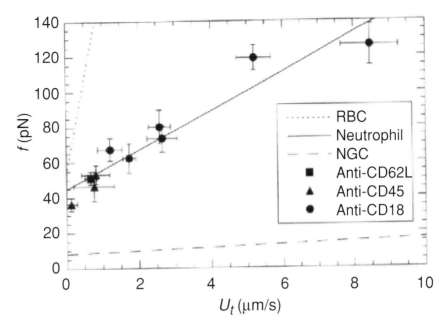

Figure 2.32

Results of experiments in which human neutrophils were pulled away from coated microspheres. It can be seen that there is a linear relationship between the force applied to the neutrophil (f) and the resulting neutrophil speed, U_t. Different symbols represent different antibodies immobilized on the surface of the microsphere, as described in the text. Also shown are results for tethers created by aspiration of red blood cells (RBC; from Waugh and Bauserman [66]) and from neuronal growth cones (NGC; from Dai and Sheetz [67] and Hochmuth et al. [68]). From Shao and Hochmuth [44] with kind permission of the authors and the Biophysical Society.

This formula assumes that the cell diameter is only slightly less than the micropipette internal diameter, and it neglects several effects, including the small pressure drop along the length of the pipette and the detailed fluid mechanics in the thin region between the cell and the micropipette wall. However, for typical sized micropipettes, this leads to an error of less than 5% (see Appendix D of Shao and Hochmuth [44]) and therefore we can use Equation (2.6) directly. Note that if the cell is fixed in place ($U_t = 0$) then the applied force is just the cell's projected area, πR_p^2, times the aspiration pressure drop.

Figure 2.32 shows a plot of applied force (calculated from Equation [2.6]) versus neutrophil velocity for experiments in which tethers were observed. It is interesting that data obtained from three different antibodies against neutrophil surface molecules lie on the same line, suggesting that the slope of the fitted line is representative of the inherent stiffness of the tether. Is tether formation just a laboratory

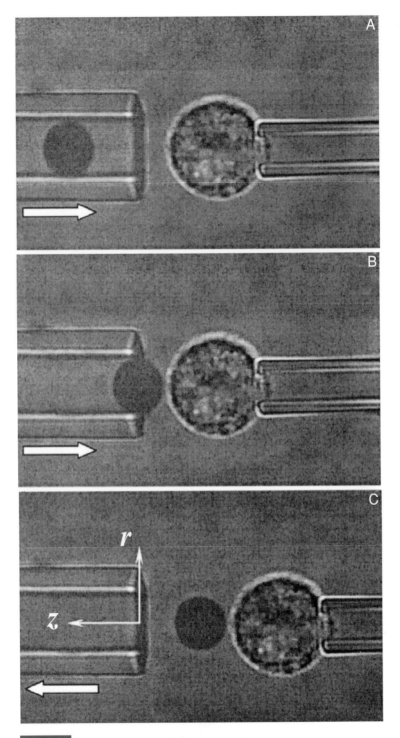

Figure 2.33

Photographs showing an extension of the two pipette aspiration technique for studying cellular and molecular biomechanical properties. (A) A neutrophil is immobilized by the pipette at right while a second pipette at left forces a coated microsphere towards the neutrophil. (B) The microsphere and cell are in contact. (C) The microsphere has adhered to the cell, which is moved slightly away from the left pipette. An aspiration pressure is then applied to the left-hand pipette to draw fluid into it. The white arrows show the direction of fluid flow in the left pipette. Reproduced with permission from Shao [48].

curiosity? No, it is not. One of the important functions that neutrophils have is to enter the vessel wall at sites of inflammation, especially in the postcapillary venules. They do this by adhering to specific molecules on the endothelium (primarily P-selectin glycoprotein ligand-1, which binds the neutrophil surface molecule P-selectin) and rolling along the endothelium until they are activated and increase the affinity of their binding to the endothelium [69]. This causes them to come to arrest (stop rolling), whereupon they can enter the vessel wall and participate in mediating the immune response. The process of rolling along the vessel wall and then arresting seems to depend critically on tether formation, and the forces measured in Fig. 2.32 are consistent with observed neutrophil rolling velocities [44]. The interaction between neutrophils and platelets also involves tethers [70].

One of the limitations of the above approach is that the force resolution may be too coarse to resolve molecular binding events adequately. Shao and co-workers [48,49] have proposed a further extension of the two pipette technique: a coated bead is once again manipulated by a second micropipette and allowed to come in contact with, and bind to, an immobilized cell. However, after binding, the bead is not aspirated into the second pipette but instead is positioned outside, but in the vicinity of the tip of, the second micropipette (Fig. 2.33). Suction pressure is then applied to the second micropipette, inducing fluid inflow into the mouth of the second pipette. This fluid flow exerts a net hydrodynamic force on the bead, and by modulating the magnitude of the suction pressure (and hence the flow and hydrodynamic force) and observing when the bead detaches from the cell, the magnitude of the cell–bead binding force can be deduced.

The magnitude of the force exerted by the flowing fluid on the bead cannot be calculated by a simple closed-form expression of the form of Equation (2.6). However, these forces can be computed, for example, by finite element modeling, which shows that the magnitude of the forces is in the range of tens to hundreds of femtonewtons (1 femtonewton = 10^{-15} N).

2.6 Models of cellular biomechanical behavior

Experimental data are clearly necessary for characterizing the mechanical behavior of cells. However, it is often desirable (and sometimes even essential) to develop models of the mechanical behavior of cells. Such mechanical models help us to understand and interpret experimental observations. They are also useful for predicting the response of a cell to a variety of inputs and for comparing responses of different cells.

In the following sections, we consider four models of the mechanical behavior of cells. The first model considers the cell as a viscoelastic body essentially made up of viscous protoplasm surrounded by a cell membrane. This approach considers the mechanics of the whole cell, without considering the contribution of individual cellular components, such as the cytoskeleton. The second and third models focus on how the mechanical properties of a cell are determined by the structure and composition of its cytoskeleton. Each of the first three models considers the behavior of the cell in isolation. In reality, cells reside within an extracellular matrix (as discussed in Section 2.4) and the properties of the matrix and its interaction with the cell will affect the mechanical behavior of the cell in response to force applied to a tissue. The fourth model we present demonstrates this point using experimental data and finite element analyses.

2.6.1   Lumped parameter viscoelastic model of the cell

The data from the magnetic bead rheometry experiments indicate that the cell's response to a step application of force is viscoelastic, with an immediate elastic response (phase I in Fig. 2.24) followed by gradual creep (phases II and III). The viscoelastic behavior of materials (including cells and many biological tissues) can be modeled using lumped parameter models constructed from arrangements of two simple elements: the linear spring and the dashpot. In this section, we construct a viscoelastic model of the cell and compare its predictions with those obtained experimentally (Fig. 2.24). It is important to understand that the elements in these models cannot be unambiguously associated with specific cellular components; instead, we represent the action of all elastic components of the cell by a linear spring (or springs) and the action of all viscous components by a dashpot (or dashpots). This "lumping" of the response of a complex biomechanical system into a small number of elements is obviously a simplification. Nonetheless, such lumped parameter models are easy to work with and represent a good first approach to quantifying cellular biomechanics.

The two basic elements used in lumped parameter viscoelastic models (also know as equivalent circuits) are the linear spring and the dashpot. When a force, $F(t)$, is applied to a linear spring, it responds instantaneously with a deformation, x, that is proportional to the load:

$$F_{spring}(t) = k_0 x_{spring}(t), \tag{2.7}$$

where k_0 is the spring constant. The reader is reminded that x_{spring} is the spring length, *measured from the equilibrium (or resting) length of the spring*, which is the length that the spring takes when no forces are applied to it. A dashpot is a viscous element, similar to a shock absorber. When a force is applied to a dashpot,

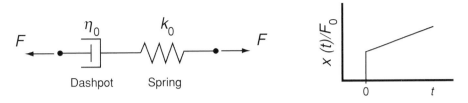

Figure 2.34

Lumped parameter model (or equivalent circuit) of a Maxwell body and the corresponding creep response to a step change in the applied force, F, from zero to F_0 at $t = 0$.

the rate of deformation is linearly related to the force:

$$F_{\text{dashpot}}(t) = \eta_0 \dot{x}_{\text{dashpot}}(t), \qquad (2.8)$$

where the dot represents differentiation with respect to time and η_0 is a constant of the dashpot called the damping coefficient. Notice the sign convention implied by the above equations: the direction for positive force must coincide with the direction of positive displacement.

These elements are obviously idealized, but their force–displacement characteristics are able to describe essential features of linear viscoelastic behavior, particularly when combined in specific arrangements. One common arrangement is a series combination of a spring and dashpot, known as the Maxwell body (Fig. 2.34). The total deformation of the Maxwell body, $x(t)$, is the sum of the deformation of the spring and that of the dashpot:

$$x(t) = x_{\text{spring}}(t) + x_{\text{dashpot}}(t). \qquad (2.9)$$

By differentiating Equation (2.9) with respect to time, it is clear that the velocity of deformation is the sum of the velocities of the two elements. Furthermore, since the elements are in series, the force in the spring, $F(t)$, is transmitted to the dashpot, i.e., $F(t) = F_{\text{spring}}(t) = F_{\text{dashpot}}(t)$. By substituting the force–displacement relationships from Equations (2.7) and (2.8) into the time derivative of Equation (2.9), we can get a force–displacement relationship for the Maxwell body:

$$\dot{x}(t) = \frac{1}{\eta_0} F(t) + \frac{1}{k_0} \dot{F}(t). \qquad (2.10)$$

This represents a differential equation that can be solved for the displacement, $x(t)$, if the force history, $F(t)$, is known.

Before we can integrate this equation, we need an initial condition, which is obtained by thinking about how the Maxwell body responds to the applied force.

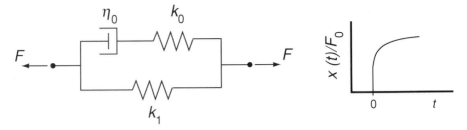

Figure 2.35

Lumped parameter model of a Kelvin body and its creep response in response to a step change in force F from zero to F_0 at time $t = 0$.

A common force history is to impose a step change in force, for example an increase in force from zero to a constant value F_0. In this case, the resulting response of the viscoelastic body is called its creep response. When a force is applied in this way, the spring element will deform instantaneously according to Equation (2.7). However, the dashpot will not deform immediately; it acquires a finite *velocity* immediately, but the displacement at time $t = 0$ is zero because a finite velocity acting over an infinitesimal duration produces no displacement. Therefore, if the force is applied at time zero, the initial condition for Equation (2.10) is $x(0^+) = F(0^+)/k_0 = F_0/k_0$.[7] If we restrict attention to times $t > 0$ then $F(t) = F_0$ and $\dot{F} = 0$ in Equation (2.10), so that the differential equation is trivially solved to obtain the displacement of the Maxwell body per unit applied force as:

$$\frac{x(t)}{F_0} = \frac{1}{k_0} + \frac{1}{\eta_0}t. \tag{2.11}$$

The contributions of the spring and the dashpot are evident from this response (Fig. 2.34), with an immediate elastic response from the spring followed by linear deformation of the dashpot with time. Unfortunately, the Maxwell body clearly does not accurately represent the cell response shown in Fig. 2.24, indicating that this simple model is insufficient.

A slightly more complex model is the Kelvin body, which consists of a Maxwell body in parallel with a spring (Fig. 2.35). In this case, because the elements are in parallel, the deformation of the entire Kelvin body is equal to the deformation of each of the two parallel paths. Therefore, the deformation of the Kelvin body is equivalent to that of the Maxwell body and is also given by Equation (2.10), with the understanding that $F(t)$ in Equation (2.10) should now be interpreted not as the total applied force, but just the force supported by the Maxwell body,

---

[7] The notation $t = 0^+$ means that we should take the limit as t tends towards zero from positive times. In other words, this is the time immediately after application of the force at time $t = 0$.

$F_{\text{Maxwell}}(t)$. $F_{\text{Maxwell}}(t)$ can be computed from

$$F(t) = F_{\text{Maxwell}}(t) + F_{\text{spring}}(t), \tag{2.12}$$

where F_{spring} is the force across the spring with stiffness k_1 and $F(t)$ is the total force applied to the Kelvin body.

By rearranging Equation (2.12) and substituting a suitably modified form of Equation (2.7) for $F_{\text{spring}}(t)$, the analogue to Equation (2.10) for a Kelvin body is:

$$\dot{x}(t) = \frac{1}{\eta_0}(F(t) - k_1 x(t)) + \frac{1}{k_0}(\dot{F}(t) - k_1 \dot{x}(t)) \tag{2.13}$$

or, after some rearranging,

$$F(t) + \frac{\eta_0}{k_0}\dot{F}(t) = k_1\left(x(t) + \frac{\eta_0}{k_1}\left(1 + \frac{k_1}{k_0}\right)\dot{x}(t)\right), \tag{2.14}$$

where $x(t)$ is the total displacement of the body.

As for the Maxwell body, we solve for the creep response by imposing a step force and considering times $t > 0$, in which case $F(t) = F_0$ and $\dot{F} = 0$. For this case, Equation (2.14) simplifies to:

$$F_0 = k_1\left(x(t) + \tau \dot{x}(t)\right), \tag{2.15}$$

where we have defined τ, the relaxation time, as:

$$\tau = \eta_0 \frac{k_0 + k_1}{k_0 k_1}. \tag{2.16}$$

To solve the differential Equation (2.15) for $x(t)$, we need an initial condition. Once again we recognize that the displacement of the dashpot is zero at time $t = 0^+$, so that the force must be carried by the springs only, i.e., $F_0 = k_0 x(0^+) + k_1 x(0^+)$. In this case the solution of Equation (2.15) is:

$$\frac{x(t)}{F_0} = \frac{1}{k_1}\left(1 - \frac{k_0}{k_0 + k_1}e^{-t/\tau}\right). \tag{2.17}$$

The Kelvin body does a much better job of representing the creep behavior of the cell subjected to magnetic bead traction than does the Maxwell body (compare Figs. 2.24 and 2.35). However, there is still a discrepancy between the model predictions and the experimental data for cellular deformation during phase III (Fig. 2.24).

The model predictions can be further improved by adding a dashpot in series with the Kelvin body (Fig. 2.36), giving a total of four components: two dashpots and two springs. The creep response for this model is simply the superposition of the response of the Kelvin body and the series dashpot, and therefore is given

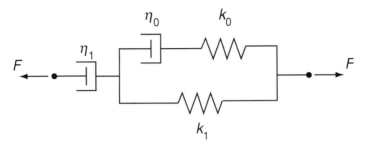

Figure 2.36

Lumped parameter model proposed by Bausch *et al.* [42] for the viscoelastic creep response of a cell subjected to a sudden force by magnetic bead rheometry. This model is used to interpret the data shown in Fig. 2.24.

by:

$$\frac{x(t)}{F_0} = \frac{1}{k_1}\left(1 - \frac{k_0}{k_0 + k_1}e^{-t/\tau}\right) + \frac{t}{\eta_1}. \tag{2.18}$$

As can be seen in Fig. 2.37, the model provides an excellent fit to the experimental data obtained by Bausch *et al.* [42]. The quality of the fit suggests that there must be (at least) two components of the cell that are responsible for generating viscous responses and (at least) two components that generate an elastic response. Of course, this model tells us nothing about what those components might be. It is even possible that all of the viscous behavior could come from a single component that relaxes with two different time scales.

From such fits, viscoelastic constants that describe the creep response of a cell can be estimated and used to compare the response of one cell with that of another cell, or to determine the response of a cell to a variety of inputs. Bausch *et al.* [42] did these comparisons and found that while the viscoelastic constants measured at different sites on an individual fibroblast cell are approximately equal (for forces up to 2000 pN), the constants can vary by up to an order of magnitude from cell to cell. Why such a variation from cell to cell? The viscoelastic constants estimated by the lumped parameter model are indicative of the properties of the whole cell, but those properties are determined in part by the properties of the cell membrane, in part by the properties of the cell cytoplasm, and in part by the structure of the cytoskeleton. As we discussed in Section 2.3, the cytoskeleton is a dynamic structure that is constantly changing its composition and organization, meaning the cytoskeletal architecture in one cell can differ substantially from that in another. We will show in the next two sections that these structural differences can have profound effects on the mechanical properties of the whole cell, explaining some of the variations seen in the viscoelastic properties from one cell to another.

Up to this point we have considered only step increases in force. An alternate way of forcing magnetic beads attached to a cell is to subject them to an oscillating

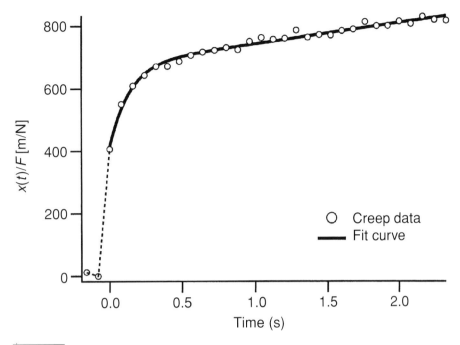

Figure 2.37

Comparison of experimentally obtained creep data (similar to that of Fig. 2.24) with model predictions from the lumped parameter model in Fig. 2.36. The applied force (*F*) was 1100 pN. The dashed line represents the instantaneous elongation of the body ($1/(k_0 + k_1)$) and the solid line is fit to the data using Equation (2.18). From Bausch *et al.* [42] with kind permission of the authors and the Biophysical Society.

force of the form

$$F(t) = F_0(1 + \sin \omega t). \tag{2.19}$$

For a Kelvin body, the initial condition is still $x(0^+) = F_0/(k_0 + k_1)$, but the response of the bead is slightly more complex. It can be shown that the displacement has both exponential and harmonic components, being given by

$$x(t) = \frac{F_0}{k_1} \left(1 - \frac{k_0}{k_0 + k_1} e^{-t/\tau} \right.$$

$$\left. + \frac{\tau\omega\left(1 - \frac{\eta_0}{k_0\tau}\right)(e^{-t/\tau} - \cos \omega t) + \left(1 + \frac{\eta_0\tau\omega^2}{k_0}\right)\sin \omega t}{1 + (\tau\omega)^2} \right) \tag{2.20}$$

where the constant τ is given by Equation (2.16).

2.6.2 Tensegrity model of the cytoskeleton

In this section, we consider an alternative model to describe the mechanical properties of a cell. It is based on the somewhat controversial theory of *tensegrity* and is largely a mechanical theory. (We will discuss some of the biological implications of tensegrity when we consider mechanotransduction in Section 2.7.) We start by defining the term tensegrity, which is a contraction of *tensional integrity*. This is a building technique in which the mechanical integrity of a structure is maintained by internal members, some of which are under tension and others of which are under compression. More formally, "tensegrity structures can be defined as the interaction of a set of isolated compression elements with a set of continuous tension elements [with] the aim of providing a stable form in space" [71]. Examples of tensegrity structures include geodesic domes and our bodies, where the muscles play the role of the tension elements and the bones are in compression.

At the level of the cell, it has been proposed by Ingber and co-workers [72] that actin microfilaments play the role of tension elements and microtubules are the compression elements. There is some experimental evidence that supports this general idea [72]:

- actin microfilaments can generate tension
- there are interconnections between actin microfilaments and microtubules
- microtubules may be under compression.

One difficulty in evaluating the implications of the tensegrity model is that the topology of the interconnected filaments within the cell is very complex. However, insight into the tensegrity mechanism can be obtained if we consider a very simple model, first introduced by Stamenovic and Coughlin [71]. In this approach, the cell is assumed to have only six compression elements (*struts*), oriented as shown in Fig. 2.38. Why six? This is the smallest number of struts that can provide a nontrivial tensegrity system that is spatially isotropic. These six compression elements are joined by 24 tension elements, as shown in Fig. 2.38.

Using this model, we ask a simple question. If the cell is placed under tension, what effective modulus does it have? To answer this question, we apply a tension $T/2$ to each end of the compression members A–A in the x direction (Fig. 2.38). We will assume that the compression members are perfectly rigid and of equal length L_0. Further, we will assume that the tension elements act as linear springs, so that the tension force that they generate can be written as

$$F = k(l - l_r),$$
(2.21)

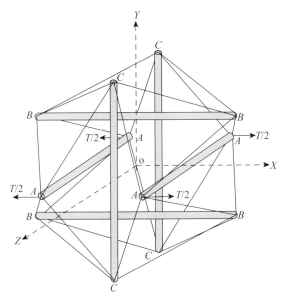

Figure 2.38

Simple tensegrity model of a cell, consisting of six compressive elements (struts) and 24 tension elements. Note the X, Y, and Z coordinate directions. From Stamenovic and Coughlin [71]. Reproduced with kind permission of ASME.

where l_r is the relaxed length of the element, l is its actual length, and k is a constant. We will assume that the length of the tension elements is l_0 (for all elements) when the cell is at rest. Note that even when the cell is at rest, there is tension in the actin filaments; that is, the tension elements are not relaxed when the cell is at rest. This implies that $l_0 > l_r$.

If we define coordinate axes as shown, then everything is symmetric about the origin. We can therefore characterize the strut positions by s_x (total distance between the two struts A–A), s_y (total distance between the two struts B–B) and s_z (total distance between the two struts C–C). From the geometry of the problem, we can express the lengths of the tension elements A–B as:

$$l_{AB}^2 = \underbrace{\left(\frac{L_{BB} - s_x}{2}\right)^2}_{\Delta x^2} + \underbrace{\left(\frac{s_y}{2}\right)^2}_{\Delta y^2} + \left(\frac{L_{AA}}{2}\right)^2. \tag{2.22}$$

It follows that

$$l_{AB} = \frac{1}{2}\sqrt{(L_{BB} - s_x)^2 + s_y^2 + L_{AA}^2}. \tag{2.23}$$

Similar expressions can be written for the lengths of the tension elements A–C and B–C:

$$l_{AC} = \frac{1}{2}\sqrt{(L_{AA} - s_z)^2 + s_x^2 + L_{CC}^2} \tag{2.24}$$

$$l_{BC} = \frac{1}{2}\sqrt{(L_{CC} - s_y)^2 + s_z^2 + L_{BB}^2}. \tag{2.25}$$

Now we consider a force balance (at equilibrium) for each of struts A–A, B–B, and C–C. Balancing forces in the x direction for strut A–A we write:

$$\xrightarrow{\;+\;} \Sigma F_x = m a_x = 0.$$

$$T + \underbrace{4F_{AB}}_{\substack{\text{4 tension} \\ \text{elements} \\ \text{AB}}} \underbrace{\left[\frac{L_{BB} - s_x}{2} \frac{1}{l_{AB}} \right]}_{\substack{\text{Projection of tension} \\ \text{in A–B onto the} \\ x\text{-axis}}} - \underbrace{4F_{AC}}_{\substack{\text{4 tension} \\ \text{elements} \\ \text{AC}}} \underbrace{\left[\frac{s_x}{2} \frac{1}{l_{AC}} \right]}_{\substack{\text{Projection of} \\ \text{tension} \\ \text{in A–C} \\ \text{onto the } x\text{-axis}}} = 0. \tag{2.26}$$

This equation can be simplified, and when Equation (2.21) is substituted into the simplified form we obtain:

$$\frac{T}{2k} + \left(1 - \frac{l_r}{l_{AB}}\right)(L_{BB} - s_x) - \left(1 - \frac{l_r}{l_{AC}}\right)s_x = 0. \tag{2.27}$$

Similar expressions can be written in the y and z directions for struts B–B and C–C, respectively, yielding:

$$0 + \left(1 - \frac{l_r}{l_{BC}}\right)(L_{CC} - s_y) - \left(1 - \frac{l_r}{l_{AB}}\right)s_y = 0 \tag{2.28}$$

$$0 + \left(1 - \frac{l_r}{l_{AC}}\right)(L_{AA} - s_z) - \left(1 - \frac{l_r}{l_{BC}}\right)s_z = 0. \tag{2.29}$$

Together Equations (2.23) to (2.25) and (2.27) to (2.29) represent six equations for seven unknowns: $l_{AC}, l_{AB}, l_{BC}, s_x, s_y, s_z$ and T. If we specify T, we can then solve the system for the displacements and lengths of the tension elements, although it is algebraically messy.

Special case: $T = 0$

If $T = 0$, then by isotropy we have $s_x = s_y = s_z$. Call this length s_0. Also, the lengths of the tension elements are all the same, $l_{AB} = l_{BC} = l_{AC}$. Call this length l_0. If we put all of this information into Equations (2.27) to (2.29) we get $L_{BB} = 2s_x$, $L_{CC} = 2s_y$ and $L_{AA} = 2s_z$. Therefore, we obtain $s_0 = L_0/2$. If we now substitute this into Equations (2.23) to (2.25), we obtain $4l_0^2 = 2(L_0^2/4) + L_0^2 = 3L_0^2/2$, so that $l_0 = \sqrt{3/8}L_0$. This represents the resting (unloaded) state of the cell. We will look for perturbations away from this base state.

Small strain case

We want to compute Young's modulus, E, for the cell. More specifically, we want to compute Young's modulus for the case of infinitesimally small values of the strain; in this case the modulus is denoted as E_0. How can we do this? The easiest way is to consider the energy stored in the cell owing to an incremental extension δs_x caused by a small applied tension T acting in the x direction.

When an object experiences elongation from length s_0 to length s_x caused by a uniaxial force T acting in the x direction, then the work done on that object may be written as $\int_{s_0}^{s_x} T\,\mathrm{d}x$. We introduce the *strain energy* per unit mass, W, which represents the amount of energy stored in the body (per unit mass) owing to deformation. If the undeformed density of the body is ρ_0 and the undeformed volume is V_0, then the product $\rho_0 W V_0$ is the total elastic energy stored in the body, and we can write for a non-dissipative system:

$$V_0 \rho_0 W = \int_{s_0}^{s_x} T\,\mathrm{d}x. \tag{2.30}$$

Differentiation of Equation (2.30) with respect to s_x shows that

$$\frac{\mathrm{d}(\rho_0 W)}{\mathrm{d}s_x} = \frac{T}{V_0}. \tag{2.31}$$

Now, for infinitesimal displacements, the stress acting on a body is given in terms of the strain energy function [73] by:

$$\sigma_{ij} = \frac{\partial\,(\rho_0 W)}{\partial \varepsilon_{ij}} \tag{2.32}$$

where σ_{ij} and ε_{ij} are the ij^{th} components of the stress and strain tensors respectively. It is now convenient to use the fact that $\varepsilon_{xx} = (s_x - s_0)/s_0$ for the special case of uniaxial tension to rewrite Equation (2.32) as:

$$\sigma_{xx} = s_0 \frac{\mathrm{d}(\rho_0 W)}{\mathrm{d}s_x}. \tag{2.33}$$

By combining Equations (2.31), (2.33) and Hooke's law for a body under uniaxial tension in the x direction, we can write

$$E\varepsilon_{xx} = \frac{s_0 T}{V_0}. \tag{2.34}$$

We once again use $\varepsilon_{xx} = (s_x - s_0)/s_0$ to rewrite Equation (2.34) as:

$$E = \frac{s_0^2}{V_0}\frac{T}{s_x - s_0}. \tag{2.35}$$

Finally, for the case of small displacements (and correspondingly small forces T), we can replace E by the incremental Young's modulus E_0, and the ratio $T/(s_x - s_0)$

by the derivative dT/ds_x to obtain:

$$E_0 = \frac{s_0^2}{V_0} \frac{dT}{ds_x}\bigg|_0 \tag{2.36}$$

where the subscript 0 reminds us that we are considering the small deformation case.

We will now apply this result to the tensegrity model of the cell. The resting cell volume (at $T = 0$) can be computed as $V_0 = 5L_0^3/16$, and the reference length is $s_0 = L_0/2$. Treating the cell as a continuum, we can then express the incremental Young's modulus as

$$E_0 = \frac{4}{5L_0} \frac{dT}{ds_x}\bigg|_0 \tag{2.37}$$

Alternatively, recalling that the resting lengths of the tension and compression elements are related by $l_0 = \sqrt{3/8}L_0$, we have

$$E_0 = \frac{2\sqrt{3}}{5\sqrt{2}l_0} \frac{dT}{ds_x}\bigg|_0. \tag{2.38}$$

The last task is to compute the derivative appearing in Equation (2.38), which is algebraically messy and will not be presented in detail here. Briefly, it involves differentiation of Equations (2.23) to (2.25) and (2.27) to (2.29) with respect to s_x, followed by algebraic manipulation. The net result is that the incremental Young's modulus is

$$E_0 = 5.85 \frac{F_0}{l_0^2} \frac{1 + 4\varepsilon_0}{1 + 12\varepsilon_0}, \tag{2.39}$$

where $\varepsilon_0 = l_0/l_r - 1$ is the initial strain in the tension elements (under resting conditions) and $F_0 = k(l_0 - l_r)$ is the corresponding force.

One of the essential features of the tensegrity model is the existence of a non-zero force in the cytoskeletal tension elements, even when the cell is in the resting state. This generates an effective stress within the cell, which we call the *prestress*.[8] The prestress, P, is defined as the net tensile force transmitted by the actin filaments (tension members) across a cross-sectional area of the cell, divided by that area. In computing the prestress, we only consider the component of force in the direction perpendicular to the cross-sectional area. If all the actin filaments in the cell were aligned and under uniform tension F_0, then the prestress in the direction of the actin filaments would be

$$P = \frac{nF_0}{A} \tag{2.40}$$

---

[8] Note that at equilibrium the prestress is balanced by other forces, e.g., compressive forces in other cytoskeletal elements (microtubules), forces from extracellular attachments, etc., as discussed in more detail below.

where n is the number of actin filaments crossing the area A. In the more realistic case of randomly oriented actin filaments, the right-hand side of the above equation has to be divided by a factor of three to account for the component of the force F_0 that is normal to the area A [74]. Furthermore, it is convenient to write F_0 as $a\sigma_c$, where a is the cross-sectional area of a tension member and σ_c is the stress in that member, to obtain

$$P = \frac{na\sigma_c}{3A}. \tag{2.41}$$

Finally, na/A is recognized as the fractional area occupied by actin filaments, or, for a randomly oriented collection of filaments, the volume fraction of tension members, φ. Therefore the prestress can be written as

$$P = \frac{\varphi\sigma_c}{3}. \tag{2.42}$$

Now, for the model shown in Fig. 2.38, $\varphi = 24al_0/V_0$, and the resting volume V_0 can be expressed in terms of l_0 as $V_0 = \frac{5\sqrt{2}}{3\sqrt{3}}l_0^3$. Substituting these expressions into Equation (2.42), we obtain an expression for the prestress, as $P = 5.85\,F_0/l_0^2$ [74]. If we assume that $\varepsilon_0 \ll 1$ (which is likely the case for small applied strain since actin yields at 0.9% strain), then Equation (2.39) shows that an estimate for the incremental Young's modulus is simply

$$E_0 = P. \tag{2.43}$$

Thus, this tensegrity analysis predicts that the Young's modulus of a cell varies linearly with prestress, with an approximate 1:1 correspondence. How do these predictions agree with what is actually observed in living cells?

To answer this question, we first need an estimate of prestress. We could calculate an upper bound on P from an estimate of the yield stress of actin for F_0 and assumptions on the characteristic lengths of the actin filaments (l_0). However, this estimate would be rather crude because of uncertainty in these quantities. Recently, experimental estimates of the prestress in human airway smooth muscle cells cultured on flexible substrates have been made [75,76] and compared with experimentally measured stiffnesses in these cells (Fig. 2.39). From Fig. 2.39, we immediately note that the linear relationship between E_0 and P predicted by the tensegrity model is observed experimentally. However, the slope of the experimental E_0 versus P curve ($\sim$0.4) is less than that predicted analytically (1.0 from Equation [2.43]), and therefore the values for the modulus, in this case measured by magnetic twisting cytometry, are less than those predicted by the tensegrity model. What about modulus measurements from other cell systems and with other devices? Data from several experiments show a high degree of scatter in measured Young's moduli (over five orders of magnitude!), with strong dependence on the

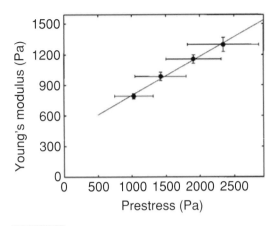

Figure 2.39

Plot of Young's modulus, E_0, versus prestress, P, for human airway smooth muscle cells cultured on flexible substrates. Shear modulus values (G) from the original paper were converted to Young's modulus values by assuming the cell is isotropic and incompressible and therefore, $E = 3G$. Graph adapted from Wang *et al.* [75]. Copyright 2001 National Academy of Sciences, U.S.A.

experimental method used (Fig. 2.40). So what does all this mean, and what are the implications for the tensegrity model?

First, the linear relationship between E_0 and P observed experimentally agrees with that predicted by the analytical model, suggesting that the tensegrity model correctly describes at least some of the physics of the small stress response of the cytoskeleton.

Second, the analytical predictions for the value of Young's modulus (based on measured values of the prestress, P) are in rough agreement with the moduli measured by certain techniques, although the moduli measured by AFM and magnetic bead rheometry are up to two orders of magnitude larger. Unfortunately, it is difficult to assess further the tensegrity model predictions because of at least two major sources of uncertainty:

1. The model predicts that the Young's modulus is equivalent to the prestress in the cell. As can been seen from the wide standard error bars in Fig. 2.39, there is a large degree of uncertainty in the measured values of P and therefore a significant degree of uncertainty in the absolute value of the predicted modulus. This raises the question: What dictates the prestress in the cell? More specifically, can the range of values for experimentally measured Young's modulus be caused in part by variability in prestress? The answer seems to be that it can: prestress in a cell is the net result of forces generated by active contractile forces in the actinomyosin apparatus, and it is balanced by forces

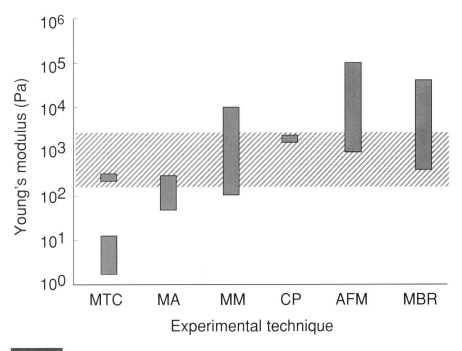

Figure 2.40

Young's moduli measured by various experimental techniques compared with the range predicted by the tensegrity model (shaded region). The range predicted analytically is based on the simple tensegrity model and experimental measurements of prestress [75,76]. Measurements were made by magnetic twisting cytometry (MTC) of endothelial cells, epithelial cells, fibroblasts, and smooth muscle cells; micropipette aspiration (MA) of endothelial cells; micromanipulation (MM, a compression method) of fibroblasts; cell poking (CP) of fibroblasts; atomic force microscopy (AFM) of myocytes, endothelial cells, and cardiocytes; or magnetic bead rheometry (MBR) of fibroblasts. The higher value of Young's modulus measured by MTC is likely more accurate than the lower values, as it was based on an improved technique that minimized experimental artefacts in earlier MTC measurements. Reprinted from Stamenovic and Coughlin [77], with permission from Elsevier.

from adhesion of the cell to its substrate, contact with other cells, internal cell elements under compression (e.g., microtubules), and possibly intracellular cytoplasmic pressure [76]. Changing any of these parameters will change the prestress, and it is likely that these parameters varied significantly between experiments performed by different investigators using different substrates, cell types, cell densities, and experimental conditions. Comparisons of experimental data from different sources are further complicated by the fact that the measured stiffness of a cell depends on the applied stress or strain because cells exhibit prestress-dependent strain stiffening (Fig. 2.41). Again, these parameters typically vary between experiments.

2. The mechanical loading applied to a cell during an experiment depends rather strongly on the measurement technique used. The tensegrity analysis

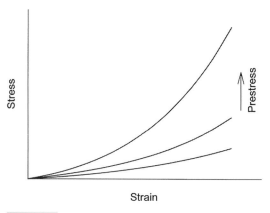

Schematic diagram of stress versus strain plots at different levels of prestress based on magnetic twisting measurements on endothelial cells. Note that for a given prestress, the cell stiffness (slope of the stress–strain curve) increases with increasing strain (strain stiffening). Also, for a given strain, cell stiffness increases with increasing prestress (prestress-induced stiffening). Reprinted with permission from Stamenovic and Wang [78].

presented here does not account for some cellular components and phenomena that may make important contributions to the cell's mechanical behavior under certain loading conditions. For instance, under large applied compressive stresses, actin filaments may actually bend; this feature is not taken into account in the tensegrity model, although it is in the foam model presented in the next section. The tensegrity model also does not include intermediate filaments as load-bearing structures, although we know they must bear forces since they deform in response to fluid shear stress [79]. The cell membrane and the cytosol may also provide mechanical resistance to deformation under certain loading conditions, such as large compressive loading, and these components of the cell are not included in the simple tensegrity model. Therefore, the lack of agreement between the tensegrity model and the measured properties using certain techniques indicates that while tensegrity can reasonably predict cell properties for certain loading conditions, it may not be appropriate for all loading conditions [77]. This is consistent with the observation that the measured moduli seem to depend on the experimental method; that is, each technique is measuring slightly different aspects of the cell's mechanical behavior.

Finally, it is important to realize that the tensegrity model is only one analytical model of cell behavior. As our understanding of the cytoskeleton and cell mechanics improves, we expect that so too will our abilities to model cellular biomechanics and predict the cell's biomechanical behavior analytically.

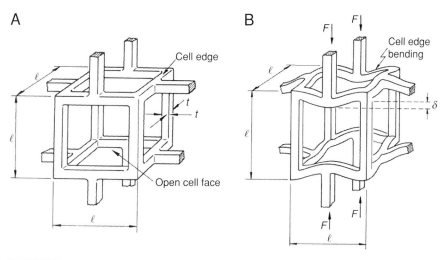

Simplified "unit cell" model of fibrous cross-linked material. (A) The undeformed unit cell has side length *l* and fiber thickness *t*. (B) The deformation of the same unit cell in response to an imposed loading *F* that acts to produce a deformation *δ* in the fibers. From Gibson and Ashby [80], reproduced with permission of the authors and publisher.

2.6.3 Modeling actin filaments as a foam

The tensegrity model that we have presented above represents one possible model of how the cytoskeleton works, but it is certainly not the only one. Here we present a different approach to analyzing the biomechanics of the cytoskeleton, in which the cross-linked actin filament network is treated as a porous random solid matrix with open pores (an *open-celled foam*). The theory of the mechanical behavior of open and closed cell foams is summarized in an excellent book by Gibson and Ashby [80], and most of the following derivation is ultimately based on that source.

It is clear that the cytoskeleton consists of a large number of filaments that are connected to one another with a complex topology (Fig. 2.5). Many naturally occurring and synthetic materials have a similar cross-linked fibrous microstructure, such as paper, felt, cotton wool, and trabecular bone (see Section 9.2). The mechanical behavior of such materials can be described by a simplified model in which connections between fibers are idealized to the form shown in Fig. 2.42. When the material is mechanically loaded, the fibers transmit forces and undergo deformation, as shown in Fig. 2.42B.

Unit cell model of the cytoskeleton
We will use this model to make estimates of the effective Young's modulus and shear modulus of the cytoskeleton based on mechanical properties of individual fibers. We first define the density of the solid fibers as ρ_s and the average density

of the entire network as ρ^*. Note that in defining ρ^* we do not consider the density of cytoplasm filling the interstices between the fibers. The ratio ρ^*/ρ_s is known as the relative density, and from the geometry of the unit cell it can be seen that this ratio is given by

$$\frac{\rho^*}{\rho_s} \propto \left(\frac{t}{l}\right)^2. \tag{2.44}$$

If we think of each fiber as a small beam mechanically loaded by a force F, then the deflection, δ, that the beam undergoes is given by elementary strength of materials as

$$\delta \propto \frac{F\,l^3}{E_s\,I}, \tag{2.45}$$

where E_s is Young's modulus for the fiber material and I is the moment of inertia for the beam. Here we have assumed that the fibers are homogeneous materials undergoing small deformation.

From the geometry of the unit cell, the force F is proportional to the externally imposed stress, σ, multiplied by cross-sectional area of a unit cell, l^2: $F \propto \sigma l^2$. Similarly, the strain, ε, is proportional to δ/l. We can now compute the Young's modulus for the entire cytoskeletal network, E^*, as

$$E^* = \frac{\sigma}{\varepsilon} \propto \frac{F}{l\delta}. \tag{2.46}$$

Substituting Equation (2.45) into Equation (2.46) then allows us to write $E^* \propto E_s I/l^4$. Using the fact that I is proportional to the fiber thickness to the fourth power, t^4, we can use Equation (2.44) to write

$$\frac{E^*}{E_s} = C_1 \left(\frac{\rho^*}{\rho_s}\right)^2, \tag{2.47}$$

where C_1 is a constant of proportionality. Obviously, there were many assumptions made in deriving this equation, especially with regard to the geometry of the unit cell. This makes it very difficult to know a suitable value for the constant C_1. Fortunately, experimental data gathered on a wide variety of different fibrous porous solids show that Equation (2.47) fits the data rather well if $C_1 = 1$ (Fig. 2.43). This gives us some confidence in using this model to study the cytoskeleton.

A similar development can be used to obtain the shear modulus of the network. It is assumed that a shear stress τ is externally applied to the network, causing a global shearing strain γ. Once again, there will be a force F transmitted by the fibers of the network, but it will now act to shear the fibers. From the geometry of the unit cell, we once again have that $\tau \propto F/l^2$; the microscopic deformation of

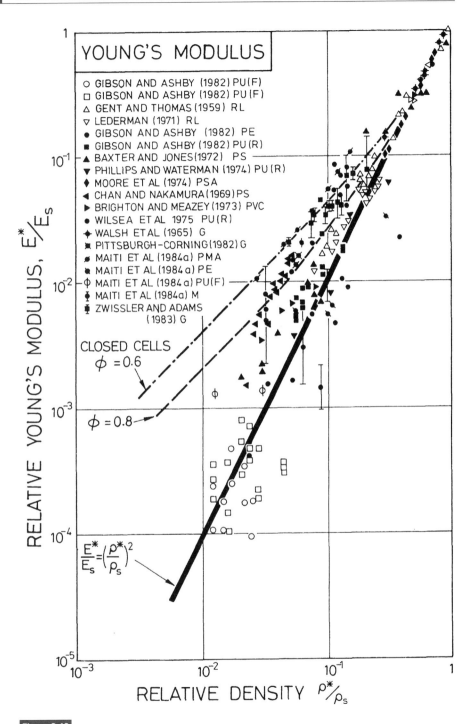

Figure 2.43

Plot of relative Young's modulus (E^*/E_s) versus relative density (ρ^*/ρ_s) for a wide variety of fibrous porous networks (foams). The heavy solid line is a plot of Equation (2.47) for the case of $C_1 = 1$. From Gibson and Ashby [80] in which the original references can be found. Reproduced with permission of the authors and publisher.

individual fibers, δ, is related to the strain by $\gamma \propto \delta/l$, where δ is once again given by Equation (2.45). We compute the network shear modulus, G^*, by

$$G^* = \frac{\tau}{\gamma} \propto \frac{F}{l\delta}. \tag{2.48}$$

Substitution and some algebraic simplification give

$$\frac{G^*}{E_s} = C_2 \left(\frac{\rho^*}{\rho_s}\right)^2 \tag{2.49}$$

where C_2 is a constant, which is empirically equal to $3/8$.

Predictions of actin network modulus

The above results can be used to estimate the mechanical properties of the actin network within a cell. The density of pure F-actin (ρ_s) is 730–850 mg/ml [81] and the density of the F-actin network in an endothelial cell is estimated as $\rho^* = 10$–20 mg/ml. The relative actin density is, therefore, about 1%. Young's modulus for pure actin is estimated to be about 2 GPa (Table 2.1; note that Satcher and Dewey used a value about 10-fold smaller than this in their original paper [81]). Plugging these values into Equations (2.47) and (2.49) gives estimates for E^* and G^* of order 10^5 Pa each. These values are significantly larger than most measurements of cytoplasmic elasticity for whole cells (Fig. 2.40), and even if smaller values for ρ^* are used, the resulting modulus values are at the very upper end of the measured ranges [74]. It is somewhat surprising that the foam model seems to markedly overpredict cellular stiffness, since the ultrastructural architecture of the amorphous actin network within many cells looks like it should be well suited to being treated as a foam. Compounding this difficulty is the fact that the modulus derived from this model should be a lower bound for the measured Young's modulus, since the actin foam model does not include other cytoskeletal elements that can carry stresses, such as microtubules and actin stress fibers. We will not go into the details here, but Satcher and Dewey [81] presented a model that included these other elements, the net effect of which is to increase the effective modulus by a factor of 2–10. The foam model, as currently formulated, also suffers from the drawback that it does not include the effects of prestress [74], although this could be conceivably included as an extension to the basic model.

2.6.4 Computational model of a chondrocyte in its matrix

The preceding three models consider the mechanical behavior of the cell in isolation. As we discussed in Section 2.4, most cells reside in an extracellular matrix

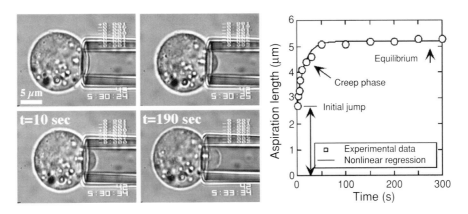

Figure 2.44

Viscoelastic behavior of chondrocytes determined by micropipette aspiration. A step pressure was applied to an isolated chondrocyte using a glass micropipette (left) and measurements of the length of the cell aspirated into the pipette over time were made. Based on a non-linear regression of the aspirated length versus time curve (right) and a lumped parameter viscoelastic model (similar to those in Section 2.6.1), estimates of the instantaneous and equilibrium Young's moduli and the apparent viscosity of the cell were made. Courtesy of Professor Farshid Guilak. Reprinted from Guilak [83], with permission from IOS Press.

(ECM), a composite of biopolymers that determine in large part the mechanical properties of a tissue. When determining how a cell responds when a force is applied to the entire tissue, the properties of the ECM and how it interacts with the cell are important. This is demonstrated in the following example of a chondrocyte in articular cartilage.

Chondrocytes are the cells responsible for synthesizing and maintaining cartilage. They are embedded in an extensive ECM. Chondrocytes experience appreciable strains resulting from the sizeable loads applied to articular cartilage during day-to-day activities. Articular cartilage in the hip, for instance, is cyclically loaded with peak pressures up to 20 MPa during stair climbing [82]. The chondrocytes are surrounded by an "envelope" of specialized connective tissue; together, the cell and this envelope are called the *chondron*. Force transmission from bulk tissue to the cell is "filtered" through the surrounding tissue envelope.

Direct mechanical testing of chondrocytes using a micropipette technique shows that chondrocytes have viscoelastic properties (Fig. 2.44). From these measurements, it is possible to estimate an initial Young's modulus for the cell (E_0), a final steady-state Young's modulus (E_∞), and a cellular viscosity (μ). If the same test is carried out with isolated nuclei, it is seen that the nuclei are much stiffer than entire cells (Fig. 2.45). This implies that the cell will differentially deform in response to an imposed load. Presumably this differential deformation is taken up by the cytoskeleton.

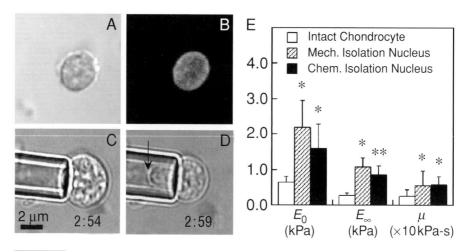

Figure 2.45

Viscoelastic properties of the chondrocyte nucleus. Nuclei from chondrocytes were isolated from the cells by either mechanical (Mech.) or chemical (Chem.) means. Isolated nuclei (shown by [A] differential interference contrast microscopy and [B] fluorescent confocal microscopy) were tested by micropipette aspiration. (C) A step pressure was applied to the resting nucleus at time $t = 0$ resulting in (D) deformation of the nucleus, which reached an equilibrium state 300 s after application of pressure. (E) Summary of the instantaneous (E_0) and equilibrium (E_∞) Young's moduli and the apparent viscosity of properties of the nucleus, which were two to four times greater than those of whole intact chondrocytes. This difference was statistically significant. There were also significant differences in the properties of mechanically and chemically isolated nuclei. Asterisk indicates significance at $P < 0.05$. Courtesy of Prof. Farshid Guilak. Reprinted from Guilak [83] with permission from IOS Press.

These data can then be used as input into a finite element model that predicts how a cell will respond when the entire tissue undergoes a defined compression. Figure 2.46 shows a finite element mesh describing the geometry of interest, and Fig. 2.47 shows some results of the calculations. These data demonstrate that the strain experienced by the chondrocyte depends very strongly on the ratio of cell modulus to tissue modulus, H_A^*. The point is that the mechanical behavior of the cell must be considered in the context of the surrounding extracellular material.

Now that we know a little about the mechanical properties of individual cells in isolation and in their native ECM, let's proceed in Section 2.7 to discuss how cells sense and respond biologically to the forces they experience. We will see that most cells are very sensitive to mechanical forces and respond in a variety of ways, often culminating in modification of their function and changes in the composition, structure, and function of the surrounding tissue. Not surprisingly, the interactions between cells, their ECM, and the forces applied to them are very complex and difficult to study in vivo. This has motivated researchers interested in the effects of forces on cells to develop several devices to mechanically stimulate cells in vitro,

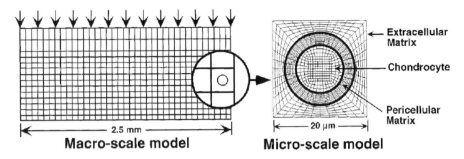

Macro-scale model   **Micro-scale model**

Figure 2.46

The mechanical environment of a single chondrocyte within a cartilage extracellular matrix was predicted using a biphasic multi-scale finite element method. First, a macro-scale model of a piece of cartilage under unconfined compression was analyzed (left). This yielded kinematic boundary conditions within a microscopic region around the cell. These boundary conditions were then applied to a micro-scale finite element model of the cell in its pericellular and extracellular matrices (right) to predict the deformation of the cell resulting from the compressive force applied at the macro-level. Reprinted from Guilak [83], with permission from IOS Press.

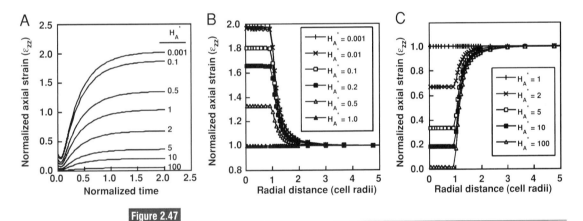

Figure 2.47

Predictions of axial strain in the chondrocyte with compression of the extracellular matrix (ECM). Using the model in Fig. 2.46, the mechanical behavior of a chondrocyte during compression of a cartilage explant was predicted. In these plots, the axial strain in the cell is normalized to the far-field strain in the ECM and the elasticity of the cell is characterized by H_A^*, the normalized aggregate modulus, where $H_A^* = H_{A,\text{cell}}/H_{A,\text{ECM}}$. The aggregate modulus, H_A, is an equilibrium modulus determined from a compression test, as described in Section 9.10.4. (A) The strain experienced by the cell is significantly dependent on its elastic properties relative to those of the matrix. (B) If the chondrocyte is assumed to be less stiff than the surrounding ECM ($H_A^* < 1$), the cell experiences strains up to twice those of the matrix. (C) In contrast, if the cell is stiffer than its matrix ($H_A^* > 1$) (as may be the case for cells embedded in the soft synthetic materials used for tissue engineering), the model predicts the strain of the cell will be significantly less than that of the extracellular matrix. Courtesy of Professor Farshid Guilak. Reprinted from Guilak [83], with permission from IOS Press.

allowing for better control over the experimental conditions. In Section 2.8, we will review some of these devices. We will then conclude this chapter with a brief summary of some of the effects of mechanical stimuli on cells from the vascular system, cartilage, and bone, three tissues that are constantly exposed to mechanical loading during normal functioning.

2.7 Mechanotransduction: how do cells sense and respond to mechanical events?

Up to this point, we have focussed on the mechanical properties of cells, describing how a cell deforms when a mechanical force is applied to it. However, that is only part of the story. In living cells, application of a mechanical stimulus causes not only a mechanical response but also a biological response. Using complex networks of sensors, transducers and actuating mechanisms, cells are able to respond and adapt to their mechanical environment. As we will discuss later in the text, these effects can be profound and critical to normal physiology and, in some cases, pathology. Here we consider some of the general principles of mechanotransduction. Because this is such a voluminous and complex topic, we have focussed here on the generalities, rather than details of how a specific cell type responds to mechanical stimulation. The reader is encouraged to consult the references provided and textbooks in cellular and molecular biology (e.g., [1,2]), which cover the biological aspects of this material in greater detail than is possible here.

To begin, it is useful to consider what components a cell needs to respond to a mechanical stimulus.

Mechanoreception. First, a cell must detect the stimulus and relay the message from outside the cell (where the stimulus acts) to inside the cell (where a response will ultimately be generated). To do so, cells use mechanoreceptors.

Signal transmission. Once sensed, the mechanical signal then needs to be relayed within the cell to various targets throughout the cell; cells appear to use both biochemical pathways and the cytoskeleton to transmit this signal.

Target activation. When the signal reaches its target (usually a protein), the target is activated. This causes alterations in cell behavior through a variety of molecular mechanisms.

In the following sections, we will discuss each of these components: how cells sense mechanical stimuli, how mechanical signals are transmitted intracellularly, and what effects those signals have on cell function. We focus our discussion on anchorage-dependent cells, although similar mechanisms are utilized by all mechanoresponsive cells.

2.7.1 Mechanoreceptors

Since mechanoreceptors must respond to extracellular signals and relay the signals from outside the cell to inside the cell, it makes sense that mechanoreceptors are physically located in the plasma membrane, at the junction of extracellular and intracellular spaces. Several mechanoreceptors have been identified in this location, including integrins, stretch-activated ion channels, and other cell-surface receptor proteins (Fig. 2.48, color plate).

Integrins

As discussed in Section 2.4, integrins are transmembrane proteins that link the ECM to the cytoskeleton via focal adhesion proteins in the cytoplasm (Fig. 2.14). Because of the physical connection between the ECM and cytosolic components, a mechanical stimulus applied to integrins can alter the structure of the cytoskeleton directly. Deformation of the cytoskeleton can have numerous consequences:

- the physical properties of the cell will change, as predicted by the analytical models presented in Section 2.6
- other receptors in the cell, including ion channels and other cell-surface receptors, can be activated (as discussed below)
- biochemical and molecular events within the cell may be regulated directly, as discussed in Section 2.7.2.

Stretch-activated ion channels

Ion channels are proteins that span the plasma membrane, connecting the cytosol to the cell exterior. Unlike other membrane pores, which are relatively large and permissive, ion channels are highly selective, allowing diffusion of specific inorganic ions across the lipid bilayer [2]. These ions, which include Na^+, K^+, Ca^{2+}, and Cl^-, are involved in a multitude of cellular activities, including intracellular signaling, gene expression, transcription, translation, and protein synthesis. Ion channels are further specialized in that they are not always open – instead they are gated, meaning a specific stimulus can cause them to open briefly, thereby allowing the flow of ions either into or out of the cell depending on the electrochemical gradients. Opening of ion channels typically involves an alteration in the channel's physical configuration. In the case of mechanically gated channels, it is not entirely clear how this occurs. One possibility is that physical deformation of the plasma membrane causes conformational changes in the embedded channel protein, leading to its activation. Also likely, however, is that cytoplasmic extensions of stretch-activated ion channels are attached to the cytoskeleton, and therefore deformation of the actin cytoskeleton can regulate gating of the channel. Interestingly, while certain channels are activated by stretch, others are actually inactivated by stretch.

Cell-surface receptor proteins

In order to respond to cues from their environment, cells rely on cell-surface receptors that bind signaling molecules to initiate an intracellular response. These cell-surface receptors are broadly classified as either G protein-linked or enzyme-linked (see for instance, Alberts *et al.* [1]). Typically, receptors respond to soluble extracellular signal molecules, such as proteins, small peptides, steroids, or dissolved gases. However, there is evidence that certain cell-surface receptors are responsive to, or are at least involved in, the sensing of mechanical signals. Again, the mechanisms are not clear, but like stretch-activated ion channels, the conformation of cell-surface receptors may be altered by membrane deformation, switching them from an inactive to an active state. Additionally, the cytoskeleton and focal adhesions may play roles in activation of these receptors. For instance, subunits of G proteins have been shown to be localized to sites of focal adhesions, in close proximity to integrins and the cytoskeleton [84]. When G protein-linked and enzyme-linked receptors are activated, they initiate several intracellular signaling pathways that distribute the signal throughout the cell, ultimately altering its behavior. These biochemical transduction mechanisms are discussed below.

2.7.2   Intracellular signal transduction

Once a mechanical stimulus is sensed and transferred from outside the cell, the signal needs to be transmitted to other points within the cell where a molecular response can be generated. It appears that cells rely on both physical and biochemical mechanisms to transmit mechanical signals (Figs. 2.49 and 2.50, both color plates).

Cytoskeleton-mediated signal transduction

Transmission of mechanical signals via integrins can lead to deformation of the cytoskeleton, which, in turn, can affect the biochemical state of the cell. For instance, because the cytoskeleton is a continuous, dynamic network that provides mechanical connections between intracellular structures, deformation of the cytoskeleton at one location may lead to deformations of connected structures at remote locations (Fig. 2.49A, color plate). This "hard-wiring" within the cell means a perturbation applied locally to an integrin can lead to movement of organelles [75] and distortion of the nucleus [86], possibly influencing gene expression. As discussed previously, cytoskeletal deformation can also activate other receptors, such as ion channels and G protein-linked receptors. This "decentralization" mechanism, by which a locally applied stimulus results in mechanotransduction at multiple, mechanically coupled sites, allows for greater diversity in the cellular response than is possible with a single uncoupled receptor, since different receptors will have different sensitivities and response times and will thus respond to different local environmental cues [85].

Another possible role for the cytoskeleton in mechanotransduction is based on the observation that many proteins and enzymes involved in protein synthesis and biochemical signal transduction appear to be immobilized on the cytoskeleton [72]. It has been proposed that these regulatory molecules will experience the mechanical load imposed on the cytoskeleton as a consequence of their binding to it. The imposed load could alter the conformation of the regulatory molecules, which, in turn, would change their kinetic behavior and biochemical activity (Fig. 2.49B, color plate). Thus, the cytoskeleton and its associated regulatory molecules might serve as a scaffold for the transduction of mechanical signals to biochemical signals within the cell.

Biochemically mediated signal transduction

The general principle behind biochemically mediated signal transduction is that activation of a receptor initiates a cascade of events mediated by a series of signaling molecules (Fig. 2.50, color plate). Ultimately these molecules interact with target proteins, altering the target proteins so they elicit changes in the behavior of the cell. These signaling pathways are utilized by the cell to respond to a variety of extracellular stimuli, including soluble signals (e.g., growth factors), cell–cell contact, and mechanical signals.

The signaling molecules involved in relaying the signal intracellularly are a combination of small intracellular mediator molecules (also known as second messengers) and a network of intracellular signaling proteins. Second messengers are generated in large numbers in response to activation of a receptor. Owing to their small size, they are able to diffuse rapidly throughout the cytosol and, in some cases, along the plasma membrane. By binding to and altering the behavior of selected signaling proteins or target proteins, second messengers propagate the signal "downstream" from the receptor. Similarly, intracellular signaling proteins relay the signal downstream by activating another protein in the chain or by generating additional small-molecule mediators (which will, in turn, propagate the signal). These are the primary mechanisms by which signals received by G protein-linked and enzyme-linked receptors are transmitted.

Interestingly, in addition to their mechanical transmission role, integrins are also able to induce biochemical responses. For instance, clustering of integrins at focal adhesion sites leads to recruitment and activation of signaling molecules (e.g., focal adhesion kinase or FAK), thereby initiating biochemical signal transduction [87]. Ultimately, the biochemical signaling pathways interact with target proteins, which are responsible for altering the behavior of the cell. Potential targets are discussed in the next section.

2.7.3 Cellular response to mechanical signals

Mechanical signals, like other extracellular signals, can influence cellular function at multiple levels, depending on the targets of the signaling pathways initiated by the stimulus (Fig. 2.51, color plate). For instance, a signaling pathway activated by a mechanical stimulus might target proteins that regulate gene expression and the transcription of mRNA from DNA (e.g., transcription factors). Additionally, the signaling targets might be molecules involved in protein production, so that alteration of those molecules will affect translation of mRNA to proteins or post-transcriptional assembly or secretion of proteins. Because cell shape and motility are dependent on the cytoskeleton, its deformation by a mechanical stimulus can alter these cytoskeleton-dependent processes. Finally, the production of proteins and their secretion from a cell can affect the function of neighboring cells (or even the secreting cell itself), thereby propagating the effect of the mechanical signal from one cell to several.

It is important to realize that the cellular response to a single type of stimulus can be quite complex, since activation of a single type of receptor usually activates multiple parallel signaling pathways and therefore can influence multiple aspects of cell behavior. Furthermore, at any one time, cells are receiving hundreds of different signals from their environment and their response is determined by integration of all the information they receive. Clearly, this makes things rather complicated, particularly if one wants to understand the response of a cell to a particular mechanical stimulus. As a result, efforts to understand the response of cells to mechanical stimuli often rely on experiments performed under controlled conditions in the laboratory. In the next section, we present some devices used to mechanically stimulate groups of cells in culture and briefly review what has been learned about the response of certain cell types from these sorts of experiments.

2.8 Techniques for mechanical stimulation of cells

It is often the case that we know that cells within a tissue respond to mechanical stimulation, yet we are not interested in the biomechanical properties of the resident cells per se. Instead we want to know what effect mechanical stimulation has on the biology of the resident cells and, by extension, on the biology of the whole tissue.

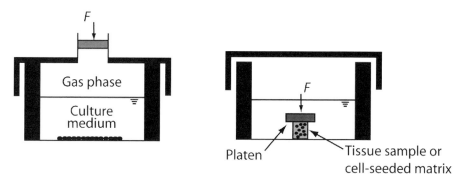

Figure 2.52

Devices for compressive loading of cultured cells and tissues. In the device on the left, hydrostatic compression of cells is achieved by pressurizing the gas phase above the culture medium. In the device on the right, three-dimensional specimens are compressed by direct loading using a platen. The specimens can be intact tissue samples or extracellular matrices (e.g., collagen or a polymer sponge) seeded with cells of interest. The specimen compressed by direct platen loading will be strained not only in the direction of loading but also in the lateral direction because of Poisson's effect (see text).

Because of the complexity of cellular biomechanics, it is not possible to predict this information from theoretical models or measurements of the properties of single cells. Therefore, we culture the tissue (or its resident cells) and mechanically stimulate the cells, observing their ensuing behavior. Here we describe some of the techniques used to stimulate cells in this way.

A wide variety of devices have been developed to apply mechanical stimuli to cells (and tissues) in culture. The choice of device and the mechanical stimulus it applies is dependent on which cells are being studied. For instance, chondrocytes, the cells in cartilage, are typically compressed within their ECM because cartilage tissue is primarily subjected to compressive loads in vivo. Smooth muscle cells are usually stretched, since these cells normally experience tensile forces in vivo, while endothelial cells, which line the inner surface of blood vessels, are typically subjected to fluid flow. These three loading modes – compression, stretching, and fluid flow – are the most common and are reviewed below. Reviews that provide more details of the advantages and disadvantages of various devices are available in Brown [88] and Frangos [89].

2.8.1 Compressive loading

Hydrostatic compression

One method to compress cultured cells is to increase the gas pressure in the culture system (Fig. 2.52, left). This results in a hydrostatic pressure being applied to cells within the liquid medium below the gas phase. Unfortunately, according to Henry's law [90], the solubility of a gas in liquid increases as its pressure increases, and since cells are sensitive to the concentrations of dissolved gases in the culture medium (particularly O_2 and CO_2), it is difficult to determine whether the effects observed result from the mechanical or chemical stimulus. Furthermore, it is unlikely that most cells experience a pure hydrostatic pressure in vivo at the pressure magnitudes shown to invoke a biological response in vitro.

It is important to understand that if cells are being cultured on a rigid substrate they will not experience a net deformation from such a hydrostatic pressure increase, since this increased isotropic stress will presumably be transmitted into the cytoplasm, resulting in zero net force change on molecular components. It is therefore difficult to see how hydrostatic pressure variations alone (in the absence of associated deformation, e.g., of a flexible substrate) can be sensed by the cell. It has been proposed that various components of the cell could differentially compress in response to hydrostatic pressure variations, but this has yet to be unequivocally demonstrated. Therefore, the physiological relevance of pure hydrostatic compression is presently unclear.

Platen compression

A common alternative for compressive loading is direct compression by a platform or platen. This method is normally used with tissue specimens or with cells that have been seeded into a natural or synthetic ECM, yielding a three-dimensional specimen that can be compressed (Fig. 2.52, right). Although this method is conceptually simple, the resulting tissue (and therefore cellular) strains can be quite complex because of Poisson's effect,[9] viscoelasticity of the matrix, the internal architecture of the matrix, and fluid flow that results from compression (much like how fluid exudes from a sponge when it is squeezed). Nonetheless, the similarity of this mode of loading with that which occurs in vivo for cartilage make it a useful method to study the response of chondrocytes to compressive forces.

2.8.2 Stretching

The most common approach to stretch cells is to grow them on a flexible surface and to deform the surface once the cells are adhered to it. This technique has been used in various configurations to apply uniaxial and biaxial strain to cells in culture (Figs. 2.53–2.55). Both static and cyclic straining can be applied using these devices. Of course, care must be taken in selecting a surface that is biocompatible

---

[9] Poisson's effect describes the strain that is generated in the direction perpendicular to the direction of loading. The strain in the direction of loading (ε_x) and the strain in the perpendicular direction (ε_y) are related by Poisson's ratio, ν, where $\nu = -\varepsilon_y/\varepsilon_x$ for a linearly elastic material. For most materials, ν is between 0 and 0.5.

A

B

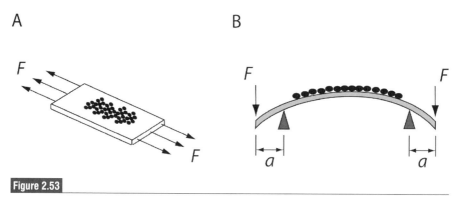

Figure 2.53

Uniaxial cell stretching devices. On the left, cells are attached to a membrane that is stretched longitudinally. On the right, cells are attached to a substrate that is deformed in a four-point bending configuration. In both cases, the substrate is strained both in the longitudinal direction and the lateral direction (see text).

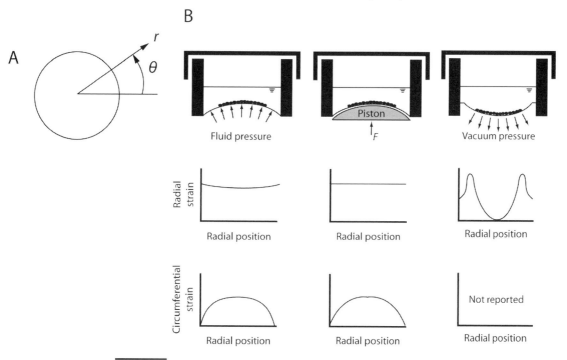

Figure 2.54

(A) The strain in a circular membrane can be described in cylindrical coordinates where r is the radial direction and θ is the circumferential direction. (B) Schematics of three actual devices that have been used to apply biaxial strain to a membrane to which cells are attached. These methods all produce non-uniform strain profiles, meaning cells attached to the membrane in different locations experience different strains. The unusual radial strain profile for the vacuum-driven device on the right is caused by the thick membrane used in this device, and therefore compressive bending strains contributed to the net strain on the surface of the membrane [91]. The membranes used in the fluid and piston-driven devices were thin, and therefore bending strains were negligible. A vacuum-driven device with a thinner membrane would produce strain profiles similar to those of the fluid-driven device (left). For details on the theoretical derivation of these strain profiles, refer to [89]. Adapted from Schaffer *et al.* [92]. Reprinted with permission of John Wiley & Sons, Inc.

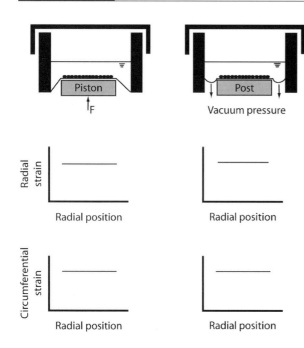

Figure 2.55

Cell-stretching devices that produce an equi-biaxial strain. Both the radial and circumferential components of the membrane strain are constant across the culture surface and are of equal magnitude.

with the cells and allows them to attach and adhere firmly. The ability to select the surface for adhesion can be an advantage, however, allowing for greater experimental control. For instance, by coating the surface with defined matrix molecules (e.g., collagen or fibronectin), one can investigate whether specific cell–matrix interactions are important for mechanotransduction in a particular type of cell.

Uniaxial stretch

One method to stretch cells uniaxially (or longitudinally) is to grow them on a flexible membrane, grip the membrane at either end, and elongate the membrane (Fig. 2.53A). Another approach is to grow the cells on a substrate that is then bent or flexed in a four-point bending configuration (Fig. 2.53B). The latter method results in a tensile strain on the convex surface. The longitudinal strain, ε, in this case is:

$$\varepsilon = \frac{6Fa}{bh^2 E} \tag{2.50}$$

where F is the applied force, a is the distance between the support and point of force application, E is the elastic modulus of the substrate, and b and h are the width and thickness of the substrate, respectively. In both the membrane elongation and substrate flexion cases, there is not only longitudinal deformation of the substrate but also deformation in the lateral direction owing to Poisson's effects.

Biaxial stretch

Rather than pull the membrane in one direction only, the outer edges of a circular membrane can be fixed and the membrane can be deformed to produce a biaxial deformation, meaning a circular membrane is strained in both the radial and circumferential directions (Fig. 2.54A). This mode of loading has been implemented with several devices that either push the membrane up from the bottom (using a piston or fluid) or pull the membrane down using vacuum pressure. Schematics of these devices and their theoretical strain profiles are show in Fig. 2.54B. From the strain profiles, it is apparent that the strain experienced by cells depends strongly upon their location on the membrane. In some cases, this can be an advantage since a range of strains can be studied in a single experiment. In practice, however, the strain input is often not well characterized, biological assays are difficult to perform on cells from an isolated area of the membrane, and communication between cells might mask any differences resulting from strain variations from one area to another. Therefore, the inhomogeneity of the strain stimulus can make interpretation of experimental results difficult and is a fundamental limitation of these devices.

To address the strain inhomogeneity, devices that apply uniform biaxial strain have been developed. Examples of two devices and their strain profiles are shown schematically in Fig. 2.55. For these cases, in which the membrane is confined on its periphery and is stretched so that the membrane remains "in-plane," the theoretical strain profile can be computed by first realizing that the stresses will be two dimensional or planar for a thin membrane. For an isotropic linear elastic material in a state of plane stress, the strains (ε) and stresses (σ) are related by:

$$\varepsilon_r = \frac{(\sigma_r - \nu\sigma_\theta)}{E} \tag{2.51}$$

$$\varepsilon_\theta = \frac{(\sigma_\theta - \nu\sigma_r)}{E} \tag{2.52}$$

where r and θ refer to the radial and circumferential components,[10] respectively, and E and ν are the Young's modulus and Poisson's ratio of the membrane material, respectively.

---

[10] Formal notation of stress components requires two subscripts, where the first subscript identifies the surface on which the stress acts and the second subscript specifies the force component from which the stress component is derived. Thus, a normal stress that acts on the x surface in the x direction is τ_{xx} and a shear stress that acts on the x surface in the y direction is τ_{xy}, where it is understood that the orientation of a surface is defined by its normal. However, because the two subscripts for normal stresses are always the same, it is common practice to denote normal stresses by σ and use only one subscript, e.g., σ_x for τ_{xx} and σ_r for τ_{rr}. Since the reader may encounter both notations in the literature, we have used them interchangeably throughout this text.

For a circular membrane such as those used in the devices shown in Fig. 2.55, the stress distribution is symmetrical about an axis that passes through the center of the membrane (i.e., axisymmetric) and it can be shown that $\sigma_r = \sigma_\theta$ (see Box 2.1 for a proof). Rearranging Equations (2.51) and (2.52) and equating the two stress components gives $\varepsilon_r = \varepsilon_\theta$. In other words, the majority of the membrane (the part that stays "in-plane"), and presumably the cells attached to it, experience a strain that is spatially constant (not a function of position) and isotropic (equal in all directions). This theoretical derivation assumes the membrane slides without friction over the underlying piston or post; to approximate this in a stretching device, a lubricant is applied to the bottom of the membrane and to the top of the post it slides over.

Although the primary stimulus applied to the cells in any stretch device is deformation of the substrate, it is important to realize that movement of the membrane can cause motion of the liquid medium within the culture dish. Therefore, the cells might not only respond to the stretch but also to the shear and pressure forces generated by the moving liquid [88]. The same is true for platen compression devices: during compression of a three-dimensional specimen containing cells, fluid can be expelled from the matrix and the fluid flows that are generated by this process might affect the cells within the matrix.

2.8.3 Fluid flow

Because there are a number of circumstances in which cells are subjected to shear stresses in vivo (e.g., endothelial cells in blood vessels and osteocytes in bone tissue), mechanical stimulation of cells using fluid flow is an important approach and one that is used frequently. The devices that have been developed to apply well-characterized shear stresses to cultured cells fall into two categories: viscometers and flow chambers (Fig. 2.56).

Viscometers

Viscometer systems for mechanical stimulation of cells were adapted from systems originally used to study the rheological properties of fluids. Two configurations have been used: the cone-and-plate viscometer and the parallel disk viscometer.

In the cone-and-plate viscometer, cells are either attached to a stationary plate or suspended in the medium between the plate and a rotating disc (Fig. 2.56A). A second plate, the cone, rotates causing the fluid between the two plates to move in the circumferential direction with a velocity v_θ. The cone is not parallel to the stationary plate, but instead makes a small angle, α, with the stationary plate; therefore, the distance between the plates, h, varies as a function of the radial

Box 2.1 **Proof that the stress is equi-biaxial for the circular membranes in the devices shown in Fig. 2.55**

For a membrane in a state of plane stress, the equilibrium equations (assuming no body forces) are given in cylindrical coordinates as [93]:

$$\frac{\partial \sigma_r}{\partial r} + \frac{1}{r}\frac{\partial \tau_{r\theta}}{\partial \theta} + \frac{\sigma_r - \sigma_\theta}{r} = 0 \tag{2.53}$$

$$\frac{1}{r}\frac{\partial \sigma_\theta}{\partial \theta} + \frac{\partial \tau_{r\theta}}{\partial r} + \frac{2\tau_{r\theta}}{r} = 0. \tag{2.54}$$

Here we have used the same nomenclature as in Equations (2.51) and (2.52), with the addition of the shear stress component, $\tau_{r\theta}$.

To show that the membrane stress field is equi-biaxial, we will introduce a stress function, Φ, which is a function of r and θ. We define Φ by setting:

$$\sigma_r = \frac{1}{r}\frac{\partial \Phi}{\partial r} + \frac{1}{r^2}\frac{\partial^2 \Phi}{\partial \theta^2}, \tag{2.55}$$

$$\sigma_\theta = \frac{\partial^2 \Phi}{\partial r^2}, \tag{2.56}$$

$$\tau_{r\theta} = \frac{1}{r^2}\frac{\partial \Phi}{\partial \theta} - \frac{1}{r}\frac{\partial^2 \Phi}{\partial r \partial \theta} = -\frac{\partial}{\partial r}\left(\frac{1}{r}\frac{\partial \Phi}{\partial \theta}\right) \tag{2.57}$$

You can verify that this definition for Φ satisfies the equilibrium equations by direct substitution into (2.53) and (2.54). Using the stress function, we can write a single partial differential equation (the equation of compatibility) that can be used to solve the two-dimensional stress field for various boundary conditions. The compatibility equation in cylindrical coordinates is [93]:

$$\nabla^4 \Phi = \left(\frac{\partial^2}{\partial r^2} + \frac{1}{r}\frac{\partial}{\partial r} + \frac{1}{r^2}\frac{\partial^2}{\partial \theta^2}\right)\left(\frac{\partial^2 \Phi}{\partial r^2} + \frac{1}{r}\frac{\partial \Phi}{\partial r} + \frac{1}{r^2}\frac{\partial^2 \Phi}{\partial \theta^2}\right) = 0. \tag{2.58}$$

In the axisymmetric case, where the stress is dependent on r only, the compatibility Equation (2.58) becomes:

$$\left(\frac{\partial^2}{\partial r^2} + \frac{1}{r}\frac{\partial}{\partial r}\right)\left(\frac{\partial^2 \Phi}{\partial r^2} + \frac{1}{r}\frac{\partial \Phi}{\partial r}\right) = \frac{\partial^4 \Phi}{\partial r^4} + \frac{2}{r}\frac{\partial^3 \Phi}{\partial r^3} - \frac{1}{r^2}\frac{\partial^2 \Phi}{\partial r^2} + \frac{1}{r^3}\frac{\partial \Phi}{\partial r} = 0. \tag{2.59}$$

This is now an ordinary differential equation with a general solution of the form:

$$\Phi = A \log r + Br^2 \log r + Cr^2 + D. \qquad (2.60)$$

With this general solution, the axisymmetric stress components (Equations (2.55) to (2.57)) are

$$\sigma_r = \frac{1}{r}\frac{\partial \Phi}{\partial r} = \frac{A}{r^2} + B(1 + 2\log r) + 2C, \qquad (2.61)$$

$$\sigma_\theta = \frac{\partial^2 \Phi}{\partial r^2} = -\frac{A}{r^2} + B(3 + 2\log r) + 2C, \qquad (2.62)$$

$$\tau_{r\theta} = 0. \qquad (2.63)$$

You will notice that Equations (2.61) and (2.62) predict infinite stresses at the center of the membrane where $r = 0$. Since this is physically impossible, the constants A and B must be zero. Therefore, $\sigma_r = \sigma_\theta = 2C = \text{constant}$ and the membrane is in a condition of uniform tension in all directions in its plane.

distance from the center, r, according to $h = r \sin \alpha$. For laminar flow in a thin gap, the fluid velocity in the gap depends on the angular velocity of the rotating plate, ω, the separation of the plates, h, the distance from the axis of rotation, r, and the distance from the stationary plate, z, according to:

$$v_\theta = \frac{\omega r z}{h} = \frac{\omega z}{\sin \alpha} \approx \frac{\omega z}{\alpha}, \qquad (2.64)$$

where we have approximated $\sin \alpha$ by α, which is valid for small angles when α is expressed in radians. Because the circumferential velocity is independent of radial position, the shear stress in a cone-and-plate viscometer is independent of position:

$$\tau = \mu \frac{dv_\theta}{dz} = \mu \frac{\omega}{\alpha}, \qquad (2.65)$$

where μ is the viscosity of the fluid. This calculation of the shear stress assumes the fluid is an incompressible Newtonian fluid, which is valid for many fluids (including cell culture medium) but not entirely accurate for some biological fluids, such as blood (see Section 3.1 for a discussion of blood rheology).

The parallel disk viscometer differs from the cone-and-plate viscometer in that the two plates are parallel to one another, separated by a constant distance, h (Fig. 2.56B). The circumferential velocity in this case is given by:

$$v_\theta = \frac{\omega r z}{h}, \qquad (2.66)$$

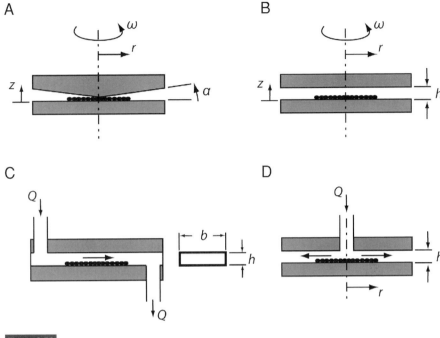

Devices used to apply fluid shear forces to cultured cells. The cone-and-plate (A) and parallel plate (B) viscometers generate circumferential fluid flow by rotation of the top plate relative to the fixed lower disc, creating shear stresses on cells either suspended in the culture medium between the discs or on cells attached to the bottom disc (as shown in the figure). In the parallel plate (C) and radial (D) flow chambers, flow of the culture medium is generated by a pressure gradient (by either hydrostatic head or a pump) and cells adherent to the bottom surface are sheared.

and the shear stress, τ, exerted by the fluid is a function of the radial position, r:

$$\tau = \mu \frac{\omega r}{h}. \tag{2.67}$$

From Equation (2.67), it is apparent that the shear stress varies from zero at the center of the plate to a maximum at the edge. As with the stretching devices that have inhomogeneous strain fields, the inhomogeneity in shear stresses with parallel disk viscometers can complicate interpretation of experimental results in many instances. In some cases, however, such as investigations of the effect of shear stress on cell shape, a range of shear stresses might be desirable.

Flow chambers

The other class of fluid shear devices, flow chambers, differs from the viscometers in that fluid motion is caused not by movement of one surface of the device but by imposing a pressure gradient. Flow chambers come in two configurations: parallel plate and radial.

The most commonly used device is the parallel plate flow chamber, in which fluid is driven by a pump or hydrostatic head through a rectangular channel (Fig. 2.56C). The chamber is designed such that the flow is fully developed and laminar (approximating plane Poiseuille flow). The shear stress on cells adhered to the bottom plate can be approximated by:

$$\tau = \frac{6Q\mu}{bh^2} \tag{2.68}$$

where Q is the volumetric flow rate through the chamber, b and h are the width and height of the chamber, respectively, and μ is the viscosity of the fluid. This approximation is valid if $b \gg h$. Therefore, the shear stress is constant for a given flow rate. The constant shear stimulus, plus the ease of use of this system – the equipment is relatively simple and the cells can be attached to a microscope slide and viewed while being sheared – make the parallel plate flow chamber an attractive option for flow studies.

An alternative configuration for a flow chamber is the radial flow chamber, in which fluid flows from an inlet in the center and moves out between two plates in the radial direction, exiting at the edge (Fig. 2.56D). In this case, the shear stress on cells adhered to the bottom plate is not constant but is a function of the radial position, r:

$$\tau = \frac{6Q\mu}{2\pi r h^2}. \tag{2.69}$$

Again, the inhomogeneous stimulus must be considered carefully when designing an experiment and interpreting results from a radial flow chamber.

In all these devices, the flow is steady, fully developed and laminar. In the body, however, fluid flow is often unsteady and in some cases, such as at branch points in large arteries, the flow pattern will be complicated, with flow reversals and separation. Investigators have therefore introduced modifications to the devices described above, such as pulsatile flow in parallel plate flow chambers [94] and flow chamber geometries that cause flow disturbances [95].

2.9 Summary of mechanobiological effects on cells in selected tissues

In this final section, we consider how cells from selected tissues respond biologically to mechanical forces. Much of what is known in this area has come from experiments using the devices covered in Section 2.8, and we present some of these data in this section. Observations from animal models have also contributed to our understanding of how mechanical forces regulate cell and tissue function, and some

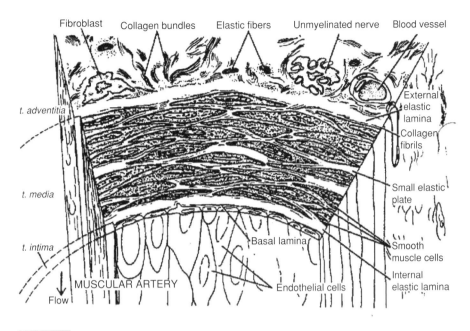

Fibroblast   Collagen bundles   Elastic fibers   Unmyelinated nerve   Blood vessel

t. adventitia

t. media

t. intima

MUSCULAR ARTERY

Flow

External elastic lamina

Collagen fibrils

Small elastic plate

Basal lamina

Smooth muscle cells

Internal elastic lamina

Endothelial cells

Figure 2.57

Schematic diagram of the wall of a muscular artery. From this figure the multilayered structure of the artery wall – consisting of the *tunica* (t) *intima*, *media*, and *adventitia* – can be appreciated. Circumferentially oriented smooth muscle cells within the media can be seen. Compare also with Fig. 4.15, color plate. From Ross and Reith [96]. Reprinted with permission of Lippincott Williams & Wilkins.

of those results are presented here too. We will limit our discussion to vascular tissue, cartilage, and bone; these tissues are particularly interesting to biomechanical engineers because they are constantly subjected to mechanical loading, and mechanical factors are associated with growth, maintenance, and pathology in these tissues.

2.9.1 Endothelial cells in the vascular system

Endothelial cells line all blood-contacting surfaces in the body (Fig. 2.57). They have a number of functions, including regulating the passage of water and solutes across the vessel wall. They are also sensitive transducers of the mechanical forces imposed on them. To make this more concrete, consider cells lining a large artery. These cells are subjected to a shear stress from the passage of blood adjacent to their surfaces. They are also subjected to a cyclic distension from the pulsation of the artery during the cardiac cycle. This combination of shear stress and stretch affects the endothelium in several ways.

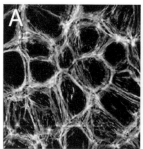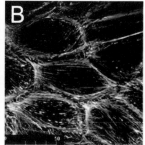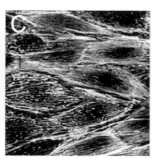

Figure 2.58

Fluorescent micrographs of F-actin in cultured vascular endothelial cells. (A) Endothelial cells 2 days after they had grown to confluence, at which time dense peripheral bands of F-actin assemble primarily at the cell–cell junctions. (B) Endothelial cells 2 days post-confluence were exposed to a shear stress of 15 dynes/cm^2 for 16 h. The shear stress caused the cells to reorganize in order to elongate and align with the direction of flow (left to right), with associated partial loss of the junctional actin and enhancement of basal actin stress fibers. (C) After 48 h of shear stress in these 2-day postconfluent cells, alignment of the actin filaments with the flow direction and changes in cell shape are complete. Scale bar (B) is in micrometers. Adapted from [97], with permission from the American Society for Investigative Pathology and the authors.

1. Endothelial cells rearrange their cytoskeleton and change their shape in response to elevated levels of shear stress (Fig. 2.58). Specifically, the stress fibers of the actin cytoskeleton at the base of the cell extend and align with the direction of flow. This occurs by regulated actin polymerization at the ends of the fibers, which drives protrusion of the cell membrane in the flow direction [97]. This presumably acts to stabilize the cell and improves resistance to shear forces (although see also comments in Satcher and Dewey [81]).

2. Arteries change their caliber in response to long-term changes in blood flow brought about by alterations in the distal vascular beds. This process, known as arterial remodeling, is a physiological response that ensures that the artery is the "right" caliber to deliver the required blood flow rate. It is known that the endothelium is required for this remodeling to take place [98]. Further, a given artery will remodel itself so as to maintain a target value of shear stress, equal to 15 dynes/cm^2 for many (but not all) arteries [99]. Presumably this remodeling process is driven by the endothelium sensing a local time-averaged shear stress and then inducing the artery wall to remodel itself to maintain a "target" value of this long-term shear stress.

3. In cell culture, vascular endothelial cells will alter their production of many substances (prostacyclins, nitric oxide, etc.) in response to imposed shear stress. This effect can also be elicited by cyclic stretch, and in fact these two stimuli seem to be synergistic in some cases. Secretion of such molecules is almost

certainly how the endothelium signals the artery wall to remodel in response to changes in time-averaged shear stress.

2.9.2 Smooth muscle cells in vascular tissue

Smooth muscle cells are found in several tissues in the body including the genitourinary, respiratory, and digestive tracts, and in blood vessels (which are the focus of this section). Vascular smooth muscle cells make up the cellular component of the medial layer of blood vessels (Figs. 2.57 and 4.15). The cells are aligned circumferentially around the blood vessel, an orientation that allows them to control the diameter of the vessel by contracting or relaxing. Like endothelial cells lining a large artery, smooth muscle cells are also subjected to cyclic stretching as the artery distends during the cardiac cycle. This stretching can affect smooth muscle cells in a variety of ways.

1. Mills *et al.* [100] used the biaxial stretch device shown on the right in Fig. 2.54 to cyclically stretch smooth muscle cells isolated from the aortas of cows. They used the non-uniform strain profile to their advantage in these experiments by seeding cells only in the center or on the periphery of the membrane to determine the effect of different levels of strain magnitude. The cells in the center of the membrane (where the strains were 0–7%) grew more quickly than did smooth muscle cells on the periphery of the membrane or cells that were grown under static conditions. The cells on the periphery of the membrane, which experienced much higher strain levels (up to 24% higher), also grew more quickly than cells grown statically. Additionally, the cells on the periphery aligned themselves circumferentially in response to the strain stimulus, assuming an orientation similar to that observed in blood vessels.
2. Cyclic stretch also increases the production of extracellular matrix proteins (such as collagen and elastin) and makes the smooth muscles more "contractile" [101].

Kim *et al.* [102] took advantage of the responsiveness of smooth muscle cells to mechanical signals to engineer smooth muscle tissue. They seeded vascular smooth muscle cells into three-dimensional collagen sponges and stretched them longitudinally (using a principle similar to that shown on the left in Fig. 2.53) at 7% strain (1 Hz) for several weeks. They found that the strain stimulus increased the proliferation of the smooth muscle cells, increased the production of collagen and elastin, and in the long term led to organization of the tissue matrix. Most importantly, the ultimate tensile strength and elastic modulus of the resulting tissue were increased 12- and 34-fold, respectively, over those for tissues grown without mechanical stimulation.

2.9.3 Chondrocytes in articular cartilage

In Section 2.6.4, we examined how chondrocytes (cartilage cells) deform in their matrix. This is important because articular cartilage is subjected to a wide range of static and dynamic loads, with peak stresses as high as 20 MPa and compressive strains of up to 45% [103]. Those loads can have profound effects not only on the mechanical integrity of the tissue matrix but also on the biological response of the chondrocytes; some of those responses are summarized here. Interested readers are also encouraged to read the review by Grodzinsky *et al.* [103], which makes reference to many original studies in this area.

As with many cells, it is difficult to study the response of chondrocytes to mechanical stimuli in vivo. Instead, in vitro testing is performed and a platen compression device (similar to that shown on the right in Fig. 2.52) is often used. The test samples are usually either small cartilage explants (e.g., cylinders cored out from articular cartilage of the knee) or synthetic systems comprising chondrocytes embedded in three-dimensional hydrogels. These types of samples are used because they emulate the native cell–ECM structure.

Chondrocytes respond very differently to compressive forces, depending on the nature of the stimulus.

1. Static compression of as little as 15% significantly inhibits the synthesis of cartilage matrix components, such as proteoglycans and proteins [104]. After the static compression is released, biosynthetic levels can return to normal, with the time to recovery dependent on the duration and amplitude of the original static compression.

2. Dynamic (cyclic) compression can stimulate matrix production considerably. However, the response to dynamic compression is very dependent on the frequency and amplitude of the compression. For instance, low-amplitude compression ($< 5\%$) at a low frequency (< 0.001 Hz) had no effect on proteoglycan or protein synthesis by chondrocytes in articular cartilage implants from calf knee joints [104]. At higher frequencies of stimulation (0.01–1 Hz), however, low-amplitude compression resulted in up to a 40% increase in proteoglycan and protein synthesis.

3. The rate at which compressive strain is applied is also an important parameter. For instance, Kurz *et al.* [105] showed that a 50% strain applied to cartilage explants at a rate of 0.01 s^{-1} had no effect on chondrocyte activity or matrix properties, whereas higher rates of 0.1–1 s^{-1} caused significant decreases in proteoglycan and protein synthesis and increased cell death. Interestingly, the

cells that survived the injurious compression of high strain rates were no longer able to respond normally to low-amplitude cyclic compression (as described above).

On first glance, uniaxial compression seems like a relatively simple loading condition. However, compression not only causes deformation of the chondrocytes and matrix (Section 2.6.4) but also leads to hydrostatic pressure gradients, fluid flow, electrical streaming potentials (from movement of ions), and other physicochemical changes. Each of these mechanical, chemical, and electrical stimuli likely plays some role in affecting the biological response of chondrocytes. In an attempt to determine which stimuli are most important, several researchers have developed theoretical models to predict these various signals. By correlating spatial patterns of matrix synthesis observed in cartilage explants during compression with patterns of mechanical, chemical, or electrical stimuli predicted by theoretical models, conclusions regarding the most important stimuli can be made (Fig. 2.59). Studies like these suggest that fluid flow within the tissue and deformation of the ECM are critically important at regulating ECM synthesis by chondrocytes in response to dynamic compression [103,106].

In addition to experiencing compression, cartilage in many joints is subjected to shear deformation. Devices similar to the compression device shown on the right in Fig. 2.52 have been built to apply shear forces to cartilage explants [107]. This type of loading also affects chondrocyte function; for instance, a dynamic shear deformation of 1% at 0.1 Hz for 24 h increases proteoglycan and protein synthesis by 41% and 25%, respectively [103].

Many of the studies on the effects of mechanical forces on cartilage have been motivated by a desire to understand the role of biomechanics in cartilage deterioration, such as occurs in osteoarthritis. However, the understanding gained from these basic studies might also have application for the repair and replacement of damaged cartilage tissue. Several researchers are currently investigating how mechanical stimulation might be used in the laboratory to grow or "engineer" better replacement cartilage tissue. Initial studies with this approach indicate that significant improvements in the mechanical properties and composition of engineered cartilage can be obtained with dynamic compression [108]. This mixture of biomechanics and the relatively new field of tissue engineering is exciting and might be critical to obtaining functional engineered tissues that are useful clinically.

2.9.4   Osteoblasts and osteocytes in bone

As we will discuss in Chapter 9, bone is an amazingly dynamic tissue that responds to its mechanical environment. This adaptive behavior allows bones to withstand repeated loading and alterations in loading without breaking or sustaining extensive

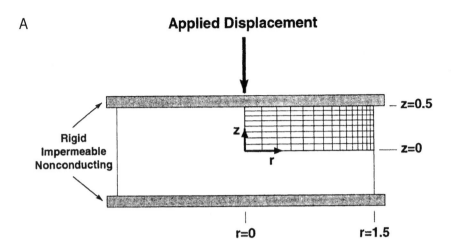

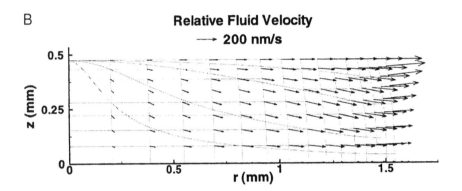

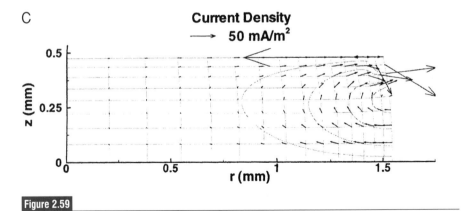

Figure 2.59

Finite element analysis of a cartilage explant subjected to dynamic compression. (A) The cartilage disc was modeled with an axisymmetric finite element model that considered the coupled mechanical, electrical, and chemical/osmotic phenomena resulting from compression. The model predicted the resulting intra-tissue profiles of fluid velocity (B) current density (C), and axial strain (not shown). These spatial profiles can be compared with patterns of biosynthetic activity in the cartilage tissue to determine which stimuli might be the most important in regulating chondrocyte function. From Levenston *et al.* [106] with permission of the ASME.

damage. The mechanisms by which bone senses and responds to mechanical loads is an area of active research with many unresolved questions. A review by Burr *et al.* [109] summarized some of what is known about bone adaptation to mechanical stimulation, with focus on the tissue level responses. Here we focus on in vitro studies of bone cells, with particular attention to osteoblasts and osteocytes, two types of bone cell that are believed to play important roles in maintaining bone tissue.

Osteoblasts are responsible for making new bone. Osteocytes are cells that were formerly osteoblasts but have stopped synthesizing bone tissue and are now embedded directly in the mineralized matrix of bone in pores called lacunae. Osteocytes are not completely isolated from other cells, however; they are connected to other osteocytes and to osteoblasts on the surface of the bone matrix through cell processes that travel through channels (canaliculi) in the bone matrix (Fig. 9.3). These intercellular connections will be important to our discussion later on.

As with cartilage, the loads applied to bones at the tissue level are experienced by the cells residing in the bone matrix. One consequence of the mechanical loading of bones is that the bone matrix deforms. This type of mechanical stimulation (deformation of the substrate to which the bone cells are attached) has been emulated in vitro with a variety of devices, including the uniaxial and biaxial stretch devices presented in Section 2.8.2. The general conclusions from these studies are that substrate deformations of approximately 1–10% influence a variety of measures of bone cell function, including DNA synthesis, enzyme production, synthesis of collagen and non-collagenous proteins, and mineral production. The molecular mechanisms for these effects are also being investigated. For instance, Ziros *et al.* [110] found that mechanical stimulation of human osteoblasts causes an increase in the expression and DNA-binding activity of a transcription factor known to be critical to osteoblast differentiation. In this study, the osteoblasts were stretched with an average strain of 2.5% using a device similar to the one illustrated in the middle panel of Fig. 2.54B.

It turns out, however, that because of the relative stiffness of mineralized bone tissue, peak strains on the surface of most bones do not usually exceed 0.3% [111]. It is possible that in certain circumstances matrix deformations might exceed 0.5%, for example, in pre-mineralized bone tissue during fracture repair or in regions where the microstructure of the bone results in local strain concentrations. In general, however, the physiological relevance of the in vitro studies that apply strains in excess of 1% is unclear.

More recent studies have focussed on the response of bone cells to fluid flow rather than substrate deformation. Mechanical loads applied to bones not only deform the matrix, but also cause movement of extracellular fluid through the lacunae and canaliculi in the bone matrix. This movement of fluid is hypothesized to stimulate osteocytes and osteoblasts through fluid shear stress effects, streaming potentials, or chemical transport mechanisms [112]. Osteocytes, owing to their location with the lacunae, may therefore act as mechanosensors, sensing fluid flows and signaling to osteoblasts (through their cell processes, for instance) to cause alterations in bone formation. Consistent with this theory, bone cells are more sensitive to physiological levels of fluid flow than they are to physiological levels of stretching, as shown in two studies.

1. Smalt *et al.* [113] measured the biological response of a variety of types of osteoblast to stretch and shear stress by monitoring the production of nitric oxide and prostaglandins, two cellular products known to be mechanically responsive in vivo. Using the device shown on the left in Fig. 2.53, they showed that the osteoblasts did not respond to uniaxial strains of 0.05–0.5% but did produce nitric oxide and prostaglandins when subjected to a period of steady shear stress of 148 dynes/cm^2 (applied using a parallel plate flow chamber; Fig. 2.56C).

2. You *et al.* [114] reached the same conclusion but used more physiologically relevant fluid flows. Theoretical models of fluid flow in the lacunar–canalicular network predict the flow will be oscillatory (i.e., reversing), with wall shear stress of the order of 8–30 dynes/cm^2 [115]. Uniaxial stretching of 0.5% strain at 1 Hz had no effect on the expression of osteopontin mRNA by the bone cells. (Osteopontin is a non-collagenous protein important in bone formation and shown to be modulated by mechanical stimulation.) When they subjected the cells to oscillatory (1 Hz) fluid flow in a parallel plate flow chamber, with shear stresses of 20 dynes/cm^2, osteopontin mRNA expression went up significantly.

As with cartilage biomechanical research, the motivating factors for understanding how bone responds to its mechanical environment are bone diseases and failures. And as with cartilage, recent developments in tissue engineering have spurred interest in using mechanical stimulation to engineer replacement bone tissue. Initial studies have shown promising results. Using three-dimensional titanium mesh matrices seeded with osteoblast precursor cells, Bancroft *et al.* [116] showed that mineralized bone tissue formed more rapidly in matrices perfused with cell culture media than in matrices not subjected to fluid flow. Interestingly, this effect occurred with flow rates as low as 0.3 ml/min (corresponding to an estimated shear stress of 1 dyne/cm^2). Although it is not clear from these experiments how a stimulus of such low magnitude could cause an increase in bone formation, the results do suggest that mechanical stimulation will be valuable for bone tissue engineering.

2.10 Problems

2.1 Write a short essay (about 300–500 words) on a eukaryotic cellular organelle of your choice, describing its structure, function, etc. (This may require a quick trip to the library; a good place to look would be under cell biology in the QH 581 section of your Science and Medicine library).

2.2 A 75 kg person doing light work requires about 3000 kcal of food energy per day, 40% of which is actually used by the body's cells. (The other 60% is lost as heat and in waste products.) Before being used by the cells, effectively all of this energy is stored in ATP, which is then cleaved into ADP and PO_3^{-2}, with the release of 12.5 kcal per mole of ATP.
   (a) How many moles of ATP are turned over per day in this fashion? What mass of ATP does this correspond to? (The molecular weight of ATP is 507 g/mol.)
   (b) The body actually contains approximately 5 g ATP. Estimate the average recycle time for an ATP molecule. You see that it is much more efficient to reuse ADP rather than to synthesize it *de novo* (from scratch).

2.3 We have discussed how ATP is produced by the mitochondria and used by the other organelles within the cell. Transport of ATP to these organelles is passive (i.e., by diffusion), and the purpose of this question is to find out how rapidly that transport can occur.
   (a) Consider a small cube of sides δx by δy by δz. The cube is filled with a solution of concentration $c(x, y, z)$. The fluid is quiescent, so that solute transport occurs by diffusion only. Fick's law (which is analogous to Fourier's law for heat transfer) states that the flux of solute is proportional to $D\nabla c$, where D is a constant known as the diffusivity. By referring to Fig. 2.60, show that the accumulation of solute within the cube must be proportional to $\nabla \cdot (D\nabla c)\, \delta x\, \delta y\, \delta z$. Hence deduce that the differential equation describing the solute distribution must be:

$$\frac{\partial c}{\partial t} = \nabla \cdot (D\nabla c). \tag{2.70}$$

   (b) Now consider a simple one-dimensional system where $c = c(x, t)$. This is an acceptable model for treating diffusion between a row of mitochondria and a contractile set of actin–myosin fibers, such as is typically found in muscle cells (see Fig. 2.60). Call the distance between the fibers and the mitochondria L and assume that initially the concentration of ATP is uniformly 5 mM within the cell. At $t = 0$ the muscle contracts, causing the concentration of ATP in the cytoplasm immediately

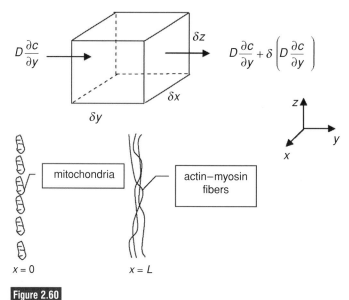

Figure 2.60

For Problem 2.3.

adjacent to the fibers to fall rapidly to zero (or at least approximately to zero.) The concentration of ATP in the surrounding cytoplasm then falls in an unsteady fashion; to a good approximation you can assume that the mitochondria act to maintain the local ATP concentration at 5 mM. Write down the boundary and initial conditions implied by this scenario.

Next, show that the function

$$\frac{c(x,t)}{c_{init}} = 1 - \frac{x}{L}$$
$$+ \sum_{n=1}^{\infty} [a_n \sin(2\pi nx/L) + b_n \cos(2\pi nx/L)] \, e^{-(2\pi n)^2 Dt/L^2}$$

$$(2.71)$$

satisfies Equation (2.70) and the boundary and initial conditions, for appropriate choice of a_n and b_n. c_{init} is the initial concentration of ATP in the cytoplasm. *Note: do not solve for* a_n *and* b_n.

(c) The second term in equation (2.71) is obviously the unsteady term, while the expression $1 - x/L$ is the steady-state solution. Express the slowest decaying mode in the unsteady solution as $e^{t/\tau}$, and identify τ. If $D = 4.5 \times 10^{-10} \, m^2/s$ for ATP, and $L = 0.5 \, \mu m$, what is τ? This represents the "lag time" needed for diffusion to begin transporting ATP to the organelles efficiently.

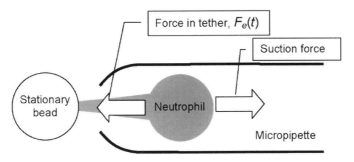

Force in tether, $F_e(t)$

Suction force

Stationary bead

Neutrophil

Micropipette

Figure 2.61

For Problem 2.5.

(d) One way of overcoming the lag mentioned above is to store a small amount of ATP in the organelle that uses it. Muscle can produce approximately 40 W/kg at high exertion, and cleaving 1 mole ATP (which has a mass of 507 g) yields 104 J. How much ATP should be stored (per kilogram of muscle) to be safe?

Note: This problem assumes that the synthesis and uptake of ATP by mitochondria and organelles is instantaneous, which is not true.

2.4 Consider a cell that has a magnetic bead attached to it. We will treat the cell as a Kelvin body. Suppose that the magnetic field has been turned on for a very long time and is producing a constant force on the bead F_0. At time $t = 0$, the force is suddenly switched off. Derive an expression for the resulting displacement of the bead as a function of time, $x(t)$. Note that you do not need to re-derive any formulae given in the text.

2.5 Consider a neutrophil in a micropipette that is being pulled away from a stationary bead by a suction force (Fig. 2.61). The neutrophil is attached to the bead by an elastic tether that lengthens over time, so that the separation between bead and cell obeys the relationship:

$$x(t) = x_{\text{final}}(1 - e^{-t/\tau}) \tag{2.72}$$

where t is time and τ is a constant.

(a) Show that the initial velocity of the neutrophil is x_{final}/τ.

(b) Using Equation (2.6) for the suction force, compute the force in the tether as a function of time, $F_e(t)$. Note that U_∞ in Equation (2.6) is the same as the initial velocity computed in (a).

(c) Show that a condition to neglect the inertia of the cell in this analysis is that the ratio $\pi R_p^2 \Delta p \tau^2 / m x_{\text{final}}$ be $\gg 1$, where m is the mass of the cell.

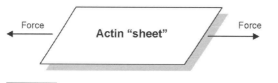

Force ⟵   Actin "sheet"   Force ⟶

Figure 2.62

For Problem 2.7.

(d) When the cell's inertia is small, what is the effective spring constant of the tether?

2.6 Consider the actin network described in Section 2.6.3.

(a) If a uniaxial force F is applied to the network, show that the energy stored in the network per unit volume is proportional to $F^2/E_s I$, where E_s is the modulus of the actin filaments and I is the area moment of inertia of a single actin filament.

(b) For a linearly elastic material subjected to uniaxial stress, the energy per unit volume stored in the material is given by $1/2E^*\varepsilon^2$, where E^* is the effective modulus of the material and ε is the strain. Derive Equation (2.47) using this fact, as well as the results of part (a) and Equation (2.44). You do not need to derive the value of the coefficient C_1. Do not repeat the derivation in the text; instead, formulate a derivation based on energy considerations.

2.7 In the text we described how the actin cytoskeleton can be modeled as a three-dimensional foam. However, in many cells, the actin forms a very thin layer adjacent to the cell's membrane. In such a case, it may be more appropriate to model the actin network as essentially two dimensional.

(a) Suppose that the actin forms a two-dimensional "sheet" that is only one fiber thick (see Fig. 2.62). Draw the two-dimensional version of Fig. 2.42A and clearly label all dimensions.

(b) Following a derivation similar to that given in the text, compute the dependence of relative elastic modulus E^*/E_s on relative network actin density, ρ^*/ρ_s. Here ρ^* is the actin density within the actin sheet.

2.8 The density of actin filaments (and other cytoskeletal elements) is not spatially uniform within a cell (e.g., Helmke *et al.* [117]). Assume that the density of actin varies with position in a cell according to

$$\rho^*(x) = \rho_0 e^{-kx} \qquad (2.73)$$

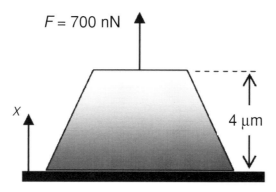

$F = 700$ nN

X

4 μm

Figure 2.63

For Problem 2.8

with $\rho_0 = 25$ mg/ml and $k = 0.1$ μm^{-1}. Further, assume that the cross-sectional area of the cell (measured normal to the x axis); varies as

$$A(x) = A_0 e^{-cx} \tag{2.74}$$

with $A_0 = 350$ μm^2 and $c = 0.05$ μm^{-1}. Neglecting the nucleus, calculate the vertical extension of a 4 μm tall cell as a result of a 700 nN force being applied to its top, assuming the bottom of the cell is anchored to a substrate (Fig. 2.63). Use values given in the text as appropriate. Note that 1 dyne/cm$^2 = 0.1$ Pa.

2.9 A magnetic bead microrheometry system will be used to measure the mechanical properties of adherent cells.

(a) The first step is to calibrate the system by observing how quickly beads are dragged through a glycerol solution by the magnet. For this particular magnet, we expect the magnetic force on a bead, F_{mag}, will be given by an expression of the form $F_{mag} = \beta/z$, where β is a constant and z is the distance between the magnet pole piece tip and the bead. Knowing that the fluid drag force on a bead of diameter D moving at velocity V is given by Stokes' law, $F_{Drag} = 3\pi\mu DV$, and neglecting the mass of the bead, derive an expression for the constant β. Your expression will involve the starting position of the bead, z_0, as well as the bead position $z(t)$ and time t.

(b) The magnet was designed to have a coefficient $\beta = 200$ nN μm. Calibration data in Table 2.3 were gathered to test magnet performance using 4.5 μm diameter beads immersed in glycerol with viscosity $\mu = 1.2$ g/(cm s); distance from the magnet to the bead was measured as a function of time. Are these data compatible with the design

Table 2.3. For Problem 2.9

Time (s)	z (μm)
0.000	100
0.095	80
0.158	60
0.214	40
0.246	20

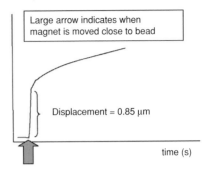

Bead displacement (μm)

Large arrow indicates when magnet is moved close to bead

Displacement = 0.85 μm

time (s)

Figure 2.64

For Problem 2.9.

specifications? There is some noise in the data, so you should consider all the data points somehow.

(c) Assume that the cell can be modeled as a Kelvin body plus an extra dashpot (Fig. 2.36). If the magnet pole piece tip is suddenly moved to within 50 μm of the bead and the bead moves 0.85 μm (see Fig. 2.64), use Equation (2.18) to determine the value of $k_0 + k_1$. For purposes of this question take $\beta = 200$ nN μm.

(d) Can you think of any experimental difficulties with this approach, making reference to the graph in Fig. 2.64?

(e) Cells are treated with an agent called Latrunculin-B. Here is a description of this agent from the Calbiochem catalogue: "A structurally unique marine toxin. Inhibits actin polymerization *in vitro* and disrupts microfilament organization as well as microfilament-mediated processes." What do you think will happen to the magnitude of $k_0 + k_1$ in cells treated with Latrunculin-B? Briefly justify your answer.

2.10 A cartilage sample consists of cells (chondrocytes, with effective Young's modulus E_{cell}) and ECM (effective Young's modulus E_{ECM}), as shown in the left panel of Fig. 2.65. The volume fraction of cells (cell volume/total tissue volume) is ϕ. The tissue sample is subjected to a uniaxial tension, T, and has unstretched length L and cross-sectional area A. We wish to determine the effective Young's modulus for the tissue sample, E_{tissue}. In general, this is a complicated problem; however, we can get bounds for E_{tissue} by considering two special cases. In the first case, we replace the real tissue configuration by a "series" configuration where a cell-containing volume is in series with an ECM-containing volume (see middle panel of Fig. 2.65). In the second case, we replace the real configuration by a "parallel" arrangement. In both cases, we require the total tension, T, applied to the tissue to match that for the real case, and the total elongation to also match that occurring in the real case.

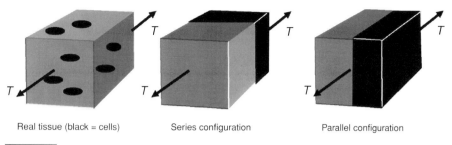

Real tissue (black = cells)   Series configuration   Parallel configuration

Figure 2.65

For Problem 2.10.

(a) Considering the work done by the tension force T, show that the energy stored in the real tissue sample is $ALE_{\text{tissue}}\varepsilon^2/2$.

(b) Consider the series configuration. You should first convince yourself that in this configuration, the stress σ is the same for both tissue components and is equal to the true stress in the tissue, but the strain ε is different for the two components. Thinking about the energy stored in each of the two tissue components, show that the overall tissue modulus can be written as $1/E_{\text{tissue}} = \phi/E_{\text{cell}} + (1 - \phi)/E_{\text{ECM}}$.

(c) Using a similar approach as for the series configuration, show that $E_{\text{tissue}} = \phi E_{\text{cell}} + (1 - \phi)E_{\text{ECM}}$ for the parallel configuration. Hint: What is the same for the two components? (It is not the stress.)

2.11 A cilium of length $L = 2\,\mu\text{m}$ and diameter $0.4\,\mu\text{m}$ sweeps through extracellular fluid ($\mu = 1\,c\text{P}$) with constant angular velocity, ω. You can model this motion as rotation of the cilium about its base, assuming that the cilium has the form of a rigid cylinder.

(a) The drag force exerted per unit length on a cylinder is $f = k_1\mu U$, where k_1 is a constant and U is the *local* fluid velocity passing over the cylinder. Show that the torque, M_0, exerted on the base of the cilium by the cell body must therefore be $k_1\mu\omega L^3/3$.

(b) Assume that the nine tubulin dimers (Fig. 2.66) are evenly spaced around the periphery of the cilium. Further assume that the perimeter to the left of the solid line is the trailing edge and the perimeter to the right is the leading edge (see cross-sectional view in Fig. 2.66). All dimers to the right of this line are under maximum tension (tension $= F_{\max}$); all to the left are under maximum compression (compression $= -F_{\max}$). Compute the value of $F_{\max}$ at the base of the cilium, assuming that the dimers transfer the entire torque M_0 from the cell body to the cilium. Take $k_1 = 2$ and $\omega = 0.2\,\text{rad/s}$. Make and state other assumptions as needed.

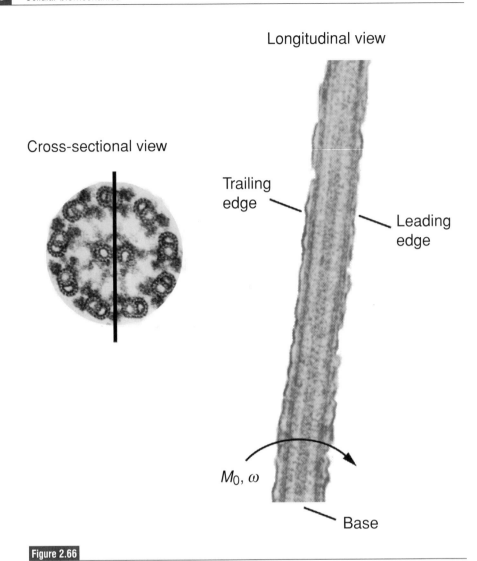

Longitudinal view

Cross-sectional view

Trailing
edge

Leading
edge

M_0, ω

Base

Figure 2.66

For Problem 2.11. Modified from Cormack [18] (with permission of Lippincott Williams & Wilkins) and Alberts *et al.* [1].

2.12   The flexural rigidity of a rod-like structure like a microtubule is a measure of its bending stiffness. Flexural rigidity is given by EI, where E is the Young's modulus and I is the geometrical moment of inertia of the rod cross-section. For a hollow cylindrical cylinder with outer diameter, D_o, and inner diameter, D_i, the moment of inertia about an axis through the middle of its cross-section is $I_z = \pi(D_o^4 - D_i^4)/64$. The flexural rigidity of microtubules can be measured in several ways. For example, Kurachi *et al.* [118] determined the flexural rigidity of microtubules grown in solution

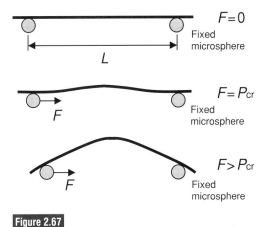

Figure 2.67

For Problem 2.12.

and stabilized by microtubule-associated proteins by applying compressive forces to both ends of the microtubule and determining the force at which the microtubule buckled. To apply the compressive end forces, microspheres were attached to either end of the microtubule, one microsphere was fixed to the surface, and the other microsphere was moved using optical tweezers to compress the microtubule until it buckled (i.e., initially deflected) (Fig. 2.67). Assuming the microtubule behaves like a slender rod of length L, the critical force, P_{cr}, at which buckling occurs is given by Euler's formula:

$$P_{cr} = \frac{\pi^2 E I}{L^2} \tag{2.75}$$

Applying this equation, the flexural rigidity, EI, is estimated by measuring the force at which a microtubule of known length is observed to buckle.

(a) Critical loads for 10 microtubules are shown in Table 2.4. Calculate the flexural rigidity for each microtubule and determine the mean flexural rigidity for all the microtubules tested.

(b) Several other studies (e.g., Felgner *et al.* [119]) have shown that flexural rigidity is constant for microtubules of different lengths. Show that the data presented here do not agree with those findings and speculate on potential sources of error in the microsphere/optical tweezer method.

(c) Use the mean flexural rigidity you calculated in (b) to estimate the Young's modulus of a microtubule. State your assumptions about the geometry of a microtubule.

Table 2.4. Critical buckling loads for 10 microtubules stabilized by microtubule-associated protein. Data approximated from graphical data presented in Kurachi *et al.* [118].

Microtubule	Length (μm)	Critical load (pN)
1	7.5	2.6
2	10.5	2.9
3	22.0	2.1
4	20.0	2.0
5	10.0	3.8
6	30.5	1.1
7	18.0	1.7
8	19.0	1.1
9	9.0	3.7
10	28.0	1.4

(d) It is believed that microtubules in the cell are loaded primarily in compression or bending, whereas actin is primarily loaded in tension. Demonstrate that microtubules are designed to withstand much higher bending loads than are actin filaments.

2.13 Recall that AFM can be used in force mapping mode to determine the local modulus of a cell. By rastering the AFM probe across the surface of the cell, an elasticity map can be obtained. As described in Section 2.5.1, if we model the AFM as a rigid, conical tip penetrating the soft, locally planar and linearly elastic cell, then the Hertz contact solution applies. For this situation, the penetration depth of the probe, u_z, is related to the contact radius, a, by:

$$u_z = \left(\frac{\pi}{2} - \frac{r}{a} \right) a \cot \alpha \quad r \leqslant a \qquad (2.76)$$

where r is the radial coordinate, α is the cone half angle, and u_z is measured relative to the undeformed surface (Fig. 2.68).

(a) Determine an expression for the contact radius in terms of the penetration depth at the end of the probe tip, δ, and the cone half angle.

(b) Express the contact radius, a, as a function of the cone half angle, the material properties of the cell, and the applied force.

(c) Plot the contact diameter (i.e., twice the radius) versus the modulus of the cell for 1, 10, 100, and 1000 pN, the range of forces that are typically applied to cells with AFM. Assume that the cone half

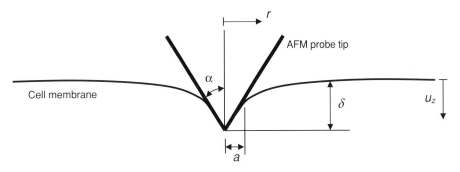

Figure 2.68

For Problem 2.13.

angle for an AFM probe tip is $30°$ and that the cell is incompressible (i.e., the cell volume does not change when it is deformed by the AFM probe tip). You will find it clearer to use a log–log plot for this question.

(d) Show on your plot the typical range of cell moduli and applied forces and use this to determine how large an area is indented by the AFM probe tip. This defines the lateral resolution that can be expected for an elasticity map. Is this resolution adequate? Why or why not?

2.14 We derived a relationship in Section 2.6.2 between the incremental modulus, E_0, of a six-strut tensegrity model of the cell and the resting force in the actin filaments, F_0, the resting length of the actin filaments, l_0, and the resting strain of the cell, ε_0 (Equation [2.39]). Recalling that the length of the actin filaments is related to the length of the microtubules, L_0, by $l_0 = \sqrt{3/8}L_0$, we can express E_0 in terms of L_0 as

$$E_0 = 15.6\frac{F_0}{L_0^2}\frac{1 + 4\epsilon_0}{1 + 12\epsilon_0} \tag{2.77}$$

In this question, you will use this equation to estimate the upper and lower bounds of E_0 as predicted by the tensegrity model.

(a) The upper bound of the prediction can be determined by assuming the actin filaments are on the verge of breaking in the resting position. Assume that actin filaments have an effective radius of 2.8 nm, a Young's modulus of 1.8 GPa, and break at an average force of approximately 400 pN. Use these values to estimate the strain at which an actin filament will break.

(b) Plot E_0 versus L_0 for $L_0 = 1$–6 μm, taking into account the estimate for resting strain you calculated in (a). You will find it clearer to use a

Table 2.5. Young's moduli for a variety of cell types and measurement techniques. Data after Stamenovic and Coughlin [71].

Cell type	Measurement technique	L_0 (μm)[a]	E_0 (dyne/cm^2)
Spread fibroblast	Atomic force microscopy	2	16 000
Sheared endothelial cell	Micropipette aspiration	3	1575
Round endothelial cell	Micropipette aspiration	3	750
Spread epithelial cell	Magnetic bead rheometry	4	75
Spread smooth muscle cell	Magnetic bead rheometry	5.5	115
Spread endothelial cell	Magnetic bead rheometry	5.5	45
Round endothelial cell	Magnetic bead rheometry	5.5	22

[a] Assumed to be the probe diameter.

log axis for the modulus. This curve represents the upper bound of the tensegrity prediction.

(c) Based on force balances on the struts, it can be shown that the force in a strut in the resting position, P_0, is related to the resting force in the actin filament by $P_0 = \sqrt{6}F_0$. The lower bound can be determined by assuming the microtubules are on the verge of buckling in the resting position. Express E_0 in terms of the microtubule flexural rigidity, initial actin filament strain, and L_0 assuming the microtubules are on the verge of buckling in the resting position.

(d) When the microtubules are on the verge of buckling as described in (c), the actin filament strains are relatively small (i.e., $\varepsilon_0 \ll 1$) and contribute negligibly to the estimate of E_0. On the same graph as you produced in part (b), plot E_0 versus L_0 for $L_0 = 1$–6 μm for the case of small strain and the microtubules on the verge of buckling. You may assume the flexural rigidity of a microtubule is 21.5 pN μm^2. This curve represents the lower bound of the tensegrity prediction.

(e) Cell stiffnesses for various cell types and measured with a variety of techniques are listed in Table 2.5. If we assume that the six-strut tensegrity model describes the local deformation of the cytoskeleton around the point at which the cell is probed, then L_0 can be approximated as the diameter of the probe. These values are also listed in Table 2.5. Plot the experimental data in Table 2.5 on your graph of the upper and lower bounds for E_0 predicted by the tensegrity model.

(f) Comment briefly on the comparison between the model predictions for E_0 and the experimentally observed values from Table 2.5.

References

1. B. Alberts, A. Johnson, J. Lewis, M. Raff, K. Roberts *et al. Molecular Biology of the Cell*, 4th edn (New York: Garland Science, 2002).

2. H. Lodish, A. Berk, S. L. Zipursky, P. Matsudaira, D. Baltimore *et al. Molecular Cell Biology*, 4th edn (New York: W. H. Freeman, 2000).

3. A. J. Vander, J. H. Sherman and D. S. Luciano. *Human Physiology: The Mechanisms of Body Function*, 4th edn (New York: McGraw-Hill, 1985).

4. A. D. Bershadsky and J. M. Vasiliev. *Cytoskeleton* (New York: Plenum Press, 1988).

5. Theoretical and Computational Biophysics Group at University of Illinois at Urbana-Champaign. Available at http://www.ks.uiuc.edu/Research/cell_motility/actin/ (2005).

6. M. Lorenz, D. Popp and K. C. Holmes. Refinement of the F-actin model against X-ray fiber diffraction data by the use of a directed mutation algorithm. *Journal of Molecular Biology*, **234** (1993), 826–836.

7. W. Kabsch, H. G. Mannherz, D. Suck, E. F. Pai and K. C. Holmes. Atomic structure of the actin: DNase I complex. *Nature*, **347** (1990), 37–44.

8. J. L. McGrath, E. A. Osborn, Y. S. Tardy, C. F. Dewey, Jr. and J. H. Hartwig. Regulation of the actin cycle in vivo by actin filament severing. *Proceedings of the National Academy of Sciences USA*, **97** (2000), 6532–6537.

9. H. Kojima, A. Ishijima and T. Yanagida. Direct measurement of stiffness of single actin filaments with and without tropomyosin by in vitro nanomanipulation. *Proceedings of the National Academy of Sciences USA*, **91** (1994), 12962–12966.

10. H. E. Huxley, A. Stewart, H. Sosa and T. Irving. X-ray diffraction measurements of the extensibility of actin and myosin filaments in contracting muscle. *Biophysical Journal*, **67** (1994), 2411–2421.

11. H. Higuchi, T. Yanagida and Y. E. Goldman. Compliance of thin filaments in skinned fibers of rabbit skeletal muscle. *Biophysical Journal*, **69** (1995), 1000–1010.

12. K. Wakabayashi, Y. Sugimoto, H. Tanaka, Y. Ueno, Y. Takezawa *et al.* X-ray diffraction evidence for the extensibility of actin and myosin filaments during muscle contraction. *Biophysical Journal*, **67** (1994), 2422–2435.

13. P. A. Janmey, J. X. Tang and C. F. Schmidt. Actin filaments, In *Biophysical Society Online Textbook of Biophysics*, ed. V. Bloomfield. (Heidelberg: Springer, 1999). Available at http://www.biophysics.org/education/janmey.pdf.

14. P. A. Coulombe and P. Wong. Cytoplasmic intermediate filaments revealed as dynamic and multipurpose scaffolds. *Nature Cell Biology*, **6** (2004), 699–706.

15. N. Wang and D. Stamenovic. Mechanics of vimentin intermediate filaments. *Journal of Muscle Research and Cell Motility*, **23** (2002), 535–540.

16. N. Marceau, A. Loranger, S. Gilbert, N. Daigle and S. Champetier. Keratin-mediated resistance to stress and apoptosis in simple epithelial cells in relation to health and disease. *Biochemistry and Cell Biology*, **79** (2001), 543–555.

17. H. Lodish, A. Berk, P. Matsudaira, C. A. Kaiser, M. Krieger *et al. Molecular Cell Biology*, 5th edn (New York: W. H. Freeman, 2004).

18. D. H. Cormack. *Ham's Histology*, 9th edn (London: Lippincott, 1987).

19. E. J. DuPraw. *Cell and Molecular Biology* (New York: Academic Press, 1968).

20. K. Sato, K. Yomogida, T. Wada, T. Yorihuzi, Y. Nishimune *et al.* Type XXVI collagen, a new member of the collagen family, is specifically expressed in the testis and ovary. *Journal of Biological Chemistry*, **277** (2002), 37678–37684.

21. A. L. Lehninger. *Biochemistry*, 2nd edn (New York: Worth, 1975).

22. E. Bertini and G. Pepe. Collagen type VI and related disorders: Bethlem myopathy and Ullrich scleroatonic muscular dystrophy. *European Journal of Pediatric Neurology*, **6** (2002), 193–198.

23. T. Gloe and U. Pohl. Laminin binding conveys mechanosensing in endothelial cells. *News in Physiological Sciences*, **17** (2002), 166–169.

24. J. L. Bert and R. H. Pearce. The interstitium and microvascular exchange. In *The Handbook of Physiology, IV: Microcirculation, Part 1: Section 2: The Cardiovascular System*, ed. E. M. Renkin and C. C. Michel. Series ed. S. R. Geiger. (Bethesda, MD: American Physiological Society: 1984), pp. 521–549.

25. E. D. Hay. *Cell Biology of Extracellular Matrix*, 2nd edn (Dordrecht, Netherlands: Kluwer Academic, 1992).

26. K. Vuori. Integrin signaling: tyrosine phosphorylation events in focal adhesions. *Journal of Membrane Biology*, **165** (1998), 191–199.

27. E. Zamir and B. Geiger. Molecular complexity and dynamics of cell–matrix adhesions. *Journal of Cell Science*, **114** (2001), 3583–3590.

28. R. M. Hochmuth. Micropipette aspiration of living cells. *Journal of Biomechanics*, **33** (2000), 15–22.

29. D. Leckband. Measuring the forces that control protein interactions. *Annual Review of Biophysics and Biomolecular Structure*, **29** (2000), 1–26.

30. A. Ikai, R. Afrin, H. Sekiguchi, T. Okajima, M. T. Alam *et al.* Nano-mechanical methods in biochemistry using atomic force microscopy. *Current Protein and Peptide Science*, **4** (2003), 181–193.

31. A. Vinckier and G. Semenza. Measuring elasticity of biological materials by atomic force microscopy. *FEBS Letters*, **430** (1998), 12–16.

32. M. Radmacher. Measuring the elastic properties of biological samples with the AFM. *IEEE Engineering in Medicine and Biology Magazine*, **16** (1997), 47–57.

33. I. N. Sneddon. The relation between load and penetration in the axisymmetric Boussinesq problem for a punch of arbitrary profile. *International Journal of Engineering Science*, **3** (1965), 47–57.

34. A. C. Fischer-Cripps. *Introduction to Contact Mechanics* (New York: Springer Verlag, 2000).

35. H. Haga, S. Sasaki, K. Kawabata, E. Ito, T. Ushiki *et al.* Elasticity mapping of living fibroblasts by AFM and immunofluorescence observation of the cytoskeleton. *Ultramicroscopy*, **82** (2000), 253–258.

36. A. Ashkin. Optical trapping and manipulation of neutral particles using lasers. *Proceedings of the National Academy of Sciences USA*, **94** (1997), 4853–4860.

37. D. G. Grier. A revolution in optical manipulation. *Nature*, **424** (2003), 810–816.

38. S. C. Kuo. Using optics to measure biological forces and mechanics. *Traffic*, **2** (2001), 757–763.

39. J. Chen, B. Fabry, E. L. Schiffrin and N. Wang. Twisting integrin receptors increases endothelin-1 gene expression in endothelial cells. *American Journal of Physiology, Cell Physiology*, **280** (2001), C1475–C1484.

40. M. Puig-De-Morales, M. Grabulosa, J. Alcaraz, J. Mullol, G. N. Maksym *et al.* Measurement of cell microrheology by magnetic twisting cytometry with frequency domain demodulation. *Journal of Applied Physiology*, **91** (2001), 1152–1159.

41. N. Wang and D. E. Ingber. Probing transmembrane mechanical coupling and cytomechanics using magnetic twisting cytometry. *Biochemistry and Cell Biology*, **73** (1995), 327–335.

42. A. R. Bausch, F. Ziemann, A. A. Boulbitch, K. Jacobson and E. Sackmann. Local measurements of viscoelastic parameters of adherent cell surfaces by magnetic bead microrheometry. *Biophysical Journal*, **75** (1998), 2038–2049.

43. J. M. Mitchison and M. M. Swann. The mechanical properties of the cell surface. 1. The cell elastimeter. *Journal of Experimental Biology*, **31** (1954), 443–460.

44. J. Y. Shao and R. M. Hochmuth. Micropipette suction for measuring piconewton forces of adhesion and tether formation from neutrophil membranes. *Biophysical Journal*, **71** (1996), 2892–2901.

45. J. S. Oghalai, A. A. Patel, T. Nakagawa and W. E. Brownell. Fluorescence-imaged microdeformation of the outer hair cell lateral wall. *Journal of Neuroscience*, **18** (1998), 48–58.

46. S. Chien and K. L. Sung. Effect of colchicine on viscoelastic properties of neutrophils. *Biophysical Journal*, **46** (1984), 383–386.

47. C. Dong, R. Skalak and K. L. Sung. Cytoplasmic rheology of passive neutrophils. *Biorheology*, **28** (1991), 557–567.

48. J. Y. Shao. Finite element analysis of imposing femtonewton forces with micropipette aspiration. *Annals of Biomedical Engineering*, **30** (2002), 546–554.

49. J. Y. Shao and J. Xu. A modified micropipette aspiration technique and its application to tether formation from human neutrophils. *Journal of Biomechanical Engineering*, **124** (2002), 388–396.

50. M. A. Tsai, R. S. Frank and R. E. Waugh. Passive mechanical behavior of human neutrophils: effect of cytochalasin B. *Biophysical Journal*, **66** (1994), 2166–2172.

51. D. E. Discher. New insights into erythrocyte membrane organization and microelasticity. *Current Opinion in Hematology*, **7** (2000), 117–122.

52. E. Evans, K. Ritchie and R. Merkel. Sensitive force technique to probe molecular adhesion and structural linkages at biological interfaces. *Biophysical Journal*, **68** (1995), 2580–2587.

53. E. A. Evans and P. L. La Celle. Intrinsic material properties of the erythrocyte membrane indicated by mechanical analysis of deformation. *Blood*, **45** (1975), 29–43.

54. R. M. Hochmuth, H. C. Wiles, E. A. Evans and J. T. McCown. Extensional flow of erythrocyte membrane from cell body to elastic tether. II. Experiment. *Biophysical Journal*, **39** (1982), 83–89.

55. A. W. Jay and P. B. Canham. Viscoelastic properties of the human red blood cell membrane. II. Area and volume of individual red cells entering a micropipette. *Biophysical Journal*, **17** (1977), 169–178.

56. D. Lerche, M. M. Kozlov and W. Meier. Time-dependent elastic extensional RBC deformation by micropipette aspiration: redistribution of the spectrin network? *European Biophysics Journal*, **19** (1991), 301–309.

57. G. B. Nash and W. B. Gratzer. Structural determinants of the rigidity of the red cell membrane. *Biorheology*, **30** (1993), 397–407.

58. T. Shiga, N. Maeda, T. Suda, K. Kon, M. Sekiya and S. Oka. A kinetic measurement of red cell deformability: a modified micropipette aspiration technique. *Japanese Journal of Physiology*, **29** (1979), 707–722.

59. W. E. Brownell, A. A. Spector, R. M. Raphael and A. S. Popel. Micro- and nanomechanics of the cochlear outer hair cell. *Annual Review of Biomedical Engineering*, **3** (2001), 169–194.

60. P. S. Sit, A. A. Spector, A. J. Lue, A. S. Popel and W. E. Brownell. Micropipette aspiration on the outer hair cell lateral wall. *Biophysical Journal*, **72** (1997), 2812–2819.

61. A. A. Spector, W. E. Brownell and A. S. Popel. A model for cochlear outer hair cell deformations in micropipette aspiration experiments: an analytical solution. *Annals of Biomedical Engineering*, **24** (1996), 241–249.

62. D. E. Discher, D. H. Boal and S. K. Boey. Simulations of the erythrocyte cytoskeleton at large deformation. II. Micropipette aspiration. *Biophysical Journal*, **75** (1998), 1584–1597.

63. R. Waugh and E. A. Evans. Thermoelasticity of red blood cell membrane. *Biophysical Journal*, **26** (1979), 115–131.

64. D. E. Discher, N. Mohandas and E. A. Evans. Molecular maps of red cell deformation: hidden elasticity and in situ connectivity. *Science*, **266** (1994), 1032–1035.

65. D. H. Boal. *Mechanics of the Cell* (Cambridge: Cambridge University Press, 2002).

66. R. E. Waugh and R. G. Bauserman. Physical measurements of bilayer–skeletal separation forces. *Annals of Biomedical Engineering*, **23** (1995), 308–321.

67. J. Dai and M. P. Sheetz. Mechanical properties of neuronal growth cone membranes studied by tether formation with laser optical tweezers. *Biophysical Journal*, **68** (1995), 988–996.

68. R. M. Hochmuth, J. Y. Shao, J. Dai and M. P. Sheetz. Deformation and flow of membrane into tethers extracted from neuronal growth cones. *Biophysical Journal*, **70** (1996), 358–369.

69. L. Klaus. Integration of inflammatory signals by rolling neutrophils. *Immunological Reviews*, **186** (2002), 8–18.

70. D. W. Schmidtke and S. L. Diamond. Direct observation of membrane tethers formed during neutrophil attachment to platelets or P-selectin under physiological flow. *Journal of Cell Biology*, **149** (2000), 719–730.

71. D. Stamenovic and M. F. Coughlin. A quantitative model of cellular elasticity based on tensegrity. *Journal of Biomechanical Engineering*, **122** (2000), 39–43.

72. D. E. Ingber. Tensegrity II. How structural networks influence cellular information processing networks. *Journal of Cell Science*, **116** (2003), 1397–1408.

73. Y. C. Fung. *Biomechanics: Mechanical Properties of Living Tissues*, 2nd edn (New York: Springer Verlag, 1993).

74. D. Stamenovic and D. E. Ingber. Models of cytoskeletal mechanics of adherent cells. *Biomechanics and Modeling in Mechanobiology*, **1** (2002), 95–108.

75. N. Wang, K. Naruse, D. Stamenovic, J. J. Fredberg, S. M. Mijailovich *et al.* Mechanical behavior in living cells consistent with the tensegrity model. *Proceedings of the National Academy of Sciences USA*, **98** (2001), 7765–7770.

76. N. Wang, I. M. Tolic-Norrelykke, J. Chen, S. M. Mijailovich, J. P. Butler *et al.* Cell prestress. I. Stiffness and prestress are closely associated in adherent contractile cells. *American Journal of Physiology, Cell Physiology*, **282** (2002), C606–C616.

77. D. Stamenovic and M. F. Coughlin. The role of prestress and architecture of the cytoskeleton and deformability of cytoskeletal filaments in mechanics of adherent cells: a quantitative analysis. *Journal of Theoretical Biology*, **201** (1999), 63–74.

78. D. Stamenovic and N. Wang. Invited review: engineering approaches to cytoskeletal mechanics. *Journal of Applied Physiology*, **89** (2000), 2085–2090.

79. B. P. Helmke, A. B. Rosen and P. F. Davies. Mapping mechanical strain of an endogenous cytoskeletal network in living endothelial cells. *Biophysical Journal*, **84** (2003), 2691–2699.

80. L. J. Gibson and M. F. Ashby. *Cellular Solids: Structure and Properties*, 2nd edn (Cambridge: Cambridge University Press, 1997).

81. R. L. Satcher, Jr. and C. F. Dewey, Jr. Theoretical estimates of mechanical properties of the endothelial cell cytoskeleton. *Biophysical Journal*, **71** (1996), 109–118.

82. W. A. Hodge, R. S. Fijan, K. L. Carlson, R. G. Burgess, W. H. Harris *et al.* Contact pressures in the human hip-joint measured in vivo. *Proceedings of the National Academy of Sciences USA*, **83** (1986), 2879–2883.

83. F. Guilak. The deformation behavior and viscoelastic properties of chondrocytes in articular cartilage. *Biorheology*, **37** (2000), 27–44.

84. C. A. Hansen, A. G. Schroering, D. J. Carey and J. D. Robishaw. Localization of a heterotrimeric G protein gamma subunit to focal adhesions and associated stress fibers. *Journal of Cell Biology*, **126** (1994), 811–819.

85. P. F. Davies. Flow-mediated endothelial mechanotransduction. *Physiological Reviews*, **75** (1995), 519–560.

86. A. J. Maniotis, C. S. Chen and D. E. Ingber. Demonstration of mechanical connections between integrins, cytoskeletal filaments, and nucleoplasm that

stabilize nuclear structure. *Proceedings of the National Academy of Sciences USA*, **94** (1997), 849–854.

87. F. G. Giancotti and E. Ruoslahti. Integrin signaling. *Science*, **285** (1999), 1028–1032.

88. T. D. Brown. Techniques for mechanical stimulation of cells in vitro: a review. *Journal of Biomechanics*, **33** (2000), 3–14.

89. J. A. Frangos (ed.) *Physical Forces and the Mammalian Cell* (San Diego: Academic Press, 1993).

90. R. B. Bird, W. E. Stewart and E. N. Lightfoot. *Transport Phenomena* (New York: John Wiley, 1960).

91. J. A. Gilbert, P. S. Weinhold, A. J. Banes, G. W. Link and G. L. Jones. Strain profiles for circular cell culture plates containing flexible surfaces employed to mechanically deform cells in vitro. *Journal of Biomechanics*, **27** (1994), 1169–1177.

92. J. L. Schaffer, M. Rizen, G. J. L'Italien, A. Benbrahim, J. Megerman *et al.* Device for the application of a dynamic biaxially uniform and isotropic strain to a flexible cell culture membrane. *Journal of Orthopaedic Research*, **12** (1994), 709–719.

93. S. Timoshenko and J. N. Goodier. *Theory of Elasticity* (New York: McGraw-Hill, 1970).

94. G. Helmlinger, R. V. Geiger, S. Schreck and R. M. Nerem. Effects of pulsatile flow on cultured vascular endothelial cell morphology. *Journal of Biomechanical Engineering*, **113** (1991), 123–131.

95. N. DePaola, P. F. Davies, W. F. Pritchard, Jr., L. Florez, N. Harbeck *et al.* Spatial and temporal regulation of gap junction connexin43 in vascular endothelial cells exposed to controlled disturbed flows in vitro. *Proceedings of the National Academy of Sciences USA*, **96** (1999), 3154–3159.

96. M. H. Ross and E. J. Reith. *Histology: A Text and Atlas* (New York: Harper and Rowe, 1985).

97. S. Noria, F. Xu, S. McCue, M. Jones, A. I. Gotlieb and B. L. Langille. Assembly and reorientation of stress fibers drives morphological changes to endothelial cells exposed to shear stress. *American Journal of Pathology*, **164** (2004), 1211–1223.

98. B. L. Langille and F. O'Donnell. Reductions in arterial diameter produced by chronic decreases in blood flow are endothelium-dependent. *Science*, **231** (1986), 405–407.

99. C. K. Zarins, M. A. Zatina, D. P. Giddens, D. N. Ku and S. Glagov. Shear stress regulation of artery lumen diameter in experimental atherogenesis. *Journal of Vascular Surgery*, **5** (1987), 413–420.

100. I. Mills, C. R. Cohen, K. Kamal, G. Li, T. Shin *et al.* Strain activation of bovine aortic smooth muscle cell proliferation and alignment: study of strain dependency and the role of protein kinase A and C signaling pathways. *Journal of Cellular Physiology*, **170** (1997), 228–234.

101. G. K. Owens. Regulation of differentiation of vascular smooth muscle cells. *Physiological Reviews*, **75** (1995), 487–517.

102. B. S. Kim, J. Nikolovski, J. Bonadio and D. J. Mooney. Cyclic mechanical strain regulates the development of engineered smooth muscle tissue. *Nature Biotechnology*, **17** (1999), 979–983.

103. A. J. Grodzinsky, M. E. Levenston, M. Jin and E. H. Frank. Cartilage tissue remodeling in response to mechanical forces. *Annual Review of Biomedical Engineering*, **2** (2000), 691–713.

104. R. L. Sah, Y. J. Kim, J. Y. Doong, A. J. Grodzinsky, A. H. Plaas *et al.* Biosynthetic response of cartilage explants to dynamic compression. *Journal of Orthopaedic Research*, **7** (1989), 619–636.

105. B. Kurz, M. Jin, P. Patwari, D. M. Cheng, M. W. Lark *et al.* Biosynthetic response and mechanical properties of articular cartilage after injurious compression. *Journal of Orthopaedic Research*, **19** (2001), 1140–1146.

106. M. E. Levenston, E. H. Frank and A. J. Grodzinsky. Electrokinetic and poroelastic coupling during finite deformations of charged porous media, *Journal of Applied Mechanics*, **66** (1999), 323–333.

107. E. H. Frank, M. Jin, A. M. Loening, M. E. Levenston and A. J. Grodzinsky. A versatile shear and compression apparatus for mechanical stimulation of tissue culture explants. *Journal of Biomechanics*, **33** (2000), 1523–1527.

108. R. L. Mauck, M. A. Soltz, C. C. Wang, D. D. Wong, P. H. Chao *et al.* Functional tissue engineering of articular cartilage through dynamic loading of chondrocyte-seeded agarose gels. *Journal of Biomechanical Engineering*, **122** (2000), 252–260.

109. D. B. Burr, A. G. Robling and C. H. Turner. Effects of biomechanical stress on bones in animals. *Bone*, **30** (2002), 781–786.

110. P. G. Ziros, A. P. Gil, T. Georgakopoulos, I. Habeos, D. Kletsas *et al.* The bone-specific transcriptional regulator Cbfa1 is a target of mechanical signals in osteoblastic cells. *Journal of Biological Chemistry*, **277** (2002), 23934–23941.

111. S. P. Fritton and C. T. Rubin. In vivo measurement of bone deformations using strain gauges. In *Bone Mechanics Handbook*, 2nd edn, ed. S. C. Cowin. (Boca Raton, FL: CRC Press, 2001), pp. 8.10–8.34.

112. E. H. Burger and J. Klein-Nulend. Mechanotransduction in bone: role of the lacuno-canalicular network. *FASEB Journal*, **13** Suppl (1999), S101–S112.

113. R. Smalt, F. T. Mitchell, R. L. Howard and T. J. Chambers. Induction of NO and prostaglandin E_2 in osteoblasts by wall-shear stress but not mechanical strain. *American Journal of Physiology*, **273** (1997), E751–E758.

114. J. You, C. E. Yellowley, H. J. Donahue, Y. Zhang, Q. Chen *et al.* Substrate deformation levels associated with routine physical activity are less stimulatory to bone cells relative to loading-induced oscillatory fluid flow. *Journal of Biomechanical Engineering*, **122** (2000), 387–393.

115. S. Weinbaum, S. C. Cowin and Y. Zeng. A model for the excitation of osteocytes by mechanical loading-induced bone fluid shear stresses. *Journal of Biomechanics*, **27** (1994), 339–360.

116. G. N. Bancroft, V. I. Sikavitsas, J. van den Dolder, T. L. Sheffield, C. G. Ambrose, J. A. Jansen *et al.* Fluid flow increases mineralized matrix deposition in 3D perfusion culture of marrow stromal osteoblasts in a dose-dependent manner. *Proceedings of the National Academy of Sciences USA*, **99** (2002), 12600–12605.

117. B. P. Helmke, D. B. Thakker, R. D. Goldman and P. F. Davies. Spatiotemporal analysis of flow-induced intermediate filament displacement in living endothelial cells. *Biophysical Journal*, **80** (2001), 184–194.

118. M. Kurachi, M. Hoshi and H. Tashiro. Buckling of a single microtubule by optical trapping forces: direct measurement of microtubule rigidity. *Cell Motility and the Cytoskeleton*, **30** (1995), 221–228.

119. H. Felgner, R. Frank and M. Schliwa. Flexural rigidity of microtubules measured with the use of optical tweezers. *Journal of Cell Science*, **109** (1996), 509–516.

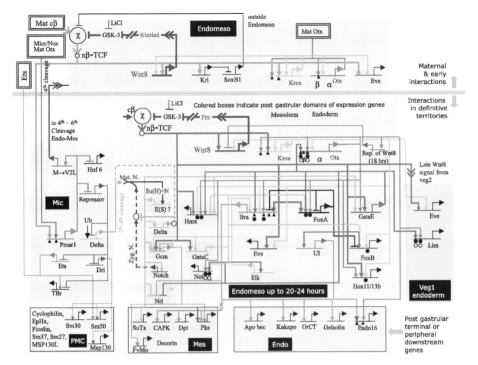

Figure 1.7

Genetic network regulating the development of endoderm and mesoderm in the sea urchin embryo. Regulated genes are shown names in black type, e.g., *FoxA*, *Notch*, *Ets*, etc. The short horizontal lines above these names from which bent arrows extend represent regulatory elements in the genome that influence expression of the genes named beneath the line. Lines with arrows represent an activation of the regulatory element, while lines ending in bars represent repression. Reprinted with permission from Davidson *et al.* [46]. Copyright 2002 AAAS. For an updated version of this figure, see http://sugp.caltech.edu/endomes/.

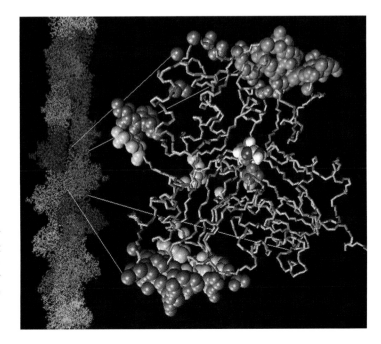

Figure 2.7

Three-dimensional structure of a single G-actin monomer (right) and an F-actin strand (left). Amino acids on the actin monomer are color coded to indicate interactions with adjacent monomers in the F-actin filament. Colored elements in the central portion of the monomer are bound ADP and Ca^{2+}. Reproduced with permission from the University of Illinois [5], based on Lorenz *et al.* [6] and Kabsch *et al.* [7].

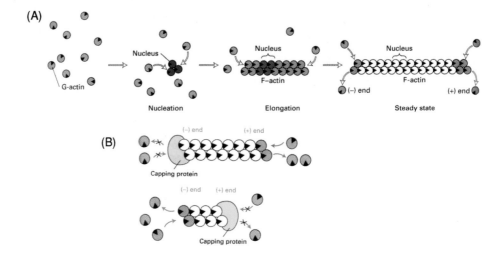

Figure 2.8
Schematic summary of actin polymerization. (A) G-actin monomers (bound to ATP) nucleate to form small stable filaments that are then elongated by addition of monomers to both ends of the filament. After being incorporated into the filament, the ATP of the G-actin subunits is hydrolyzed to become stable ADP–F-actin. After elongation, the ends of the actin filament are in steady state with the G-actin monomers. (B) In the presence of a capping protein (such as CapG), the ends are capped and elongation is markedly slowed or halted. Most of the regulation of F-actin dynamics is mediated through capping proteins. Adapted from Lodish *et al.* [2]. Reproduced with kind permission of W. H. Freeman.

Figure 2.9
Schematic diagram of the assembly of tubulin dimers into a microtubule. Note that there are two types of tubulin monomers: α and β. These associate to form short protofilaments (1). The protofilaments assemble together to form a sheet that bends (2) and eventually wraps on itself to form a hollow microtubule made up of 13 protofilaments (3). Once the microtubule is formed, tubulin dimers can exchange at both the + and − ends, although they tend to do so preferentially at the + end. From Lodish *et al.* [17]. Reproduced with kind permission of W. H. Freeman.

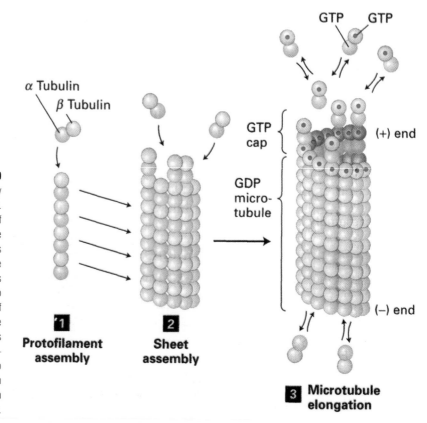

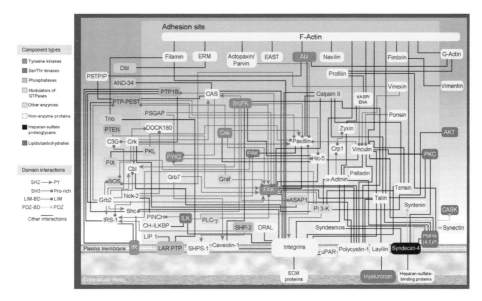

Figure 2.15

A schematic diagram summarizing known interactions between the various constituents of cell–matrix adhesion complexes (as of 2001). Components within the central green box are associated with cell–matrix adhesion sites, while the outer blue region contains additional selected proteins that affect cell–matrix adhesion but are not known to stably associate with the cell–matrix adhesion complex. The general property of each component is indicated by the color of its box, and the type of interaction between the components is indicated by the style and color of the interconnecting lines (see legend). This schematic illustrates the overwhelming complexity of cell–matrix adhesion. From Zamir and Geiger [27]. Reproduced with permission of the Company of Biologists, Ltd.

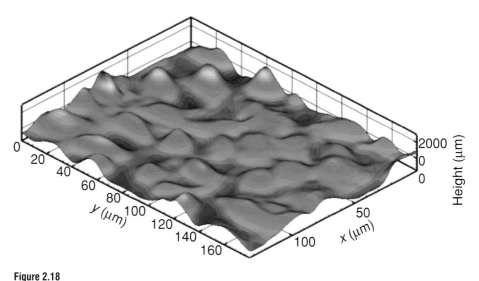

Figure 2.18

Atomic force microscopic map of cellular topography for confluent porcine aortic endothelial cells. The vertical scale has been exaggerated. Data courtesy of Dr. Shigeo Wada (Sendai University, Japan), and Mr. James Shaw and Dr. Chris Yip (University of Toronto).

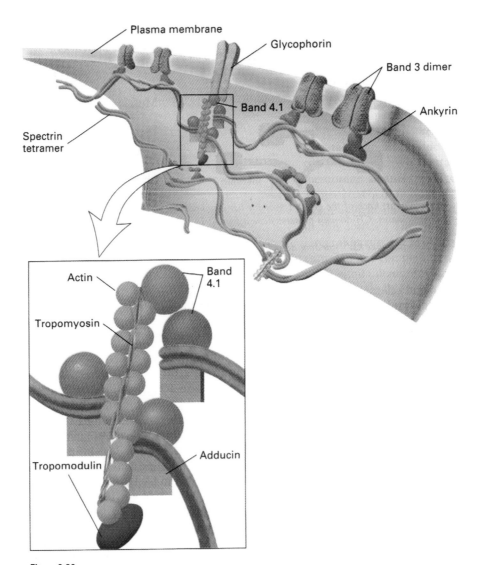

Figure 2.28

Schematic view of the erythrocyte cortical cytoskeleton, showing cross-linked spectrin fibers (green). Note the connections between spectrin and integral membrane proteins provided by ankyrin (purple). From Lodish *et al.* [2]. Reproduced with kind permission of W. H. Freeman.

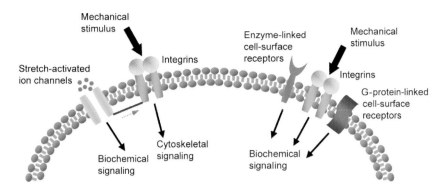

Figure 2.48

Candidate mechanoreceptors for relaying extracellular mechanical signals to the cell's interior to activate intracellular signaling pathways. Connections between receptors probably exist; for example, forces received by integrins cause deformation of the cytoskeleton (dark blue line), which "tugs" on the ion channel (as indicated by the red dotted arrow), causing the channel to open permitting flow of ions. Refer to the text for further details.

Figure 2.49

Possible roles for the cytoskeleton in mediating mechanical signals through physical means. (A) Mechanical coupling via the cytoskeleton hypothetically allows for mechanotransduction at multiple sites (After Davies [85]). (B) Transduction of mechanical signals into biochemical signals by integrins and the cytoskeleton, on which regulatory proteins and enzymes are physically immobilized. In the scheme shown, the chemical conversion of substrate 1 into product 2 is regulated by the middle protein immobilized to the microtubule. In response to a mechanical stimulus, the force balance in the cytoskeleton is altered, leading to deformation of the immobilized proteins and changes in protein kinetics (more of product 2 converted in this case). In contrast, pulling a transmembrane protein that is not physically connected to the cytoskeleton (red oval protein in cell membrane) has no effect. From Ingber [72] with permission of the Company of Bioligists Ltd.

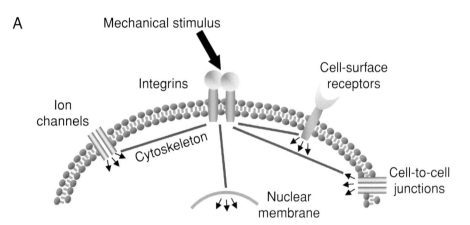

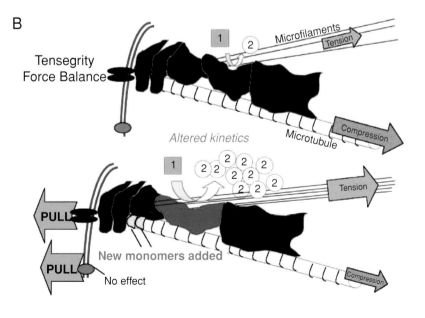

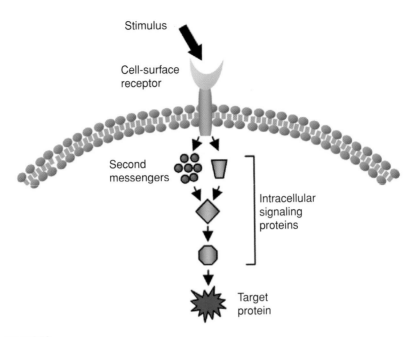

Figure 2.50

Schematic of simple intracellular signaling pathways using biochemical mediators.

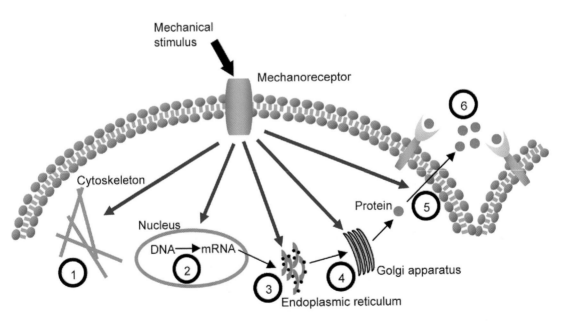

Figure 2.51

Mechanical signals transduced intracellularly can affect several cellular processes, including: (1) cytoskeletal organization; (2) transcription of mRNA from DNA (gene expression); (3) translation of proteins from mRNA; (4) assembly and post-translational modification of proteins; and (5) secretion of proteins into the extracellular space, where they can regulate the function of the stimulated cell or its neighbors through signaling by soluble factors (6). Note that secreted proteins are always enclosed in vesicles for transport from the Golgi apparatus to the cell membrane; this detail is not shown in this simplified scheme.

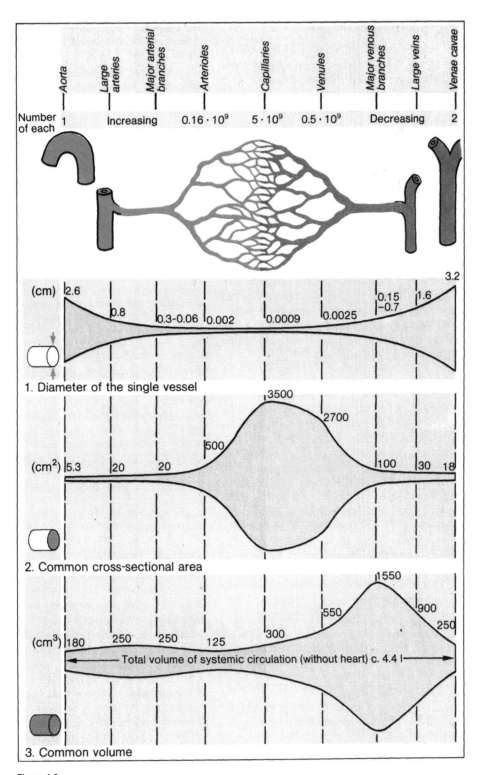

Figure 4.2
A schematic overview of the systemic vascular tree. The values of diameter etc. are for humans and are approximate. Note the large blood volume contained in the veins, and the large aggregate cross-sectional area exhibited by the capillaries. From Despopoulos and Silbernagl [3]. Reproduced with kind permission of Thieme.

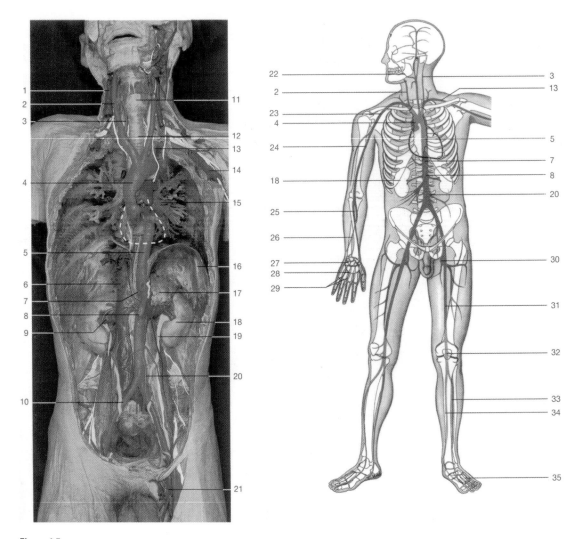

Figure 4.5

Major vessels of the trunk (left) and schematic overview of the major arteries in humans (right). In the left panel, the position of the heart is indicated by the white dashed line; arteries are colored red, while veins are colored blue.

1, internal jugular vein; 2, common carotid artery; 3, vertebral artery; 4, ascending aorta; 5, descending aorta; 6, inferior vena cava; 7, celiac trunk; 8, superior mesenteric artery; 9, renal vein; 10, common iliac artery; 11, larynx; 12, trachea; 13, left subclavian artery; 14, left axillary vein; 15, pulmonary veins; 16, diaphragm; 17, suprarenal gland; 18, kidney; 19, ureter; 20, inferior mesenteric artery; 21, femoral vein; 22, facial artery; 23, axillary artery; 24, brachial artery; 25, radial artery; 26, ulnar artery; 27, deep palmar arch; 28, superficial palmar arch; 29, common palmar digital arteries; 30, profunda femoris artery; 31, femoral artery; 32, popliteal artery; 33, anterior tibial artery; 34, posterior tibial artery; 35, plantar arch. From Rohen *et al.* [7]. Reproduced with permission from Lippincott Williams & Wilkins.

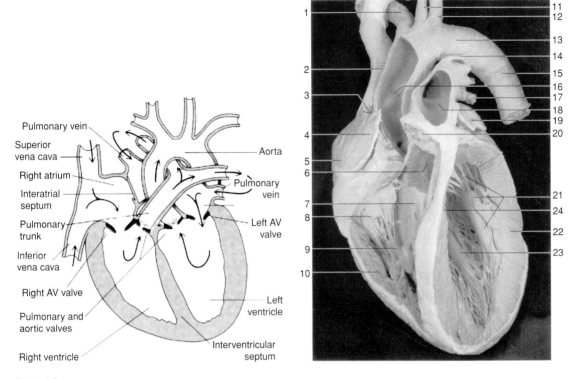

Figure 4.7

Gross anatomy of the heart. On the left, the heart is shown in section (AV, atrioventricular). Major structures are labeled and arrows represent blood flow directions. The right panel shows a human heart partially dissected. Note the thickness of the left ventricular wall and the extensive network of chordae tendineae.

1, brachiocephalic trunk; 2, superior vena cava; 3, sulcus terminalis; 4, right auricle; 5, right atrium; 6, aortic valve; 7, conus arteriosus (interventricular septum); 8, right atrioventricular (tricuspid) valve; 9, anterior papillary muscle; 10, myocardium of right ventricle; 11, left common carotid artery; 12, left subclavian artery; 13, aortic arch; 14, ligamentum arteriosum (remnant of ductus arteriosus); 15, thoracic aorta (descending aorta); 16, ascending aorta; 17, left pulmonary vein; 18, pulmonary trunk; 19, left auricle; 20, pulmonary valve; 21, anterior papillary muscle with chordae tendineae; 22, myocardium of left ventricle; 23, posterior papillary muscle; 24, interventricular septum. Left panel from Vander *et al.* [10] with kind permission of McGraw-Hill Education; right panel from Rohen *et al.* [7] with kind permission of Lippincott Williams & Wilkins.

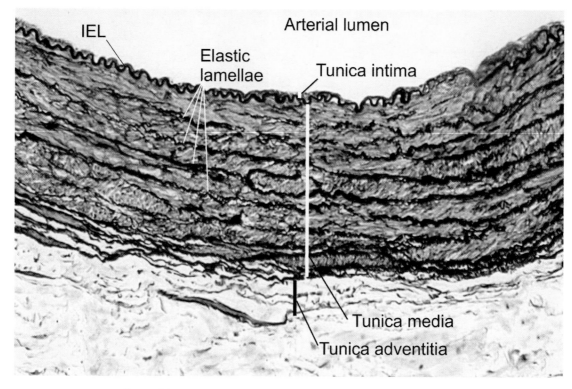

Figure 4.15
Cross-section through the wall of an artery, demonstrating the tunicas intima, media, and adventitia, as well as elastic lamellae within the media. The elastin appears black in this preparation, which has been stained with Verhoeff's stain and lightly counter-stained to make the collagen appear blue. IEL, internal elastic lamina. Modified from Vaughan [19] by permission of Oxford University Press, Inc.

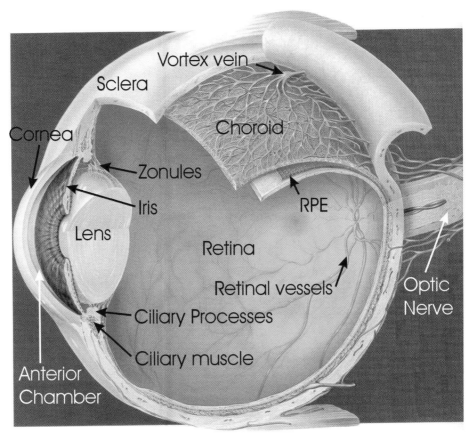

Figure 6.1
Overview of the eye, with several key structures labeled. RPE, retinal pigment epithelium. Modified from Krey and Bräuer [1] with kind permission of Chibret Medical Service.

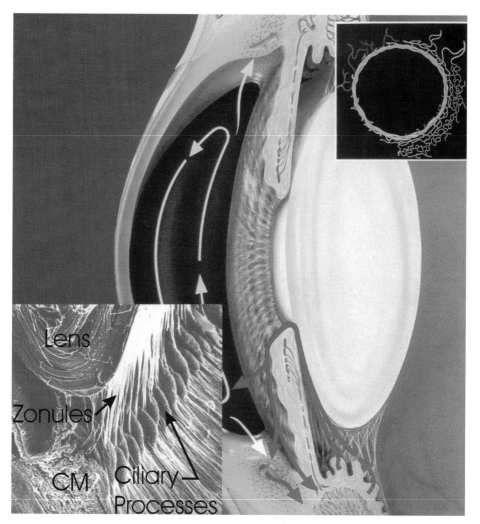

Figure 6.7
View of anterior segment of a human eye. Aqueous humor is produced by the ciliary processes and travels along the routes shown by the red arrows. The white arrows in the anterior chamber show thermal convection patterns. The lower left inset is a scanning electron micrograph of the zonular apparatus (CM, ciliary muscle). The inset at upper right shows Schlemm's canal (green) as seen face-on. Green vessels anastomosing with Schlemm's canal are collector channels; blue vessels are aqueous veins; red vessels are arterioles. Modified from Krey and Bräuer [1] with kind permission of Chibret Medical Service.

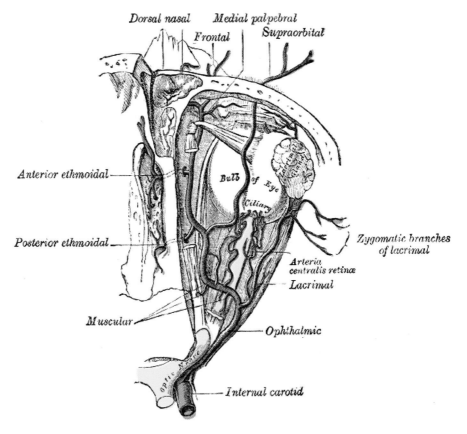

Figure 6.15

Blood supply to the eye and orbit in the human. The ophthalmic artery derives from the internal carotid artery, and supplies blood to the orbit of the eye (the cavity in which the eye sits), the lacrimal (tear) gland, the extraocular muscles and the eye globe proper. Reprinted wth permission from Gray's Anatomy [71].

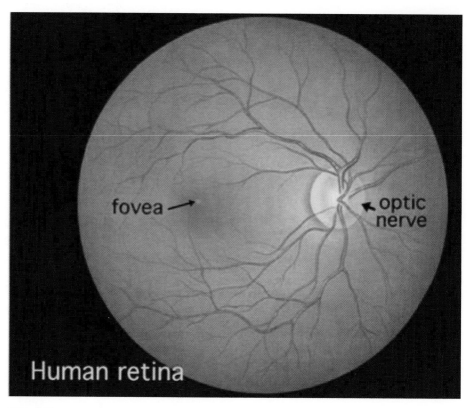

Figure 6.16
Image of a human retina, showing the retinal vasculature. The central retinal artery enters the eye through the center of the optic nerve, branching to form retinal arteries running along the surface of the retina. After passing through the retinal capillaries, blood drains back to the central retinal vein, which exits the eye through the optic nerve. This image is taken from the outside of the eye, through the cornea. The fovea is the specialized part of the retina where we do most of our seeing; it contains an especially high density of photoreceptors and has no overlying retinal circulation. From Kolb *et al.* [72] with kind permission of Dr. Helga Kolb.

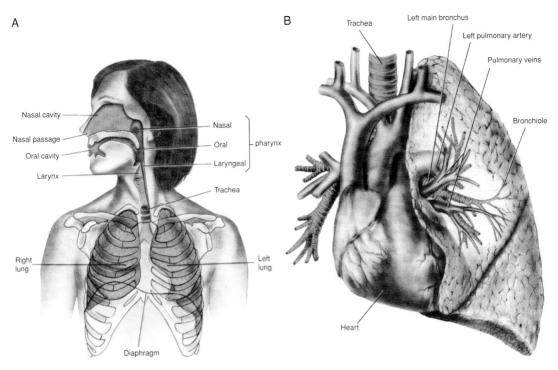

Figure 7.1

Organization of the respiratory system. (A) The structures that conduct ambient air to the lungs. (B) The large airways and main vascular structures of the left human lung and the heart. In this panel the tissue of the lungs is shown as transparent so that the internal airways and blood vessels can be better seen. From Vander *et al.* [1]. Reproduced with kind permission of The McGraw-Hill Companies.

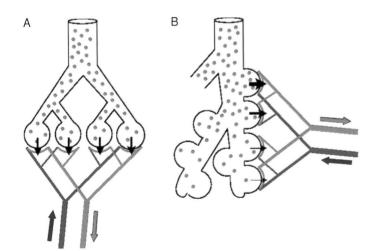

Figure 7.15

Two conceptual models of mass transfer in the acinus, showing oxygen concentration (red dots), oxygen mass transfer into pulmonary capillaries (black arrows), pulmonary artery blood flow (blue arrows,) and pulmonary vein blood flow (red arrows). (A) A "parallel ventilation/ parallel perfusion" model, where all alveoli reside at the end of airway and are effectively equivalent from an air-side mass transfer viewpoint. (B) The actual situation ("series ventilation/ parallel perfusion"), where alveoli reside in multiple airway generations so that distal alveoli see lower oxygen concentrations. From Sapoval *et al.* [11]. Copyright 2002 National Academy of Sciences, U.S.A.

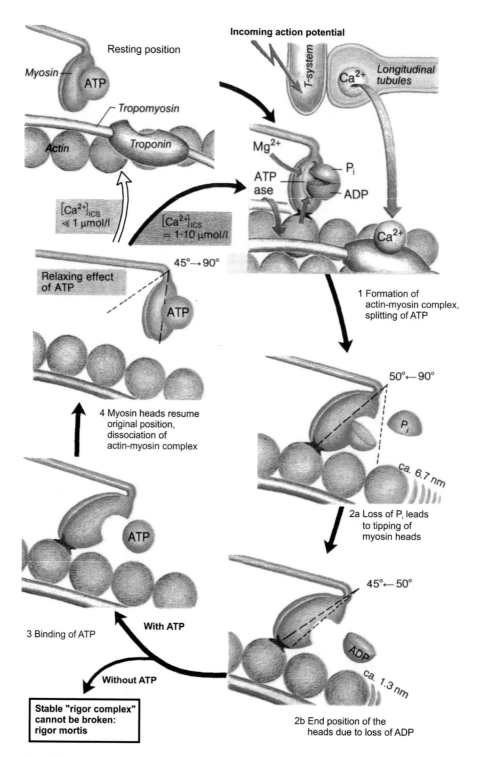

Figure 8.5

Description of sliding filament model, showing some of the molecular mechanisms involved in the sliding of actin and myosin filaments (isotonic contraction). See text for discussion of specific features. ICS, intracellular stores. From Despopoulos and Silbernagl [4]. Reproduced with kind permission of Thieme.

3 Hemodynamics

The term *hemodynamics* comes from the Greek words *haima* (blood) and *dunamis* (power) and refers to the movement and deformation (i.e., flow) of blood, and the forces that produce that flow. In this chapter we will examine this fascinating (and complex) topic.

Everyone is familiar with blood's role as a transport medium: it carries oxygen and nutrients to metabolically active tissues, returns carbon dioxide to the lungs, delivers metabolic end-products to the kidneys, etc. However, the reader should be aware that blood does much more than simply deliver substances to target tissues. For example, it:

- provides a buffering reservoir to control the pH of bodily fluids
- serves as an important locus of the immune system
- transports heat, usually from centrally located tissues to distal ones, in order to help maintain a suitable temperature distribution throughout the body.

Unfortunately, in this book we will not be able to examine all of these roles, and to a large extent we will simply view blood as a passive carrier, a fluid that transports physiologically important compounds within the body. However, within this context, it will soon become clear that something so "simple" as an analysis of blood flow as a transport mechanism is non-trivial. We begin by examining blood rheology.

3.1 Blood rheology

Rheology is the study of how materials deform and/or flow in response to applied forces. The applied forces are quantified by a quantity known as the *stress*, defined as the applied force per unit area. Students know that the applied force is a vector, meaning that its orientation is important; what may be less clear is that the surface to which the force is applied has an orientation too. This surface orientation is characterized by its normal vector. The orientations of both the force and the surface must be accounted for when computing the stress, and therefore the resulting quantity

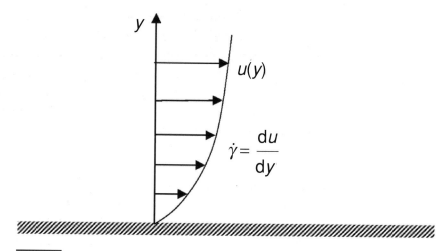

Figure 3.1

Velocity profile near a solid surface showing the definition of the velocity gradient for a simple flow with one non-zero velocity component.

is a second-order tensor whose diagonal elements are *normal stresses* and whose off-diagonal elements are *shear stresses*, as discussed in more detail in the footnote on p. 85. To simplify the problem, we consider below only one entry of this tensor, which allows us to treat the stress as a scalar.

For a solid, the deformation is quantified in terms of the fractional change in the dimensions of a small material element of the solid. This quantity, known as the *strain*, depends on both the direction of the deformation and the orientation of the material element that is being deformed, and it is therefore also a second-order tensor. For a fluid, a similar situation obtains, except that we replace the deformation by the rate of deformation to obtain a *rate-of-strain* tensor. Once again, we will consider only one element of these tensors and talk simply about a scalar strain (or rate of strain).

Rheological knowledge of a material can be expressed (in part) by a *constitutive relationship* between the applied stress and the resulting strain or rate-of-strain. Simple constitutive relationships are well known and frequently used in engineering. For example, for a Hookean solid under uniaxial stress, the constitutive relationship relating applied normal stress, σ, to normal strain, ε, is $\sigma = E\varepsilon$, where E is a material constant known as Young's modulus. Similarly, for a Newtonian fluid we may write

$$\tau = \mu\dot{\gamma}, \tag{3.1}$$

where τ is the applied shear stress, $\dot{\gamma}$ is the *rate of strain*, and μ is a material constant known as the *dynamic viscosity*. For a simple unidirectional flow in which the velocity u varies only with the transverse coordinate y, $\dot{\gamma}$ is just equal to the velocity gradient du/dy (Fig. 3.1).

Fluids that obey Equation (3.1) for constant μ are called *Newtonian*. Many fluids of engineering importance, such as gases and water, are Newtonian. However, as we shall see, blood is quite strongly non-Newtonian, at least in certain flow regimes. A detailed discussion of non-Newtonian rheology is beyond the scope of this book; however, several points are worth noting. First, we can fit the shear stress versus strain rate behavior of any fluid by using an equation of the form (3.1), so long as we replace μ by an effective viscosity μ_{eff}:

$$\tau = \mu_{eff}\dot{\gamma}. \tag{3.2}$$

Note that μ_{eff} is no longer constant but will in general depend on $\dot{\gamma}$ and numerous other factors. Nonetheless, to give an indication of fluid viscous behavior under a specified set of conditions, Equation (3.2) is useful. We can further classify the type of fluid by examining the dependence of μ_{eff} on $\dot{\gamma}$: if the effective viscosity decreases as $\dot{\gamma}$ increases, the fluid is said to be *shear thinning*, while if μ_{eff} increases with increasing $\dot{\gamma}$ the fluid is *shear thickening*.

It should be noted that many biological materials have rheological behavior that is a combination of fluid and solid; such materials are known as viscoelastic (simple models for viscoelastic materials were introduced in Section 2.6.1 and will be discussed further in Section 9.10.5). For example, there are fluids for which μ_{eff} depends not only on $\dot{\gamma}$ but also on the *history* of $\dot{\gamma}$. Strictly speaking, blood falls into this category. However, as a first approximation, we will ignore the elastic nature of blood and treat it as a purely viscous non-Newtonian fluid.

3.1.1 Blood composition

To understand why blood is a non-Newtonian fluid, we consider blood composition. There are approximately 5 liters of blood in an average human being. This complex fluid is essentially a suspension of particles (the *formed elements*) floating in an aqueous medium (the *plasma*). More specifically, blood composition can be broken down grossly as shown in Fig. 3.2, and in more detail in Table 3.1 (for plasma) and Table 3.2 (for the formed elements).

Although all blood constituents are important physiologically, for rheological purposes we can make some simplifications. For example, white cells (*leukocytes*) and platelets play a major role in the immune response and in blood clotting, respectively. However, as Table 3.2 shows, there are relatively few white cells and platelets compared with the number of red cells (*erythrocytes*); consequently, the mechanical behavior of the formed elements is usually dominated by the red cells.[1] In fact, the volume fraction of red cells is so important to the rheological and

---

[1] An exception to this rule occurs for flow in very small vessels, where the somewhat larger and much stiffer white cells can temporarily "plug" the vessel and markedly affect flow behavior [3].

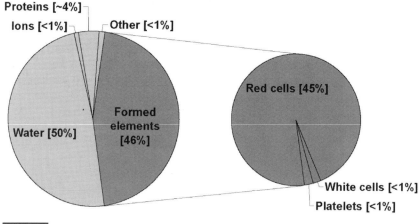

Figure 3.2

Overall composition of human blood. The light-colored pie slices (water, ions, proteins, and other) collectively make up the plasma. The composition of the formed elements is shown on the right. Numbers in brackets represent approximate volume fractions for males; the volume fraction of red cells is approximately 4% lower in females.

physiological characteristics of blood that a specific term, known as the *hematocrit*, H, is commonly used, defined by:

$$H = \frac{\text{volume of red blood cells}}{\text{total blood volume}}. \tag{3.3}$$

In addition, measurements have shown that although plasma itself has numerous constituents, it is a Newtonian fluid with a viscosity of approximately $\mu_{plasma} = 1.2$ cP at $37\,^\circ$C.[2] Therefore, plasma viscosity slightly exceeds that of water, as might be expected from the presence of proteins and other macromolecules in the plasma. Based on the above discussion, we approximate blood as a suspension of red cells in a Newtonian fluid and focus on the behavior of the red cells to explain blood's non-Newtonian rheology.

Individual red cells are shaped like "biconcave disks," with a diameter of approximately 8 μm and a maximal thickness of approximately 3 μm (Table 3.3). There are approximately 5 billion red cells per milliliter of blood, so the total red cell surface area in a normal human adult is approximately $3000\,\text{m}^2$. The cytoplasm contains large amounts of the iron-containing protein hemoglobin (Table 3.4). Hemoglobin is highly efficient at binding oxygen; hence, the presence of large amounts of hemoglobin sequestered inside red cells greatly increases the oxygen transport capacity of blood (see Section 7.5). For example, the oxygen-carrying capacity of whole blood is approximately 65 times that of plasma alone (~ 21 ml O_2 per 100 ml blood versus 0.3 ml O_2 per 100 ml plasma, at 1 atmosphere).

[2] The cgs unit of viscosity is the Poise (abbreviation: P), equal to 1 g/(cm s). For reference, the viscosity of water at $20\,^\circ$C is approximately 1 centiPoise (1 cP).

Table 3.1. Constituents of arterial plasma. From Vander et al. [1]. Reproduced with kind permission of the McGraw-Hill companies.

Constituent	Amount/concentration	Major functions
Water	93% of plasma weight	Medium for carrying all other constituents
Electrolytes (inorganic)	Total <1% of plasma weight	Keep H_2O in extracellular compartment; act as buffers; function in membrane excitability
Na^+	142 mM	
K^+	4 mM	
Ca^{2+}	2.5 mM	
Mg^{2+}	1.5 mM	
Cl^-	103 mM	
HCO_3^-	27 mM	
Phosphate (mostly HPO_4^{2-})	1 mM	
SO_4^{2-}	0.5 mM	
Proteins	Total = 7.3 g/100 mL (2.5 mM)	Provide non-penetrating solutes of plasma; act as buffers; bind other plasma constituents (lipids, hormones, vitamins, metals, etc.); clotting factors; enzymes; enzyme precursors; antibodies (immunoglobulins); hormones
Albumins	4.5 g/100 mL	
Globulins	2.5 g/100 mL	
Fibrinogen	0.3 g/100 mL	
Gases		
CO_2	2 mL/100 mL plasma	
O_2	0.2 mL/100 mL	
N_2	0.9 mL/100 mL	
Nutrients		
Glucose and other	100 mg/100 mL (5.6 mM)	
Total amino acids	40 mg/100 mL (2 mM)	
Total lipids	500 mg/100 mL (7.5 mM)	
Cholesterol	150–250 mg/100 mL (4–7 mM)	
Individual vitamins	0.0001–2.5 mg/100 mL	
Individual trace elements	0.001–0.3 mg/100 mL	
Waste products		
Urea	34 mg/100 mL (5.7 mM)	
Creatinine	1 mg/100 mL (0.09 mM)	
Uric acid	5 mg/100 mL (0.3 mM)	
Bilirubin	0.2–1.2 mg/100 mL (0.003–0.018 mM)	
Individual hormones	0.000001–0.05 mg/100 mL	

Table 3.2. Formed elements in blood: erythrocytes (red cells) are responsible for oxygen and carbon dioxide transport; leukocytes (white cells) are responsible in part for immune function; and platelets are responsible in part for blood coagulation. From Caro *et al.* [2]. Reproduced with kind permission of Oxford University Press.

Cell	Number per mm^3	Unstressed shape and dimensions (μm)	Volume concentration in blood (%)
Erythrocyte	$4-6 \times 10^6$	Biconcave disc; $8 \times 1-3$	45
Leukocytes			
Total	$4-11 \times 10^3$		
Granulocytes			
Neutrophils	$1.5-7.5 \times 10^3$		
Eosinophil	$0-4 \times 10^2$	Roughly spherical; $7-22$	1
Basophil	$0-2 \times 10^2$		
Lymphocytes	$1-4.5 \times 10^3$		
Monocytes	$0-8 \times 10^2$		
Platelets	$250-500 \times 10^3$	Rounded or oval; $2-4$	

The red cell is somewhat unusual from a biomechanical viewpoint, being more readily deformable than other cell types. This is important since red cells must often pass through very narrow openings. This large deformability can be explained by several structural features. First, the major cytoskeletal protein in erythrocytes, *spectrin*, is closely associated with the cell membrane, rather than traversing the cytoplasm (Section 2.5.4). This confers strength to the membrane without unduly constraining the cell's deformability. Second, red cells lack most organelles (including, in mammals, a nucleus).

3.1.2 Relationship between blood composition and rheology

Rheologically, there are two major effects from the presence of red cells.

1. *Rouleaux formation.* If blood is allowed to sit for several seconds, stacks of red cells (*rouleaux*) begin to form. If allowed to grow, these rouleaux can become quite large and will eventually form an interconnected network extending throughout the blood (Fig. 3.3). Rouleaux provide mechanical coupling between different fluid regions, and thereby increase the resistance to

Table 3.3. Geometric parameters of normal human red cells. Statistics are from pooled data on 1581 cells taken from 14 subjects; cells were suspended in 300 mOsm buffer. From Tsang [4] as shown in Fung [5] with kind permission of Professor Y. C. Fung.

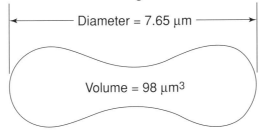

Diameter = 7.65 μm

Volume = 98 μm³

Surface area = 130 μm²

	Diameter (μm)	Minimum thickness (μm)	Maximum thickness (μm)	Surface Area (μm²)	Volume (μm³)	Sphericity index
Mean	7.65	1.44	2.84	129.95	97.91	0.792
Standard error of mean	±0.02	±0.01	±0.01	±0.40	±0.41	±0.001
Standard deviation (σ)	0.67	0.47	0.46	15.86	16.16	0.055
Minimum value	5.77	0.01	1.49	86.32	47.82	0.505
Maximum value	10.09	3.89	4.54	205.42	167.69	0.909
Skewness (G_1)	0.26	0.46	0.52	0.53	0.30	−1.13
Kurtosis (G_2)	1.95	1.26	0.24	0.90	0.30	3.27

Table 3.4. Composition of red cells: as a first approximation, the red cell is essentially hemoglobin and water enclosed in a membrane. From Caro et al. [2]. Reproduced with kind permission of Oxford University Press.

Component	Percentage of mass
Water	65
Membrane components (protein, phospholipid, cholesterol)	3
Hemoglobin	32
Inorganic	
Potassium	0.420 g per 100 ml
Sodium	0.025 g per 100 ml
Magnesium	0.006 g per 100 ml
Calcium	Small amount

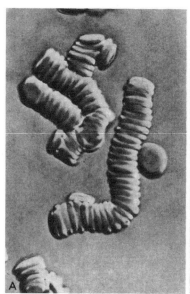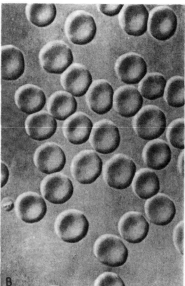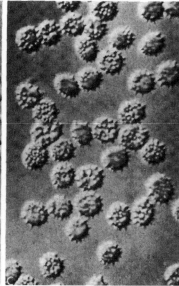

Figure 3.3

Rouleaux forming in stagnant blood. (A) Rouleaux formation under stagnant conditions. (B) Normal erythrocytes, showing the classical "biconcave disk" shape characterized by a thick rim and a central depression. (C) Under certain conditions, erythrocytes can adopt a spiny configuration, in which case they are known as *echinocytes*. This process, known as *crenation*, indicates red cell pathology or artefacts during specimen preparation. From Fawcett [6] with kind permission of Hodder Education and Springer Science and Business Media.

deformation of fluid elements. This implies that the effective viscosity μ_{eff} is increased by the presence of rouleaux.

Rouleaux are broken up by imposed shearing of the blood. Based on the above discussion, we expect that this will cause a relative decrease in μ_{eff}. Thus, the action of rouleaux formation is to make blood a shear-thinning fluid.

2. *Red cell alignment.*   Because individual red cells are disc shaped, their hydrodynamic influence depends on their orientation. There are two forces that compete in orienting red cells. First, Brownian motion, always present, tries to randomize the orientation of red cells. Second, fluid shearing forces cause red cells to align their long axes with streamlines. These forces lead to two possible extrema, as shown in Fig. 3.4. When red cells are randomly or near-randomly oriented, then individual cells will "bridge" between different streamlines, thereby providing mechanical coupling between two different fluid regions having potentially different velocities. This will tend to increase μ_{eff}. The magnitude of this effect will decrease as red cells become progressively more oriented, that is, as $\dot{\gamma}$ increases.

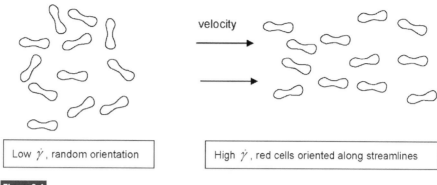

| Low $\dot{\gamma}$, random orientation | High $\dot{\gamma}$, red cells oriented along streamlines |

Figure 3.4

Two limiting orientations for red cells. On the left, the fluid shear rate is very low so Brownian forces predominate and the red cells are randomly oriented. On the right, the fluid shear rate is large, causing the red cells to line up along streamlines with very little randomization of orientation.

Table 3.5. Summary of red cell interaction effects on blood rheology

Characteristic	Low shear rate	High shear rate
Rouleaux behavior	Rouleaux formation enhanced; effective viscosity μ_{eff} is increased	Rouleaux break up; effective viscosity μ_{eff} is decreased
Individual red cell orientation	Red cells are randomly oriented; μ_{eff} is increased	Red cells are aligned with streamlines; μ_{eff} is decreased

The above effects are summarized in Table 3.5, from which we conclude that we expect blood to have shear-thinning behavior. We expect that the magnitude of the effects in Table 3.5 will depend on how many red cells are present, i.e., on the value of H. Since plasma is Newtonian, we expect Newtonian behavior for H = 0 (pure plasma), with increasing non-Newtonian rheology as the value of H increases.

How does this theoretical argument compare with reality? Quite well, as shown in Fig. 3.5. We see Newtonian behavior for H = 0, and increasing non-Newtonian (shear-thinning) rheology as the value of H increases. Closer examination of Fig. 3.5 reveals several additional interesting features. For example, note that red cells in Ringer's solution (essentially, in saline) are less non-Newtonian than is whole blood. This is because rouleaux formation is promoted by macromolecules in plasma; when these macromolecules are removed by replacing plasma by Ringer's solution, rouleaux formation and non-Newtonian effects are markedly reduced. The clotting protein fibrinogen is especially important in this regard, which explains the large fall in μ_{eff} caused by fibrinogen removal shown in Fig. 3.5. It can also be observed that for very large shear rates ($\dot{\gamma} > 100\,\text{s}^{-1}$), μ_{eff} asymptotes to a

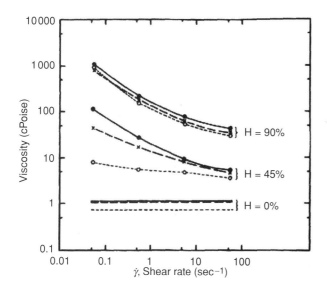

Figure 3.5

Plot of effective viscosity versus shear rate for blood of differing hematocrits (H). Note the Newtonian behavior of the fluid at zero hematocrit, and the logarithmic vertical scale. •, whole blood; ×, defibrinated blood (i.e., blood from which the clotting protein fibrinogen has been removed); ○, washed cells in Ringer's solution. The points are determined from a fifth-order polynomial curve fit to experimental data. Modified with permission of the American Physiological Society from Chien *et al.* [7].

constant. Therefore, we may say that for $\dot{\gamma} > 100\,\text{s}^{-1}$ blood acts like a Newtonian fluid with viscosity μ_{eff}. This limiting viscosity is dependent on the value of H. For normal values of hematocrit (45%), μ_{eff} is 3 to 4 cP. At the other extreme (very low $\dot{\gamma}$), μ_{eff} becomes very large.

Further information about blood rheology is contained in Fig. 3.6, which plots the *relative viscosity*, η_r, as a function of particle volume fraction for a number of different suspensions, including blood. The relative viscosity of a suspension is defined as

$$\eta_r = \frac{\mu_{\text{eff}}}{\mu_{\text{fluid}}}, \tag{3.4}$$

where μ_{fluid} is the viscosity of the fluid in which the particles are suspended, in this case, plasma. Clearly, η_r is a relative measure of how addition of particles has increased suspension viscosity. Figure 3.6 is restricted to very high shear rates, so that blood is effectively Newtonian. It can be seen that a red cell suspension flows much more easily than does a suspension of rigid spheres at the same particle volume fraction (hematocrit). This is because at high particle volume fraction, particles are in direct mechanical contact as the suspension flows. Red cells can accommodate by deforming in order to pass by one another, while solid particles cannot deform and so act to impede each other's motion. This makes the suspension harder to deform, that is, increases the value of η_r. Figure 3.6 also confirms how flexible red cells are; for example, the relative viscosity of a red cell suspension is less than that of a suspension of deformable oil droplets in water. Finally, note the curve labeled "sickle cell anemia" in Fig. 3.6. This refers to a red cell suspension obtained from patients with sickle cell anemia, a disease characterized by red cells

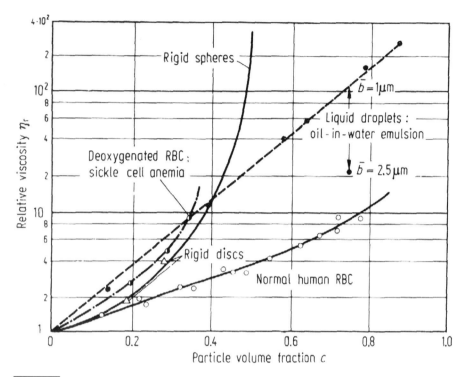

Plot of relative viscosity versus particle volume fraction for various suspensions, including red blood cells (RBC) in saline. Data for this graph were taken at high shear rates ($\dot{\gamma} > 100\,\text{s}^{-1}$), so that the effective viscosity was independent of shear rate. Note that for a suspension of red cells, the particle volume fraction is equal to the hematocrit. Note also that the vertical scale is logarithmic. Relative viscosity is defined in Equation (3.4). From Goldsmith [8]. Reproduced with kind permission of Springer Science and Business Media.

that are much less deformable than normal. It is difficult to force such red cells through the smallest vessels (the capillaries), and, consequently, patients with this disease typically have a number of circulatory system disorders.[3]

3.1.3 Constitutive equation for blood

For the remainder of this section, we will assume that the hematocrit is fixed at its physiological value, so that the effective viscosity μ_{eff} is a function of $\dot{\gamma}$ only. Recalling Equation (3.1) and the fact that blood is shear-thinning, we expect that a plot of shear stress τ versus strain rate $\dot{\gamma}$ will be concave downward.

[3] It is interesting to note that the most common form of sickle cell anemia is caused by a single mutation in the genes for the hemoglobin molecule. From an evolutionary viewpoint, it would seem that such a mutation would be unfavorable. However, the sickle cell trait seems to confer added resistance to malaria, which is why it probably remains in the general population.

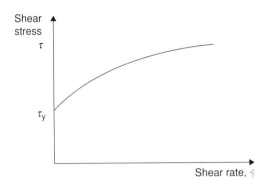

Figure 3.7

Expected qualitative behavior of shear stress vs. strain rate for blood. Note the concave downward (shear-thinning) nature of the curve, as well as the non-zero intercept at τ_y (yield stress).

Experimentally, this is observed to be the case. Interestingly, it is also observed that blood exhibits a *yield stress*: that is, there is a critical applied shear stress τ_y below which blood will not flow. Therefore, in the range $0 < \tau < \tau_y$, blood acts like a solid, while for $\tau \geqslant \tau_y$, blood acts like a fluid. This is a result of extensive rouleaux formation, which is able to form a stress-supporting network at low shear stresses. When the value of τ exceeds the *critical shear stress* τ_y, the rouleaux break up and the blood begins to flow.

Qualitatively, we expect the τ versus $\dot{\gamma}$ behavior shown in Fig. 3.7. Quantitatively, it is found that shear stress strain rate data can be fit by using the *Casson relationship*, originally derived to model the rheology of printer's ink

$$\sqrt{\tau} = \sqrt{\tau_y} + \sqrt{\mu\dot{\gamma}} \quad \text{for } \tau \geqslant \tau_y \tag{3.5}$$

$$\dot{\gamma} = 0 \quad \text{for } \tau < \tau_y. \tag{3.6}$$

This behavior is confirmed by Fig. 3.8, which shows a linear relationship between $\sqrt{\tau}$ and $\sqrt{\dot{\gamma}}$, with intercept $\sqrt{\tau_y}$. The slope is $\sqrt{\mu}$, where μ is a constant having the same dimensions as dynamic viscosity. The Casson relationship asymptotes to Newtonian behavior at high values of $\dot{\gamma}$. This can be seen by examining Equation (3.5) for $\dot{\gamma} \gg \tau_y/\mu$, in which case the constitutive relationship becomes $\tau = \mu\dot{\gamma}$.

3.2 Large artery hemodynamics

3.2.1 Physical characteristics of blood flow patterns in vivo

In vivo blood flow patterns are complex. Factors influencing fluid mechanics in large arteries include the following, arranged approximately in order of overall importance.

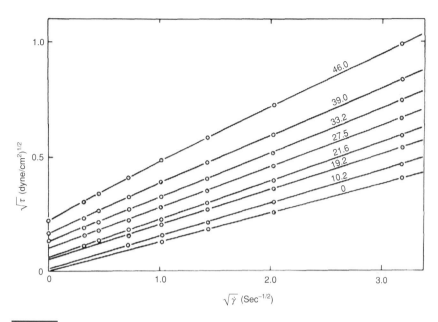

Figure 3.8

Casson plot for human blood at 25 °C and very low shear rates for hematocrit values indicated above each line. Note that plasma (hematocrit of 0, lower curve) intersects the vertical axis at the origin, confirming its Newtonian behavior. From Cokelet [9]. Adapted by permission of Pearson Education, Inc., Upper Saddle River, NJ.

1. Arteries have complex three-dimensional geometries that include one or more of the following features: bifurcation, significant changes in caliber, and compound curvature.
2. In the large arteries, flow is classified as moderately to highly unsteady. The combination of unsteadiness and highly three-dimensional geometry leads to many interesting flow features (e.g., Ku *et al.* [10], Ojha *et al.* [11], Ethier *et al.* [12]).
3. Artery walls are distensible, moving in response to the local time-varying blood pressure. However, this compliant nature is generally thought to be of only modest importance [13]. For example, it seems to have only a 10–15% effect on wall shear stress magnitudes [13,14].
4. Some arteries undergo large motions, particularly the coronary arteries (because of their position on the surface of the beating heart) and arteries that cross articulating joints. Surprisingly, this effect seems to be smaller than one would think a priori [15,16].

Typical mean Reynolds numbers in the human arterial system range from order several hundred to several thousand (Tables 3.6 and 3.7). Generally speaking, arterial flow in healthy humans is laminar, with the exception of flow in the proximal aorta and the aortic arch [2,25]. In diseased arteries, transition to turbulence occurs

Table 3.6. Typical hemodynamic parameters for selected vessels in a 20 kg dog and a 70 kg human (surface area 1.8 m^2). α is the Womersley parameter (Section 3.2.3); Re$_D$ is the Reynolds number (based on vessel diameter) given as mean with peak value in parentheses. Values collated from a variety of sources by Milnor [17]. Reproduced with kind permission of Lippincott Williams & Wilkins.

	Dog			Human		
	α	Velocity (cm/s)^a	Re$_D$	α	Velocity (cm/s)^a	Re$_D$
Systemic vessels						
Ascending aorta	16	15.8 (89/0)	870 (4900)	21	18 (112/0)	1500 (9400)
Abdominal aorta	9	12 (60.0)	370 (1870)	12	14 (75/0)	640 (3600)
Renal artery	3	41 (74/26)	440 (800)	4	40 (73/26)	700 (1300)
Femoral artery	4	10 (42/1)	130 (580)	4	12 (52/2)	200 (860)
Femoral vein	5	5	92	7	4	104
Superior vena cava	10	8 (20/0)	320 (790)	15	9 (23/0)	550 (1400)
Inferior vena cava	11	19 (40/0)	800 (1800)	17	21 (46/0)	1400 (3000)
Pulmonary vessels						
Main artery	14	18 (72/0)	900 (3700)	20	19 (96/0)	1600 (7800)
Main vein^b	7	18 (30/9)	270 (800)	10	19 (38/10)	800 (2200)

^{a} Velocities are temporal means with systolic/diastolic extremes in parentheses.
^{b} One of the usually four terminal pulmonary veins.

downstream of severe stenoses [26]. The values in Table 3.6 are averages; it can be appreciated from the range of reported values in Table 3.7 that significant variation from one person to the next is typical.

Notably absent from the above list is the non-Newtonian nature of blood. How important are the non-Newtonian characteristics of blood? Based on our discussion above, it seems reasonable to treat blood like a Newtonian fluid so long as the shear rate is greater than $100\,\text{s}^{-1}$. The question then becomes whether this shear rate is ever reached in the cardiovascular system. If it is, then we can, as a first approximation, ignore the non-Newtonian effects of blood.

From the characteristic parameters for blood flow in several clinically important large arteries, shown in Tables 3.6 and 3.7, we conclude that typical wall shear stresses are in the range 1–15 dynes/cm^2 for most arteries in humans. This corresponds to wall shear rates $\dot{\gamma}$ from approximately 30 to 450 s^{-1}. So the answer to the question as to whether we can treat blood like a Newtonian fluid is: "most of the time." In other words, for most of the large arteries, we can ignore the non-Newtonian nature of blood. More detailed analyses have confirmed that a

Table 3.7. Comparison of hemodynamic parameters in selected arteries in humans. Q, mean (cycle-averaged) flow rate in ml/s; D, cycle-averaged diameter in mm; U_0, mean (cycle-averaged) velocity in cm/s; Re_D, Reynolds number based on D, U_0 and an assumed blood kinematic viscosity of 3.5 cStokes; τ, mean wall shear stress determined from $\tau = 8\mu U_0/D = 32\,\mu Q/\pi\,D^2$, in dyne/cm² (this formula for shear stress ignores approximately 10% error from mean flow–pulsatile interactions). Reproduced from Ethier et al. [12] with permission from WIT Press, Southampton, UK.

Artery	Q	D	U_0	Re_D	τ	Comment and source
Common carotid[a]	6.0	6.3	19.3	347	8.6	Ultrasonography on young healthy volunteers (D. Holdsworth, unpublished communication, 1996)
	8.7	7.3	20.7	432	7.9	Ultrasonography on 47 healthy volunteers [18]
	7.5	7.3	17.9	373	6.9	Doppler (flow) [18] and 2 transducer (diameter) [19] ultrasonography on 35 normals
		8.2				Milnor, Table 4.3 [17]
Superficial femoral[b]	2.2	6.5	6.6	125	2.9	Color and B-mode ultrasonography, scaled to match graft flow rates [20]
	2.2	6.6	6.4	121	2.7	Ultrasonography on 4 healthy volunteers [18]
	5.2	6.5[c]	15.7	291	6.8	Doppler ultrasonography after balloon angioplasty [21]
	3.6	6.2	12.0	212	5.4	Catheter tip velocity probe (Milnor [17], Tables 4.3 and 6.3)
Thoracic aorta	45–93	23–28	8.9–18.4	640–1330	1.0–2.0	Biplanar angiography & cadaver specimens [22,23][d]

[a] Ignores flow entrance effects, but entrance length = 0.06 Re_D = 24 diameters, so this is a small effect. Compare with mean wall shear stress of 7 dyne/cm² quoted in Ku et al. [10]
[b] First flow rate is suitable for diseased patients so is a lower bound for normals.
[c] Assumed value.
[d] Compare with mean wall shear stress of 1.3 dyne/cm² quoted in Ku and Zhu [24].

Newtonian approximation is valid at the shear rates seen in medium-sized and larger arteries [27, 28].

3.2.2 Steady blood flow at low flow rates

Although the non-Newtonian rheology of blood is not of primary importance in most arteries, it is important at lower shear rates, for example such as might occur at low flow rates in veins or in extracorporeal blood-handling systems. Here we

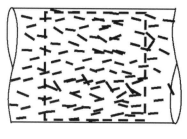

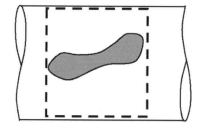

Large vessel   Small vessel

Figure 3.9

Schematic of red cells in large and small vessels. Note the relative size of red cells and control volumes (dotted lines).

consider this low shear case, and use the Casson constitutive relationship to derive the velocity profile for steady flow of blood in a large vessel or tube. It is recognized that blood flow is unsteady within the cardiovascular system, and the question may then be asked: What is the utility of studying steady flow of blood? The answer is threefold. First, steady blood flow can occur in extracorporeal blood-handling systems. Second, study of steady flow gives further insight into the importance of non-Newtonian rheology without involving a great deal of mathematical complication. Finally, pulsatile blood flow can be decomposed into a steady component and a zero mean-flow oscillatory component, and the following analysis addresses the steady component.

A second question may be posed. Why restrict attention to flow in a large vessel or tube? Implicit in the use of Casson rheology to model blood is the assumption that the fluid is a homogeneous continuum. This is true to a very good approximation in large vessels, since the red cells are very much smaller than the vessel diameter. In other words, on the scale of the vessel, the blood "looks" homogeneous, and we can therefore average red cell effects over a control volume (Fig. 3.9), much as is done in considering the continuum assumption in basic fluid mechanics. However, this is not true in, for example, capillaries, where red cells occupy a substantial fraction of the vessel.[4] We see from the above discussion that a "large" vessel in this context means a vessel with diameter "many times the size of a red cell," for example 100 times larger than a red cell.

In vivo, vessels have complex shapes, exhibiting branching and curvature. However, their essential nature is that they are tubes; consequently, as a crude first approximation, it is acceptable to treat a vessel as a uniform cylinder. Hence, we will restrict our attention to the steady laminar flow of blood in a long straight tube, and ask: What is the velocity profile in such a flow? Mathematically, we denote the

---

[4] This is another bizarre aspect of suspension rheology: flow patterns can be dependent on the particle volume fraction, shear rate, and vessel size.

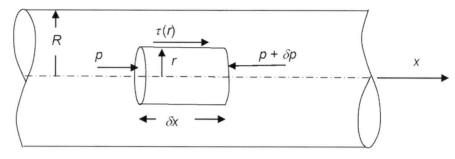

Figure 3.10

Forces acting on a small element of fluid for steady, fully developed flow in a tube of radius R.

axial velocity by u and the radial position by r, and seek an explicit representation of $u(r)$.

At this point it is worth recalling the shear stress distribution for steady, fully developed flow in a long straight tube of radius R. Considering a fluid element of length δx and radius r, we identify the following forces acting on that element: a pressure force $p\pi r^2$ on the left face, a pressure force $(p + \delta p)\pi r^2$ on the right face, and a shearing force $\tau(r)2\pi r\delta x$ on the outer face (Fig. 3.10). Taking account of directions, and noting that for steady fully-developed flow all fluid elements must experience zero acceleration and, thus by Newton's second law, zero net force, we may write

$$p\pi r^2 + \tau(r)2\pi r\,\delta x - (p + \delta p)\pi r^2 = 0. \tag{3.7}$$

Taking the limit as δx goes to zero, in which case $\delta p / \delta x$ becomes the axial pressure gradient, we obtain:

$$\tau(r) = \frac{r}{2}\frac{dp}{dx}. \tag{3.8}$$

Note that this result is valid for all types of fluid, since it is based on a simple force balance without any assumptions about fluid rheology (see Box 3.1).

Since the axial pressure gradient is a constant for steady, fully developed flow in a tube, Equation (3.8) shows that the shear stress distribution within our model vessel must take the form shown in Fig. 3.11. Evidently, $\tau = 0$ at the center line $r = 0$, and therefore, there must be a small region near the center line for which $\tau < \tau_y$. If we call R_c the radial location at which $\tau = \tau_y$, the flow can then be divided into two regions:

- $r > R_c$: fluid flows
- $r \leqslant R_c$: no flow, fluid travels as a plug.[5]

---

[5] It must be remembered that the definition of flow is "continuous deformation in response to an applied shear stress." Therefore, a lump of fluid travelling as a solid mass is not flowing, even though it is moving. For this reason, the term "plug flow" to describe such motion is somewhat of a misnomer, but since it is in common use, we will adopt it here.

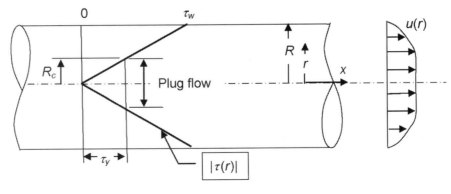

Figure 3.11

Shear stress distribution, and regime of plug flow, for flow of a Casson fluid in a long straight tube.

Box 3.1 Using Equation (3.8) to derive Poiseuille's law for a Newtonian fluid

Since Equation (3.8) is valid for any type of fluid undergoing steady, fully-developed flow, it can be used to derive Poiseuille's law for a Newtonian fluid. For such a fluid, $\tau = \mu(du/dr)$ and Equation (3.8) becomes

$$\frac{1}{r}\frac{du}{dr} = \frac{1}{2\mu}\frac{dp}{dx}. \tag{3.9}$$

Now notice that the right-hand side of Equation (3.9) can only be a function of x (the pressure must be uniform on cross-sections of the tube or a non-axial component of the velocity would be generated), while for a fully developed flow the left-hand side can only be a function of r. The only way that a function of r can equal a function of x is if the function is a constant: that is, the two sides of Equation (3.9) must be constant. This means that the pressure gradient is a constant. Integrating Equation (3.9) once with respect to r and requiring the velocity to be zero at the wall and to be symmetric about the center line, we obtain the well-known parabolic velocity profile:

$$u(r) = -\frac{dp}{dx}\frac{R^2}{4\mu}\left[1 - \frac{r^2}{R^2}\right]. \tag{3.10}$$

The last step is to compute the flow rate (Q) by integrating Equation (3.10)

$$Q = -\frac{\pi R^4}{8\mu}\frac{dp}{dx}. \tag{3.11}$$

This is Poiseuille's law. Recognizing that the pressure gradient is constant and negative (for flow in the positive x direction), we can write $-dp/dx = \Delta p/L$, where Δp is the pressure drop over the tube length L.

We first search for the velocity profile in the region $r > R_c$, where we may write:

$$\sqrt{\tau} = \sqrt{\frac{r}{2}\frac{dp}{dx}} = \sqrt{\tau_y} + \sqrt{\mu\dot{\gamma}} \tag{3.12}$$

where dp/dx is a constant, and the strain rate $\dot{\gamma} = du/dr$. By using Equation (3.12), as well as the fact that $\dot{\gamma} = 0$ at $r = R_c$, the yield stress τ_y may be written in terms of R_c as

$$\sqrt{\tau_y} = \sqrt{\frac{R_c}{2}\frac{dp}{dx}}. \tag{3.13}$$

Inserting Equation (3.13) into (3.12), rearranging, and squaring both sides we obtain:

$$\mu\frac{du}{dr} = \frac{1}{2}\frac{dp}{dx}[r - 2\sqrt{rR_c} + R_c]. \tag{3.14}$$

This is a linear first-order differential equation for $u(r)$, which can be directly integrated. The resulting expression contains one unknown constant, which is specified by imposing the no-slip boundary condition, $u = 0$ at $r = R$. The final expression for $u(r)$, valid for $R_c \leqslant r \leqslant R$, is

$$u(r) = -\frac{1}{4\mu}\frac{dp}{dx}\left[(R^2 - r^2) - \frac{8}{3}\sqrt{R_c}(R^{3/2} - r^{3/2}) + 2R_c(R - r)\right]. \tag{3.15}$$

Once the above velocity profile has been determined, it is a simple matter to determine the plug velocity in the center of the tube. Since the plug velocity must match the velocity at the inner edge of the flowing region, we set $r = R_c$ in Equation (3.15) to obtain the plug (core) velocity

$$u_{\text{plug}} = u(R_c) = -\frac{1}{4\mu}\frac{dp}{dx}\left[R^2 - \frac{8}{3}\sqrt{R^3 R_c} + 2RR_c - \frac{1}{3}R_c^2\right]. \tag{3.16}$$

The overall velocity profile is sketched on the right side of Fig. 3.11.

The final step is to obtain the total flow rate in the tube, Q, obtained by integration of $u(r)$ across the tube cross-section:

$$[L/min]$$
$$\downarrow \; Q = \int_0^R u(r)2\pi r\,dr$$

$$= \int_0^{R_c} u_{\text{plug}}2\pi r\,dr + \int_{R_c}^R u(r)2\pi r\,dr$$

$$= -\frac{\pi R^4}{8\mu}\frac{dp}{dx}F(\xi), \tag{3.17}$$

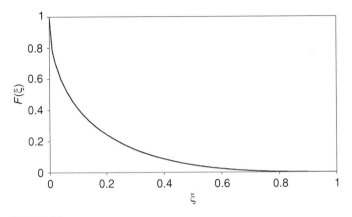

Plot of $F(\xi)$ versus ξ for a Casson fluid (Equation [3.18]). See text for the physical interpretation of $F(\xi)$ and ξ.

where the function $F(\xi)$ is given by

$$F(\xi) = 1 - \frac{16}{7}\sqrt{\xi} + \frac{4}{3}\xi - \frac{1}{21}\xi^4 \tag{3.18}$$

and ξ is given by

$$\xi = \frac{2\tau_y}{R\,|dp/dx|} = \frac{R_c}{R}. \tag{3.19}$$

The function $F(\xi)$ measures the reduction in flow rate (relative to a Newtonian fluid) experienced by the Casson fluid for a given pressure gradient and is plotted in Fig. 3.12. The parameter ξ gives an indication of what fraction of the tube is filled with plug flow. There are two limiting cases.

1. A Newtonian fluid, for which $\tau_y = 0$; therefore $\xi = 0$. In this case, $F(\xi)$ is 1, the velocity profile is the familiar parabolic shape, and Equation (3.17) reduces to Poiseuille's law. This case can be approximated by a Casson fluid with a very small yield stress τ_y, where "small" in this case means that $\tau_y \ll R|dp/dx|$.[6]
2. A Casson fluid with τ_y sufficiently large that the entire tube is filled with a plug. In this case $R_c = R$, $\xi = 1$, and $F(\xi) = 0$. No fluid moves down the tube.

3.2.3 Unsteady flow in large vessels

We now turn to the more physiologically relevant problem of unsteady flow in large vessels. Unfortunately, there is no closed-form (analytic) solution for unsteady flow

---

[6] Note that the definition of small τ_y depends on both the imposed pressure gradient and the tube radius. Consequently, the same Casson fluid may flow in a nearly Newtonian fashion or a very non-Newtonian fashion depending on R and dp/dx.

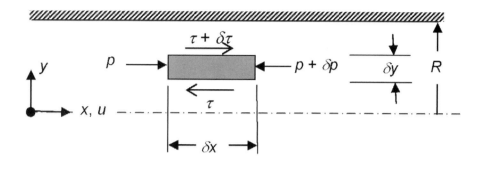

Figure 3.13

Definition sketch for analysis of unsteady flow in a two-dimensional channel. The small shaded region represents a fluid element of dimensions δx by δy by 1 unit deep into the page.

that takes account of realistic non-Newtonian blood rheology; instead, numerical solutions are required. However, as we have seen above, for most large arteries it is acceptable to treat the blood as Newtonian, and in what follows we will make this assumption. Then, the blood's motion is described by the well-known Womersley solution for oscillatory flow.[7] Derivation of the Womersley solution is slightly beyond the scope of this work, since it requires extensive use of Bessel functions. However, an analogous solution having a much simpler mathematical form can be derived for the case of a long two-dimensional channel of half-height R, for which transverse location y is directly analogous to radial location in a tube (Fig. 3.13).

Some thought will reveal that fluid particles in either a long straight tube or a long straight two-dimensional channel will travel along straight lines, never changing their radial position. Of course, they may move axially in one direction along the channel for one part of the flow cycle, then reverse direction during another part, but their paths are always straight lines. Since the fluid is incompressible, this implies that all fluid along an axial line must be moving at the same speed at each instant; otherwise fluid would be accumulating or being depleted somewhere along the line. Therefore, there can be no dependence of velocity on axial position, and the axial velocity u is a function of transverse position and time only, $u = u(y, t)$. Further, there can be no transverse pressure gradients within the channel: the pressure is uniform across the channel cross-section at every axial location. (If there were transverse pressure gradients, the fluid particles would travel in curved paths.) Considering a small chunk of fluid of thickness δy and axial length δx, we see that this slice is acted upon by pressure forces and viscous (shearing) forces. With

---

[7] Interestingly, even though this solution is known as Womersley flow, Womersley was not the first to derive it. This was done much earlier by Sexl [29].

reference to Fig. 3.13, and considering a unit depth of fluid into the page, we may write Newton's second law for the fluid chunk as:

$$\xrightarrow{+} \quad \sum F_x = ma_x \tag{3.20}$$

$$p\,\delta y + (\tau + \delta\tau)\delta x - (p + \delta p)\delta y - \tau\,\delta x = \rho\,\delta x\,\delta y \tfrac{\partial u}{\partial t}. \tag{3.21}$$

Since for this simple flow the shear stress τ is given by $\mu(\partial u/\partial y)$, we obtain the following partial differential equation for $u(y, t)$ governing the fluid motion:

$$\rho\frac{\partial u}{\partial t} = -\frac{\partial p}{\partial x} + \mu\frac{\partial^2 u}{\partial y^2}. \tag{3.22}$$

Because the pumping action of the heart is periodic, or nearly so, it is sufficient to consider oscillatory (periodic) blood flow. This can be accomplished by imposing an oscillatory pressure gradient of magnitude Π:

$$-\frac{\partial p}{\partial x} = \Pi\cos(\omega t) = \Re\{\Pi e^{i\omega t}\}, \tag{3.23}$$

where $\Re$ means "take the real part of." Now, since Equation (3.22) is linear in u and $\partial p/\partial x$, we expect that imposition of a sinusoidally shaped pressure gradient waveform will result in a velocity that also varies sinusoidally in time, with perhaps some phase shift between the pressure gradient and velocity. We therefore write the velocity as

$$u(y, t) = \frac{\Pi}{\rho\omega}\Re\{\hat{u}(y)\,e^{i\omega t}\}, \tag{3.24}$$

where $\hat{u}$ is a complex-valued function that depends only on transverse position y and $\Pi/\rho\omega$ is a scaling factor having dimensions of velocity inserted to simplify later algebra.[8] Substituting Equations (3.23) and (3.24) into Equation (3.22), and converting the resulting equation into one in the complex domain, we obtain the following ordinary differential equation for $\hat{u}(y)$

$$i\,\hat{u}(y) = \frac{\mu}{\rho\omega}\frac{d^2\hat{u}(y)}{dy^2} + 1. \tag{3.25}$$

Before obtaining the solution to this equation, it is useful to non-dimensionalize the transverse position y so that, instead of ranging from $-R$ to R, it ranges from -1 to 1. This is done by introducing the scaled coordinate $\hat{y} = y/R$, in which case Equation (3.25) becomes

$$\frac{1}{\alpha^2}\frac{d^2\hat{u}(y)}{d\hat{y}^2} - i\,\hat{u}(y) = -1. \tag{3.26}$$

---

[8] Equation (3.24) is nothing more than a good guess at a separation of variables solution for $u(y, t)$.

The parameter α in Equation (3.26) is called the *Womersley parameter* and is defined by

$$\alpha = R\sqrt{\frac{\omega\rho}{\mu}}. \qquad (3.27)$$

The solution to Equation (3.26) that satisfies the no-slip boundary conditions $\hat{u} = 0$ at $\hat{y} = \pm 1$ is

$$\hat{u}(y) = i\left[\frac{\cosh(\alpha\hat{y}\sqrt{i})}{\cosh(\alpha\sqrt{i})} - 1\right]. \qquad (3.28)$$

This expression, when combined with Equation (3.24), yields the final expression for $u(y, t)$

$$u(y, t) = \Re\left\{\frac{i\Pi}{\rho\omega}\left[\frac{\cosh(\alpha\hat{y}\sqrt{i})}{\cosh(\alpha\sqrt{i})} - 1\right]e^{i\omega t}\right\}. \qquad (3.29)$$

In order to interpret this equation, it is useful to introduce the concept of the Stokes layer. As the fluid oscillates back and forth, the action of the wall is to damp the resulting motion. However, the wall can only damp motion in its immediate vicinity, in the so-called Stokes layer. The thickness of this layer depends on the fluid viscosity, density, and the frequency of the oscillations. Dimensional analysis shows that the Stokes layer thickness δ_{stokes} is proportional to $\mu/\rho\omega$, so the Womersley parameter is simply the ratio of channel half-height to Stokes layer thickness, R/δ_{Stokes}. The shape of the velocity profiles generated by Equation (3.29) is governed by the value of the Womersley parameter α. Large values of α mean that the channel is large compared with the Stokes layer, and, consequently, the velocity profile is blunt except in a thin region near the wall. In this case, the fluid essentially sloshes back and forth in the channel. The other extreme, small values of α, means that the entire channel is filled with the Stokes layer. In the limit of vanishing α, a parabolic profile that is sinusoidally modulated in time is obtained.

This solution can be generalized to cylindrical tubes of radius R; see, for example, White [30] for details. The resulting expression for the velocity profile is

$$u(r, t) = \Re\left\{\frac{i\Pi}{\rho\omega}\left[\frac{J_0(\alpha\hat{r}i^{3/2})}{J_0(\alpha i^{3/2})} - 1\right]e^{i\omega t}\right\} \qquad (3.30)$$

where $\hat{r} = r/R$ is the dimensionless radial position, and J_0 is the Bessel function of the first kind of order zero [3]. Although the mathematical details are more complex than those of the planar case, the physics are identical. In particular, the Womersley parameter plays the same central role in determining velocity profile shape.

Application of this solution to the physiological situation is complicated by three factors.

First, the pressure gradient does not vary with time in a sinusoidal manner in the arteries. However, this difficulty can be overcome by noting the linear nature of the governing equation, (3.22), which implies that a sum of solutions is also a solution. In particular, a pressure gradient waveform can be broken down into its Fourier components, each of which produces a corresponding velocity profile. These velocity profiles can then be linearly combined to produce a net velocity profile for a given pressure gradient waveform. Usually, only the first several harmonics of the waveform are important, and therefore to get a qualitative idea about the shape of the velocity profile it is sufficient to compute the Womersley parameter for the fundamental harmonic. At rest, the range of values for α in the arteries of the human body is approximately 4 to 20, depending on location (Table 3.6 and Caro *et al.* [2]). Thus, flow can properly be described as moderately to highly unsteady throughout the arterial tree.

Second, the artery walls are elastic and deform under the action of the local pressure, which is time varying. This greatly complicates the mathematical treatment of the problem and will only be considered in a qualitative way below. However, in practice, this effect usually yields only modest perturbations on the velocity profile predicted by Equation (3.30).

Third, blood is assumed to be Newtonian in the above derivation, even though we have previously stated that it is not. However, it can be shown that for the large arteries under most flow conditions the blood acts in a Newtonian fashion. Specifically, within the Stokes layer (which is the only region where viscous effects are important), the values of $\dot{\gamma}$ are usually so high that the blood is effectively Newtonian.[9]

3.3 Blood flow in small vessels

At the opposite end of the size spectrum from the large arteries are the vessels that collectively make up the microcirculation: arterioles, capillaries, and venules. They range in diameter from 6 μm (smallest capillaries) to approximately 50 μm. In such small vessels, it is no longer possible to treat blood as a continuum with average properties such as density and viscosity. In fact, for the smallest capillaries, the red cells fit rather "tightly" within the vessel, with only a thin plasma layer between the red cell and the wall of the vessel. Therefore, the particulate nature of blood must be considered in any bioengineering treatment of the problem. This fact makes simple analytic treatment of the problem beyond the scope of this book. However,

[9] Caution must be taken in extending this result to more complex geometries.

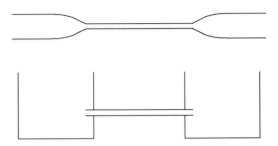

Schematic of experimental apparatus used by Fahraeus and Lindqvist in their study of blood flow in fine bore capillaries. The upper figure shows the original Fahraeus and Lindqvist apparatus; the lower figure shows a modified version used by Barbee and Cokelet [32,33]. Redrawn from Fung [5] with kind permission of Professor Y. C. Fung.

a great deal can be learned from qualitative considerations and examination of experimental data. We consider several such interesting flow phenomena.

3.3.1 Fahraeus–Lindqvist effect

In 1931, Fahraeus and Lindqvist published the results of an intriguing series of experiments [5]. They forced blood through fine glass capillary tubes connecting two reservoirs (Fig. 3.14). Capillary diameters were less than $250\,\mu\mathrm{m}$, and experiments were conducted at sufficiently high shear rates ($\dot{\gamma} \geqslant 100\,\mathrm{s}^{-1}$) so that a similar flow in a large tube would be effectively Newtonian. After correcting for entrance effects, they presented their data in terms of an effective viscosity, derived from fitting measured pressure drop and volume flow rate to Poiseuille's law for a tube of radius R

$$Q = \frac{\pi R^4}{8\mu_{\mathrm{eff}}} \frac{\Delta p}{L} \qquad (3.31)$$

where Q is the flow rate, Δp is the pressure drop across the capillary, and L is the length of the capillary.

Although Poiseuille's law is only valid for a Newtonian fluid, fitting experimental data to Equation (3.31) provides a convenient method of characterizing flow resistance by a single number, namely μ_{eff}. In general, μ_{eff} will depend on the fluid being tested, the capillary diameter, and the flow rate (or pressure drop). However, for a given fluid and a fixed pressure drop, data can be compared between capillaries of differing diameter.

Fahraeus and Lindqvist noticed two unusual features of their data. First, μ_{eff} decreased with decreasing capillary radius, R. This decrease was most pronounced for capillary diameters $< 0.5\,\mathrm{mm}$ (Fig. 3.15). Second, the "tube hematocrit" (i.e.,

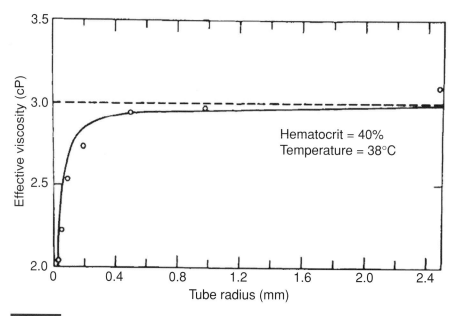

Plot of effective viscosity versus capillary tube radius for flow of ox blood. The dashed line shows the effective viscosity in a tube of very large diameter. Effective viscosity is defined in Equation (3.31); the relative viscosity can be computed from Equation (3.4) knowing that the viscosity of the suspending plasma solution was 1.3 cP. Modified from Haynes [34] with kind permission of the American Physiologial Society.

the average hematocrit in the capillary) was always less than the hematocrit in the feed reservoir. The ratio of these two hematocrits, the tube relative hematocrit, H_R, is defined as

$$H_R = \frac{\text{tube hematocrit}}{\text{feed reservoir hematocrit}} \qquad (3.32)$$

and was found to depend strongly on capillary radius and to depend weakly on feed reservoir hematocrit (Fig. 3.16).

These initially confusing results can be explained by the concept of a *plasma skimming layer*, a thin layer adjacent to the capillary wall that is depleted of red cells. Because the skimming layer is red cell-poor, its effective viscosity is lower than that of whole blood. This layer therefore acts to reduce flow resistance within the capillary, with the net effect that the effective viscosity is less than that for whole blood. Because the skimming layer is very thin (approximately 3 μm) this effect is insignificant in capillaries whose diameter is large. This explains the behavior observed in Fig. 3.15. The relative hematocrit decrease of Fig. 3.16 is also caused by the skimming layer: the central region of the tube is at hematocrit equal to that in the feed reservoir, while the skimming layer is at lower hematocrit. The average

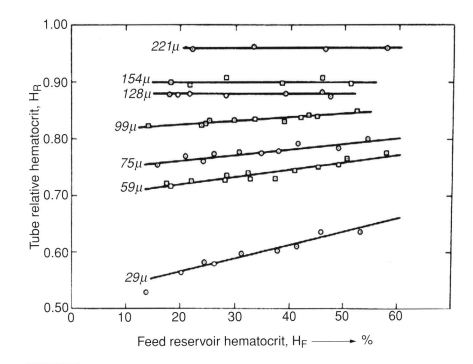

Figure 3.16

Tube relative hematocrit, for capillaries of different sizes. See the text for the definition of tube relative hematocrit. The numbers to the left of the lines are the tube diameters ($\mu = \mu$m). The existence of a relative hematocrit <1 in a capillary tube is known as the Fahraeus effect. Reprinted from Barbee and Cokelet [33], with permission from Elsevier.

of these two yields the tube hematocrit, which must be less than the feed reservoir hematocrit. This effect is more pronounced for smaller capillaries, simply because the skimming layer occupies a greater fraction of the tube cross-section for such capillaries.

This explanation, while accurate, is ultimately unsatisfying, since it fails to answer the fundamental question of *why* a plasma skimming layer exists. There are actually two factors which promote skimming layer formation.

1. For deformable particles (such as red cells) flowing in a tube, there is a net hydrodynamic force that tends to force the particles towards the center of the tube. This is known as the *Segre–Silberberg effect*.
2. It is clear that red cells cannot pass through the capillary wall, which implies that the *centers* of red cells must lie at least one red cell half-thickness away from the wall. This means that, on average, there will be more red cells near the center of the tube than very near the wall.[10]

[10] To understand this concept fully, it is convenient to think of the location of a red cell as being defined by the location of the center of the cell. Then it is clear that red cells can sample locations everywhere within the tube except close to the

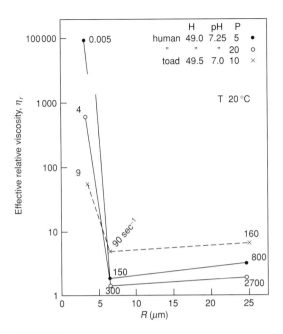

Figure 3.17

Graph demonstrating the "inverse" Fahraeus–Lindqvist effect. Human and toad blood was forced through small gaps (gap half-width R) and the effective relative viscosity, η_r, was deduced from measurements of the resulting flow rates. Note the marked increase in effective viscosity at around $R = 6$ μm. Note that the vertical scale is logarithmic and broken between 10^3 and 10^5. Numbers next to data points are the nominal wall shear rates in units of s^{-1}. Reprinted from Dintenfass [35], by permission of Macmillan Publishers Ltd.

3.3.2 "Inverse" Fahraeus–Lindqvist effect

For *very* small capillaries (diameters of 6–8 μm), a somewhat different behavior is noted: the effective blood viscosity increases rapidly as the capillary diameter decreases (Fig. 3.17). These very small capillaries are almost the same size as the diameter of a single red cell, and thus the fit of a red cell in such a capillary is a very tight one. (In fact, video analysis of red cells passing through such capillaries shows that red cells are forced to pass through the vessel in single file.) The thin gap between the edge of the red cell and the vessel wall is filled with plasma, which experiences very high shear stresses. These high shear stresses act to retard the motion of the red cell, thus increasing the effective viscosity.

tube wall. This means that red cell concentration will be reduced near the tube wall, even though it is possible for the *edge* of a red cell to be very close to the wall.

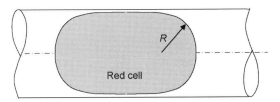

Figure 3.18
For Problem 3.2.

3.4 Problems

3.1 In the nineteenth century, Poiseuille's law was experimentally deduced by J. L. M. Poiseuille, who was interested in understanding the flow of blood in horses' arteries. We now know that Poiseuille's law is a poor way of modeling blood flow in arteries. In your own words, give at least three reasons why Poiseuille's law is not appropriate for arterial blood flow modeling, and briefly discuss why this is the case.

3.2 The red blood cell is very oddly shaped, and it is natural to wonder why it is not spherical. This question should tell you the answer. Since the body requires a certain minimum amount of hemoglobin in the blood (and hence a certain minimum red cell volume), let us consider the red cell volume to be fixed at 98 μm^3.

   (a) If the red blood cell were spherical, what is the smallest pore that it could fit through? Assume that the red blood cell membrane will rupture if stretched (a very good approximation) and remember that a sphere is the geometrical object having minimum surface area for a given volume.

   (b) Now consider the real shape of a red blood cell and allow the cell to deform as it passes through a pore of radius R (Fig. 3.18). Assume that the red blood cell is cylindrical with hemispherical ends. Taking cell membrane area as 130 μm^2, what is the minimum R value? You will get a cubic equation for R; solve it numerically.

   (c) Why is it advantageous to have non-spherical red blood cells?

3.3 At very large values of hematocrit, red cells interact with one another strongly. One model of such an interaction is shown in Fig. 3.19, in which red cells form "lines" separated by gaps. (This is an appropriate model for high shear rate flow in a planar channel.) Assume that the red cells pack together in each "line" so that the space between cells in a given "line" can be neglected. For a hematocrit of 0.75, calculate μ_{eff} for such a system in a Couette flow cell. Assume that all red cells have the same thickness (w), that red cells have

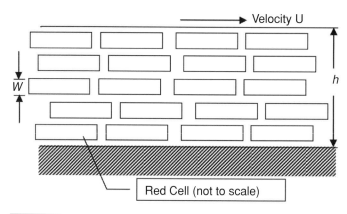

Figure 3.19

For Problem 3.3.

the shape shown in Fig. 3.19, that all fluid gaps between lines have uniform thickness (t), and that the plasma viscosity is 1.2 cP. State other assumptions. (Note that the red cells in the figure are not to scale. In a real channel there would be many such red cell "lines.") Question courtesy of Dr M. K. Sharp, based on a paper by Thurston [36].

3.4 A normal human contains 5 liters of blood, approximately 2% of which is resident in the systemic (i.e., non-pulmonary) capillaries at any given time.

(a) Assuming that the capillaries are 8 μm in diameter, estimate the total length of capillaries in the body (excluding the lungs).

(b) If an average capillary length is 1 mm, how many capillaries are there in the body?

(c) Cardiac output is 5 l/min. Assuming that this is evenly distributed throughout a parallel network consisting of the capillaries found in (b), estimate the pressure drop across the capillary bed. Assume Newtonian, laminar flow in the capillaries with μ_{eff} of 3.5 cP. What percentage of the total 85 mmHg systemic pressure drop is this?

3.5 The continually branching circulatory system can be modeled as a branching network in which each junction contains a parent tube and two daughter tubes of equal diameter. This model can be used to give a rough estimate of pressure drop as a function of position within the arterial tree. Properties of the vessels in generation n (i.e., after n branchings) are designated by the subscript n. The aorta is considered to be generation zero. Assume that Poiseuille's law holds in each generation with a (high shear) value for μ_{eff} of 3.5 cP. You may also assume that the length to diameter ratio L_n/D_n is constant and equal to

16 in all generations (i.e., for all n). The bifurcations are characterized by the area ratio α, defined by

$$\alpha = \frac{\text{area of daughter tubes}}{\text{area of parent tube}}. \tag{3.33}$$

You may assume that α is the same for each generation.

(a) Show that $D_n = (\alpha/2)^{n/2} D_0$.

(b) Show that $\Delta p_n = \Delta p_0 (2/\alpha^3)^{n/2}$, and hence that the total pressure drop from the root of the aorta to the end of the Nth generation pipe, Δp_{0-N}, is

$$\Delta p_{0-N} = \Delta p_0 \left[\frac{(2/\alpha^3)^{N+1)/2} - 1}{(2/\alpha^3)^{1/2} - 1} \right]. \tag{3.34}$$

(c) Show that the total volume of the system from the root of the aorta to the end of the Nth generation V_{0-N}, is

$$V_{0-N} = V_0 \left[\frac{1 - (\alpha^3/2)^{(N+1)/2}}{1 - (\alpha^3/2)^{1/2}} \right]. \tag{3.35}$$

(d) The aortic volume V_0 is 100 ml, and the total volume V is approximately 1200 ml. Use these facts to show that $\alpha = 1.189$ when N is very large.

(e) Our assumption of Poiseuille flow will break down in the smaller arterioles, when the diameter is approximately 20 μm. Knowing that average aortic diameter is 2 cm, estimate the number of generations N up to which the formula in part (b) is valid.

(f) Make a table showing n, the total number of vessels in the nth generation; the length from the root of the aorta to the end of the nth generation; and the pressure at the beginning of the nth generation. Take $Q = 5$ l/min and assume a pressure at the aortic root of 100 mmHg. In your table, include values of n from zero (the aorta) to the number of generations which you found above, in steps of 3; i.e., you will have entries for $n = 0, 3, 6, \ldots$

Hint: in the above you will need to use the formula for summation of a geometric series:

$$\sum_{i=0}^{n} r^i = \frac{1 - r^{n+1}}{1 - r}. \tag{3.36}$$

3.6 Use the bifurcating model for the arterial system described in the previous question, and take $Q_0 = 5$ l/min, $\mu_{\text{eff}} = 3.5$ cP, $D_0 = 2$ cm, $L_n/D_n = 16$ and $\alpha = 1.19$ to answer the following questions.

Table 3.8. For Problems 3.7 and 3.9

Vessel	Mean velocity (cm/s)	Radius (cm)
Ascending aorta	20	0.75
Abdominal artery	15	0.45
Femoral artery ·	10	0.2
Arteriole	0.3	0.0025
Inferior vena cava	12	0.5

(a) What is T_n, the time taken for a fluid particle to flow through generation n (also known as the average transit time for generation n)?

(b) Using (a), develop a formula for T_{0-N}, the total transit time from the start of the bifurcating system to the end of generation N.

(c) Calculate T_{0-N} using data given above for $N = 27$.

3.7 Consider laminar Newtonian flow in a tube of radius R.

(a) Show that the wall shear stress τ_w can be written in terms of the mean velocity V by:

$$\tau_w = \frac{4\mu V}{R}. \tag{3.37}$$

(b) Assuming Newtonian flow, with viscosity 3 cP, estimate τ_w for the vessels shown in Table 3.8.

(c) For blood at a normal hematocrit of 45%, the yield shear stress is approximately 0.05 dynes/cm^2. For each of the above vessels, can the flow of blood be approximated as Newtonian? Why or why not?

3.8 The binding strength of a single $\alpha IIb\beta 3$ integrin complex to fibrinogen has been measured to be 60 to 150 pN [37]. Suppose that a value of 100 pN is typical for all integrins. How many integrin complexes would be required to maintain adherence of a vascular endothelial cell to the wall of a 8 mm diameter artery carrying 1.4 l/min of blood? You may treat the blood flow as steady, take blood viscosity as 3.5 cP, and assume that the apical surface area of a single endothelial cell is 550 μm^2.

3.9 The purpose of this question is to estimate the importance of non-Newtonian effects on blood flow in large vessels. To do this, we will look only at the steady case (realizing all the while that blood flow is pulsatile).

(a) Starting from the expression for Q for a Casson fluid (Equation [3.17]), show that to a very good *approximation* this expression can be rewritten as:

$$Q = \frac{\pi R^3 \tau_y}{4\mu} G(\xi) \tag{3.38}$$

where

$$G(\xi) = \frac{1}{\xi} - \frac{16}{7\sqrt{\xi}} \tag{3.39}$$

valid for $\xi \ll 1$.

(b) Prepare a table showing numerical values for ξ and the ratio $dp/dx_{blood} : dp/dx_N$ for each of the vessels in Table 3.8. Take $\tau_y = 0.05$ dynes/cm^2 and $\mu = 3.5$ cP for blood. In the above, dp/dx_N is the pressure gradient that a Newtonian fluid of viscosity 3.5 cP would have.

3.10 A 35 cm long tube (internal diameter, 1 mm) is filled with blood, which has been citrated to prevent it from clotting. (Assume that the citration process does not alter the rheological properties of the blood.) Use property values for blood: $\mu = 3.5$ cP, $\tau_y = 0.05$ dynes/cm^2, $\rho = 1.06$ g/cm^3.
 (a) At what pressure difference between the tube ends (Δp) does the blood begin to flow?
 (b) What is Q when $\Delta p = 10$ Pa?

3.11 We have introduced the concept of a Casson fluid, which has a non-zero yield stress. A Bingham fluid also has a non-zero yield stress, but its constitutive relation is slightly different:

$$\tau = \tau_y + \mu\dot{\gamma} \tag{3.40}$$

where τ_y and μ are constants (the yield shear stress and the "viscosity").
 (a) Following the presentation for a Casson fluid (Section 3.2.2), derive the velocity profile in a tube of radius R in terms of τ_y, μ, and dp/dx. Integrate this to get the flow rate Q.
 (b) Will this be greater or larger than the flow rate that a Newtonian fluid of viscosity μ would have?

3.12 Blood flows in a tube of radius 1 cm because of a pressure gradient of 0.4 dynes/cm^3. Treating the blood as a Casson fluid with yield stress 0.06 dynes/cm^2, what percentage of the total volume flow rate is from blood traveling in the central non-flowing "core" of the flow?

3.13 A Casson fluid fills the space between two parallel large plates. A pressure gradient dp/dx forces the fluid from left to right. Derive an expression for

Table 3.9. For Problem 3.14

Shear rate (s^{-1})	Shear stress (dynes/cm^2)
1	1.14
2	1.24
3	1.28
4	1.32
5	1.34
6	1.35
7	1.42
8	1.46
9	1.50

the velocity of the plug flow region in this flow. Specifically, show that:

$$u_{\text{plug}} = -\frac{1}{2\mu}\frac{dp}{dx}\left[R^2 - \frac{y_c^2}{3} - \frac{8}{3}\sqrt{y_c\,R^3} + 2\,y_c\,R\right] \tag{3.41}$$

where R is the half-width between the plates, μ is the Casson "viscosity," and y_c is the location of the plug edge. Hint: this is obviously a lot like the derivation in the text, but you will have to think a little about the basic physics at the beginning.

3.14 When a Casson fluid is tested in a viscometer, the data shown in Table 3.9 are obtained. What flow rate do you expect if a pressure drop of 50 dynes/cm^2 forces this same fluid through a tube 10 cm long with radius 2 cm? Assume laminar flow. You may find that making a graph is useful as an intermediate step; before getting out your calculator, think carefully about *what* you want to plot.

3.15 Blood from two different patients is to be tested by having it flow through a long straight tube of length 10 cm and internal diameter 6 mm. A pressure drop of 50 dynes/cm^2 is imposed from one end of the tube to the other, and the resulting blood flow rate Q is measured. If the blood from patient A has a yield shear stress of 0.08 dynes/cm^2 and the blood from patient B has a yield shear stress of 0.12 dynes/cm^2, predict the ratio Q_A/Q_B. Assume that the blood from both patients follows identical Casson rheology, with the exception of the difference in yield stresses.

3.16 A Casson fluid is contained between two large parallel plates, each of area A (Fig. 3.20). The bottom plate is fixed, and the top plate has a constant

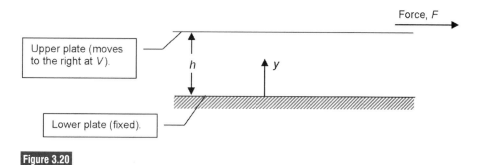

Figure 3.20

For Problem 3.16.

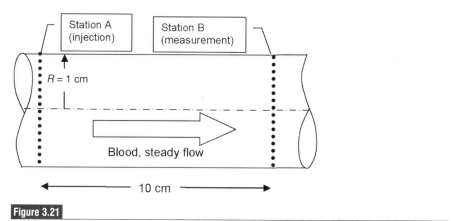

Figure 3.21

For Problem 3.17.

shearing force F applied to it. The gap between the plates is h. Assume the resulting flow is steady.

(a) Show that the shear stress at any location y is equal to F/A.

(b) Derive a relationship between the velocity of the top plate, V, and the applied force, F. Include the cases when shear stress is less than the yield stress, τ_y, and when it is greater than τ_y.

3.17 Nitric oxide (NO) is a naturally occurring vasodilator (a substance that relaxes the artery wall). Its effects are to be tested in the apparatus shown in Fig. 3.21, where NO is injected across the entire tube cross-section at station A and its effects will be measured at station B. One of the characteristics of NO is that it decays rapidly in blood, according to the relationship

$$a(t) = a_0 e^{-t/\tau} \tag{3.42}$$

where $a(t)$ is its activity at time t, a_0 is its initial activity, and τ is its decay time constant, equal to 7 s.

(a) Qualitatively sketch how the activity will vary with radial position at the measuring site. Justify your sketch, using a maximum of three sentences.

(b) If the maximum activity at the measuring site is to be 54% of the injected activity, what should the volumetric flow rate of blood be? Treat blood as a Casson fluid with $\tau_y = 0.05$ dynes/cm^2 and $\mu = 0.035$ g/(cm s). Hint: To answer this question you must numerically solve a (slightly nasty) polynomial equation. It is useful to know that the solution of this equation is in the range $0.10 \leqslant \xi \leqslant 0.11$. You should get the solution to three significant digits.

3.18 An elephant has an aorta that is approximately 8 cm in diameter and a resting heart beat of 35 beats/min. Over the cardiac cycle, do you think that velocity profiles in the aorta will be:
- fairly flat, oscillating back and forth, or
- more parabolic, oscillating back and forth?

Make and state necessary assumptions.

3.19 Consider unsteady flow of a Newtonian fluid in long two-dimensional channel of half-height R, as shown in Fig. 3.13. Suppose that the pressure gradient varies in time according to Equation (3.23). Using existing results from this chapter, derive an expression for the time-varying wall shear stress exerted by the fluid on the upper wall.

3.20 We know that the flow of blood is pulsatile, and so the velocity profile in an artery will be a function of time. This question is designed to identify the qualitative nature of this unsteady velocity profile. Consider laminar, unsteady, incompressible flow of a Newtonian fluid in a rigid infinitely long tube (Womersley flow). Suppose that the pressure gradient in the tube oscillates according to

$$\frac{\partial p}{\partial x} = \text{constant } \sin(\omega t). \tag{3.43}$$

As $\partial p / \partial x$ changes sign, it will tend to drive the fluid in the tube back and forth in an oscillatory fashion. However, the fluid in the tube will not respond to a change in sign of $\partial p / \partial x$ immediately, since it has first to slow down before it can reverse its direction. The time it takes a fluid particle to slow down depends, of course, upon its momentum.

(a) Consider two fluid "chunks" of identical mass, one near the pipe wall and one near the pipe center line. Which chunk has the greatest momentum (on average)? Why?

(b) As $\partial p / \partial x$ changes sign, which fluid chunk changes direction first?

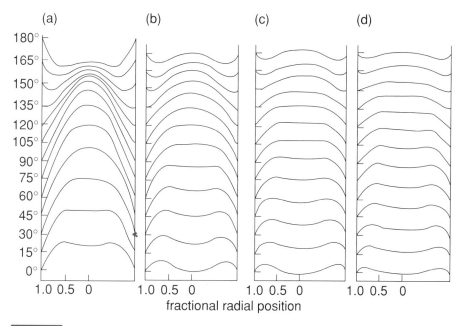

Figure 3.22

For Problem 3.20. Each panel shows velocity profiles for one value of α. Profiles within each panel are taken at 12 equally spaced times during half of an oscillatory flow cycle. From Nichols and O'Rourke [25]. Reprinted by permission of Edward Arnold.

(c) One measure of the unsteadiness of the flow is provided by the Womersley parameter

$$\alpha = \frac{D}{2}\sqrt{\frac{\omega}{\nu}} \qquad (3.44)$$

where D is the tube diameter and ν is the fluid kinematic viscosity, equal to the ratio of dynamic viscosity, μ, over fluid density, ρ. Show that α is dimensionless.

(d) Rank the four sets of velocity profiles in Fig. 3.22 in terms of increasing α. Explain your reasoning. Hint: compare fluid near the wall with fluid near the tube center line.

3.21 Using the expression for the pulsatile velocity profile in a two-dimensional channel of half-height R (Equation [3.29]), show that in the limit of small Womersley parameter, α, the pulsatile velocity profile simply becomes a parabola modulated harmonically in time, i.e.,

$$u(y, t) = \prod \frac{R^2}{2\mu}\left[1 - \left(\frac{y}{R}\right)^2\right]\cos(\omega t). \qquad (3.45)$$

You will need to remember that $\cosh(z) \rightarrow 1 + \frac{1}{2}z^2$ as $z \rightarrow 0$. (This is known as the quasi-steady limit.) For bonus marks, derive the corresponding limiting form of the velocity profile for flow in a cylindrical tube by taking the appropriate limit of Equation (3.30), using the fact [31] that $J_0(z) \rightarrow 1 - \frac{1}{4}z^2$ as $z \rightarrow 0$.

3.22 To compute the pulsatile velocity profile in a two-dimensional channel using Equation (3.29), we need to know the amplitude of the pressure gradient waveform. In practice, this is hard to measure and therefore Equation (3.29) is not very convenient. Instead, it is usually easier to measure the volumetric flow rate, $Q(t)$, and it is more convenient to derive a form of Equation (3.29) that involves the amplitude of the flow rate waveform.

(a) Integrate Equation (3.29) to show that the volumetric flow rate in a two-dimensional channel of half-height R can be expressed as

$$Q(t) = \Re \left\{ \frac{2i\Pi R}{\rho\omega} \left[\frac{\tanh(\alpha\sqrt{i})}{\alpha\sqrt{i}} - 1 \right] e^{i\omega t} \right\}. \tag{3.46}$$

(b) By eliminating Π between Equations (3.29) and (3.46), show that the unsteady velocity profile in a two-dimensional channel can be expressed in terms of the volumetric flow rate as

$$u(y, t) = \Re \left\{ \frac{\alpha\sqrt{i}\,\Lambda}{2R} \left[\frac{\cosh(\alpha\hat{y}\sqrt{i}) - \cosh(\alpha\sqrt{i})}{\sinh(\alpha\sqrt{i}) - \alpha\sqrt{i}\cosh(\alpha\sqrt{i})} \right] e^{i\omega t} \right\} \tag{3.47}$$

where Λ is the amplitude of the flow rate waveform expressed per unit depth of the channel, i.e., $Q(t) = \Re\{\Lambda\, e^{i\omega t}\}$.

(c) For the truly brave, show that the analogue of Equation (3.47) for the case of a cylindrical vessel is

$$u(r, t) = \Re \left\{ \frac{\alpha i^{3/2}\,\Lambda}{\pi R^2} \left[\frac{J_0(\alpha i^{3/2}\hat{r}) - J_0(\alpha i^{3/2})}{2J_1(\alpha i^{3/2}) - \alpha i^{3/2} J_0(\alpha i^{3/2})} \right] e^{i\omega t} \right\} \tag{3.48}$$

where $J_1(z)$ is the Bessel function of the first kind of order one, which satisfies the relationship $\int J_0(z)\, z\, dz = z\, J_1(z)$ and Λ is the 3D version of that defined in part (a).

3.23 From the text, you know that the Fahraeus–Lindqvist effect occurs when a cell-free layer forms next to the wall in a thin tube. In this question you are asked to investigate the Fahraeus–Lindqvist effect in planar channels of gap width H (see Fig. 3.23). Assume that there is a thin cell-free layer of thickness δ next to the solid wall, and that the viscosity of this thin layer is μ_1. The viscosity of the fluid in the center layer is μ_2. Look at the lower half of the channel only; the situation is the same in the upper half.

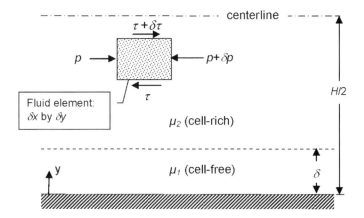

Figure 3.23

For Problem 3.23.

(a) Show by using a force balance that the shear stress in the fluid for steady flow varies according to:

$$\frac{\mathrm{d}\tau}{\mathrm{d}y} = \frac{\partial p}{\partial x}. \tag{3.49}$$

Hence show that τ is given by $\tau(y) = (y - H/2)(\partial p/\partial x)$, by using the fact that $\partial p/\partial x$ is constant and $\mathrm{d}\tau/\mathrm{d}y$ is zero at $y = H/2$ (from symmetry). Hint: consider the small fluid chunk sketched in Fig. 3.23.

(b) Assuming that the fluid in each of the two regions is Newtonian with viscosities μ_1, μ_2, show that the velocity profile in regions 1 and 2 is given by:

$$u_i(y) = -\frac{1}{2\mu_i}\frac{\partial p}{\partial x}[y(H - y) + A_i] \tag{3.50}$$

where A_i is an undetermined coefficient, and i can be 1 or 2, corresponding to regions 1 and 2.

(c) By requiring no slip at $y = 0$, and velocity matching at $y = \delta$, show that $A_1 = 0$ and that

$$A_2 = \delta(H - \delta)\left[\frac{\mu_2}{\mu_1} - 1\right]. \tag{3.51}$$

(d) Integrate the velocity profile from $y = 0$ to $y = H/2$ to get the flow rate (per unit width into the page) in the lower half channel. Then, retaining only terms of first or zeroth order in δ, show that the effective viscosity of the fluid in the gap is:

$$\mu_{\mathrm{eff}} = \frac{\mu_2}{1 + 6\,\delta(\mu_2/\mu_1 - 1)/H}. \tag{3.52}$$

Compare your results with the data of Fig. 3.15. Hint: the flow rate for a fluid of viscosity μ_{eff} is:

$$Q = -\frac{H^3}{24\mu_{\text{eff}}} \frac{\partial p}{\partial x}. \tag{3.53}$$

3.24 The purpose of this question is to investigate the implications of Fig. 3.16. Specifically, we wish to determine how the feed reservoir hematocrit changes as a function of time in the experiment shown in Fig. 3.14. Suppose that at time zero the total blood volume in the feed reservoir is V_0, and that the feed reservoir hematocrit is $H_{\text{F},0}$. Further, suppose that blood is pumped from the feed reservoir through a capillary at a constant volumetric flow rate Q.

(a) Show that the feed reservoir hematocrit changes with time, t, according to the equation

$$\frac{dH_{\text{F}}}{dt} = \frac{Q\,H_{\text{F}}}{V_0 - Qt}\,(1 - H_{\text{R}}) \tag{3.54}$$

where H_{R} is the tube relative hematocrit. Hint: start from the statement

$$H_{\text{F}} = \frac{V_{\text{RBC}}}{V_{\text{RBC}} + V_{\text{p}}} \tag{3.55}$$

where V_{RBC} is the volume of red cells in the feed reservoir, and V_{p} is the volume of plasma in the reservoir.

(b) Integration of the above equation shows that the hematocrit eventually approaches 100% in the feed reservoir. Why does this not occur in the circulatory system: that is, why does the red cell count not increase to something close to 100% at the entrance to the capillaries? You do not need to use any mathematics; a description in words (plus a simple sketch if you prefer) will do the job.

3.25 Consider blood flowing in a fine capillary of radius $R = 50\,\mu\text{m}$. Because of the Fahraeus–Lindqvist effect you may assume that the $10\,\mu\text{m}$ thick region next to the wall contains no red cells. Assume that the velocity profile has the shape sketched in Fig. 3.24, that the red blood cells travel at the same speed as the fluid in the core, and that the hematocrit in the core ($0 < r < 40\,\mu\text{m}$) is 45%. If the blood flowing in this capillary is collected in a beaker, what will the average hematocrit in the beaker be? Compare your result with Fig. 3.16. You may assume that the parabolic portion of the velocity profile joins smoothly with the blunt part; that is, that there is no slope discontinuity at the interface between the two regions. (This is not quite correct if the effective viscosity of the fluid in the two regions is different, but is an acceptable first approximation.)

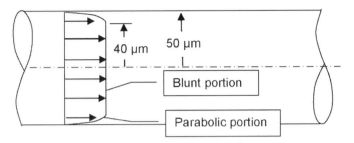

Figure 3.24

For Problem 3.25.

3.26 We have described how the effective viscosity of blood in very small capillaries increases as the capillary radius decreases ("inverse Fahraeus–Lindqvist effect"). This occurs because the red cell becomes "tight fitting" in the capillary: the red cell nearly fills the capillary and leaves only a thin plasma-filled gap between the red cell and the capillary wall. We will analyze the passage of red cells through a capillary in this case.

(a) Consider first the flow of plasma in the thin gap between the red cell and the capillary wall. To an excellent approximation, this can be treated as flow between two parallel planes separated by a gap of thickness $2h$, in which the lower plane (the red cell) travels at velocity V_{RBC}. It can be shown that the velocity profile in the gap, $u(y)$, satisfies:

$$u(y) = \frac{V_{RBC}}{2}(1 - y/h) - \frac{h^2}{2\mu_p}\frac{dp}{dx}[1 - (y/h)^2] \qquad (3.56)$$

where μ_p is the plasma viscosity, dp/dx is the axial pressure gradient in the capillary, and y is defined in Fig. 3.25.

Using this result, calculate the total shear stress acting on the surface of a red cell of length L. For the steady case, use this to derive a formula for red cell velocity, V_{RBC}. Hint: do a force balance on the red cell.

(b) If the capillary is completely filled with red cells as shown, and neglecting the small volume flow rate of fluid in the thin gaps, derive an expression for the effective viscosity of the red cell suspension based on matching the red cell velocity with the mean velocity for Poiseuille flow in a tube of radius R:

$$V_{avg} = -\frac{dp}{dx}\frac{R^2}{8\mu_{eff}}. \qquad (3.57)$$

3.27 Blood flows from a beaker through a 99 μm diameter glass capillary that is 0.1 cm long (Fig. 3.26). The pressure difference across the capillary tube is

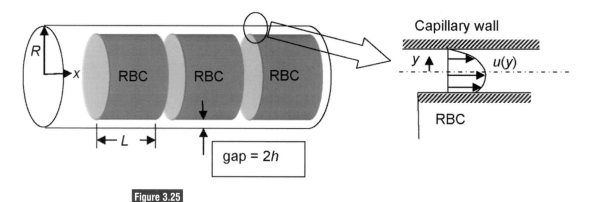

Figure 3.25

For Problem 3.26.

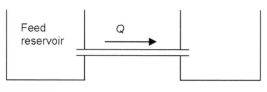

Figure 3.26

For Problem 3.27.

maintained at $60\,000$ dynes/cm^2, and the effective viscosity of the flowing blood is 2.3 cP.

(a) Show that the flow rate $Q = 6.15 \times 10^{-3}$ cm^3/s.

(b) If the volume of blood in the feed reservoir at time zero is V_0 and the feed reservoir hematocrit at time zero is $H_{F,0}$, show that the feed reservoir hematocrit, $H_F(t)$, obeys the relationship.

$$\frac{H_F(t)}{H_{F,0}} = \left(\frac{V_0}{V_0 - Qt}\right)^{1-H_R} \tag{3.58}$$

You may assume that the relative hematocrit, H_R, and the flow rate, Q, are both constant. To obtain full credit for this part you must derive the above expression without making any numerical substitutions.

(c) Using the data given above, suitable information from this chapter, and values $H_{F,0} = 30\%$ and $V_0 = 5$ ml, compute the feed reservoir hematocrit at time 3 min.

(d) For really small capillaries H_R is not constant, so the formula in part (b) is wrong. Assuming that the relationship between H_R and H_F can be expressed as $H_R = a H_F + b$, where a and b are constants, derive an expression for $H_F(t)$ analogous to that given in part (b).

References

1. A. J. Vander, J. H. Sherman and D. S. Luciano. *Human Physiology: The Mechanisms of Body Function*, 4th edn (New York: McGraw-Hill, 1985).
2. C. G. Caro, T. J. Pedley, R. C. Schroter and W. A. Seed. *The Mechanics of the Circulation* (Oxford: Oxford University Press, 1978).
3. H. Schmid-Schönbein. Rheology of leukocytes. In *Handbook of Bioengineering*, ed. R. Skalak and S. Chien. (New York: McGraw-Hill, 1987), pp. 13.1–13.25.
4. W. C. O. Tsang. The size and shape of human red blood cells. M.Sc. thesis, University of California at San Diego (1975).
5. Y. C. Fung. *Biomechanics: Mechanical Properties of Living Tissues* (New York: Springer Verlag, 1981).
6. D. W. Fawcett. *Bloom and Fawcett: A Textbook of Histology* (Philadelphia, PA: W. B. Saunders, 1986).
7. S. Chien, S. Usami, H. M. Taylor, J. L. Lundberg and M. I. Gregersen. Effects of hematocrit and plasma proteins on human blood rheology at low shear rates. *Journal of Applied Physiology*, **21** (1966), 81–87.
8. H. L. Goldsmith. The microrheology of human erythrocyte suspensions. In *Theoretical and Applied Mechanics; Proceedings of the Thirteenth International Congress of Theoretical and Applied Mechanics* (Moscow University, August 21–26, 1972), ed. E. Becker and G. K. Mikhailov. (New York: Springer Verlag, 1973), pp. 85–103.
9. G. R. Cokelet. The rheology of human blood. In *Biomechanics: Its Foundations and Objectives*, ed. Y. C. Fung, N. Perrone and M. Anliker. (Englewood Cliffs, NJ: Prentice-Hall, 1972), pp. 63–103.
10. D. N. Ku, D. P. Giddens, C. K. Zarins and S. Glagov. Pulsatile flow and atherosclerosis in the human carotid bifurcation: positive correlation between plaque location and low and oscillating shear stress. *Arteriosclerosis*, **5** (1985), 293–302.
11. M. Ojha, R. S. C. Cobbold and K. W. Johnston. Hemodynamics of a side-to-end proximal arterial anastomosis model. *Journal of Vascular Surgery*, **17** (1993), 646–655.
12. C. R. Ethier, D. A. Steinman and M. Ojha. Comparisons between computational hemodynamics, photochromic dye flow visualization and MR velocimetry. In *The Haemodynamics of Internal Organs: Comparison of Computational Predictions with in vivo and in vitro Data*, ed. X. Y. Xu and M. W. Collins. (Ashurst, UK: Computational Mechanics, 1999), pp. 131–184.
13. K. Perktold and G. Rappitsch. Computer simulation of local blood flow and vessel mechanics in a compliant carotid artery bifurcation model. *Journal of Biomechanics*, **28** (1995), 845–856.
14. D. A. Steinman and C. R. Ethier. Numerical modeling of flow in a distensible end-to-side anastomosis. *Journal of Biomechanical Engineering*, **116** (1994), 294–301.

15. J. E. Moore, Jr., E. S. Weydahl and A. Santamarina. Frequency dependence of dynamic curvature effects on flow through coronary arteries. *Journal of Biomechanical Engineering*, **123** (2001), 129–133.

16. A. Santamarina, E. Weydahl, J. M. Siegel, Jr. and J. E. Moore, Jr. Computational analysis of flow in a curved tube model of the coronary arteries: effects of time-varying curvature. *Annals of Biomedical Engineering*, **26** (1998), 944–954.

17. W. R. Milnor. *Hemodynamics*, 2nd edn (Baltimore, MD: Williams & Wilkins, 1989).

18. M. Eriksen. Effect of pulsatile arterial diameter variations on blood flow estimated by Doppler ultrasound. *Medical and Biological Engineering and Computing*, **30** (1992), 46–50.

19. S. Uematsu, A. Yang, T. J. Preziosi, R. Kouba and T. J. Toung. Measurement of carotid blood flow in man and its clinical application. *Stroke*, **14** (1983), 256–266.

20. D. A. Steinman, B. Vinh, C. R. Ethier, M. Ojha, R. S. Cobbold *et al.* A numerical simulation of flow in a two-dimensional end-to-side anastomosis model. *Journal of Biomechanical Engineering*, **115** (1993), 112–118.

21. S. T. Hussain, R. E. Smith, A. L. Clark and R. F. Wood. Blood flow in the lower-limb after balloon angioplasty of the superficial femoral artery. *British Journal of Surgery*, **83** (1996), 791–795.

22. D. N. Ku, S. Glagov, J. E. J. Moore and C. K. Zarins. Flow patterns in the abdominal aorta under simulated postprandial and exercise conditions: an experimental study. *Journal of Vascular Surgery*, **9** (1989), 309–316.

23. J. E. Moore, Jr., D. N. Ku, C. K. Zarins and S. Glagov. Pulsatile flow visualization in the abdominal aorta under differing physiological conditions: implications for increased susceptibility to atherosclerosis. *Journal of Biomechanical Engineering*, **114** (1992), 391–397.

24. D. N. Ku and C. Zhu. The mechanical environment of the artery. In *Hemodynamic Forces and Vascular Cell Biology*, ed. B. E. Sumpio. (Austin, TX: R. G. Landes, 1993), pp. 1–23.

25. W. W. Nichols and M. F. O'Rourke. *McDonald's Blood Flow in Arteries* (Philadelphia, PA: Lea & Febiger, 1990).

26. M. Ojha, R. S. C. Cobbold, K. W. Johnston and C. R. Ethier. Visualization of pulsatile flow in a modeled arterial anastomosis. In *Proceedings of the Second International Symposium on Biofluid Mechanics and Biorheology*, ed. D. Liepsch. (1989), pp. 369–379.

27. P. D. Ballyk, D. A. Steinman and C. R. Ethier. Simulation of non-Newtonian blood flow in an end-to-side anastomosis. *Biorheology*, **31** (1994), 565–586.

28. K. Perktold, R. O. Peter, M. Resch and G. Langs. Pulsatile non-Newtonian blood flow in three-dimensional carotid bifurcation models: a numerical study of flow phenomena under different bifurcation angles. *Journal of Biomedical Engineering*, **13** (1991), 507–515.

29. T. Sexl. Uber Den Von E. G. Richardson Entdeckten 'Annulareffekt.' *Zeitschrift für Physik*, **61** (1930), 349–362.

30. F. M. White. *Viscous Fluid Flow*, 2nd edn (New York: McGraw-Hill, 1991).

31. M. Abramowitz and I. A. Stegun. *Handbook of Mathematical Functions* (New York: Dover,1972).

32. J. H. Barbee and G. R. Cokelet. Prediction of blood flow in tubes with diameters as small as 29 microns. *Microvascular Research*, **3** (1971), 17–21.

33. J. H. Barbee and G. R. Cokelet. The Fahraeus effect. *Microvascular Research*, **3** (1971), 6–16.

34. R. H. Haynes. Physical basis of the dependence of blood viscosity on tube radius. *American Journal of Physiology*, **198** (1960), 1193–1200.

35. L. Dintenfass. Inversion of the Fahraeus–Lindqvist phenomenon in blood flow through capillaries of diminishing radius. *Nature*, **215** (1967), 1099–1100.

36. G. B. Thurston. Plasma release-cell layering theory for blood flow. *Biorheology*, **26** (1989), 199–214.

37. R. I. Litvinov, H. Shuman, J. S. Bennett and J. W. Weisel. Binding strength and activation state of single fibrinogen–integrin pairs on living cells. *Proceedings of the National Academy of Sciences USA*, **99** (2002), 7426–7431.

4 The circulatory system

We now turn our attention to the system that transports the blood: the heart and blood vessels. From an engineering viewpoint, the circulatory system consists of a remarkably complex branching network of tubes that convey the blood (*the vasculature*; Fig. 4.1), and two pulsatile pumps in series to force the blood through the tubes (the heart). The vasculature consists of *arteries, arterioles, capillaries, venules,* and *veins.* On average, no cell in the body is more than approximately 40 μm away from a capillary, and almost every tissue is thoroughly invested with a capillary network. A typical human contains approximately 5 liters of blood, and at rest the heart pumps approximately 6 l/min; consequently, on average, blood circulates throughout the body about once per minute. In this chapter, we emphasize the operation of the components of the circulatory system, how they interact with one another, and how they work in concert to deliver blood to target tissues.

4.1 Anatomy of the vasculature

For reasons to be described below, it is conventional to divide the vasculature into two parts: the *pulmonary* and *systemic* circulations. The loop from the right heart, through the lungs, and back to the left heart is known as the pulmonary circulation; the loop from the left heart to the body and back to the right heart is the systemic circulation. Both the pulmonary and systemic vasculature have a similar topology. Blood is supplied in large vessels (arteries), which branch to form smaller arteries, and finally arterioles and capillaries (Fig. 4.2, color plate). The capillaries then join together to form the venules, which in turn form to join larger and larger veins which eventually return the blood to the heart (Fig. 4.2). Typical vessel sizes for the vasculature are shown in Fig. 4.2 for humans and Table 4.1 for dogs. It is handy to remember that in humans the diameter of the largest artery, the aorta, is a little larger than one inch (2.54 cm). The student should keep in mind that the values in Fig. 4.2 are population averages, and, as might be expected, the sizes of the large vessels can vary appreciably from one person to another. For example, aortic

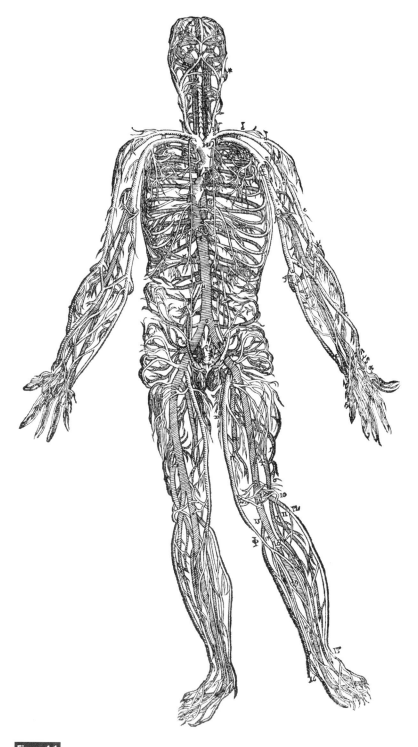

Woodcut engraving of the major veins in a human, from *De Humnai Corporis Fabrica Libri Septum* by Andreas Vesalius, published in Basel in 1543. There are a number of inaccuracies in this image (see discussion in Saunders and O'Malley [1]), but it does serve to illustrate the fantastic geometric complexity of the human vascular tree. This is reinforced when we realize that less than 0.0001% of all the vessels in the body are depicted in this image. A modern edition of this work is available [2].

Table 4.1. Characteristics of the vascular system in a 20 kg dog; this table is useful for giving a sense of the organization of the vascular tree. Similar to the human, it is clear that the majority of the blood volume is contained within the veins, while the capillaries present the largest cross-sectional area. Some entries in the table are based on direct measurements (e.g., the diameters of the largest vessels); some are based on morphometric studies of small tissue samples and extrapolation to the entire body (e.g., the capillary numbers), subject to constraints such as blood volume and percentage of blood in a given vascular region. Class is simply a way of categorizing vessels; all vessels within a given class have diameters ranging from 50 to 150% of the mean value for that class. Compiled by Milnor [4]. Reproduced with permission.

Class	Vessels	Mean diameter (mm)	Number of vessels	Mean length (mm)	Total cross-sectional area (cm²)	Total blood volume (ml)	% of total volume
Systemic							
1	Aorta	(19–4.5)	1		(2.8–0.2)	60	
2	Arteries	4.000	40	150.0	5.0	75	
3	Arteries	1.300	500	45.0	6.6	30	11
4	Arteries	0.450	6000	13.5	9.5	13	
5	Arteries	0.150	110 000	4.0	19.4	8	
6	Arterioles	0.050	2.8×10^6	1.2	55.0	7	
7	Capillaries	0.008	2.7×10^9	0.65	1357.0	88	5
8	Venules	0.100	1.0×10^7	1.6	785.4	126	
9	Veins	0.280	660 000	4.8	406.4	196	
10	Veins	0.700	40 000	13.5	154.0	208	67
11	Veins	1.800	2100	45.0	53.4	240	
12	Veins	4.500	110	150.0	17.5	263	
13	Venae cavae	(5–14)	2		(0.2–1.5)	92	
Total						1406	

Table 4.1. (continued)

Class	Vessels	Mean diameter (mm)	Number of vessels	Mean length (mm)	Total cross-sectional area (cm²)	Total blood volume (ml)	% of total volume
Pulmonary							
1	Main artery	1.600	1	28.0	2.0	6	⎫
2	Arteries	4.000	20	10.0	2.5	25	⎬ 3
3	Arteries	1.000	1550	14.0	12.2	17	⎭
4	Arterioles	0.100	1.5×10^6	0.7	120.0	8	⎫ 4
5	Capillaries	0.008	2.7×10^9	0.5	1357.0	68	⎭
6	Venules	0.110	2.0×10^6	0.7	190.0	13	⎫
7	Veins	1.100	1650	14.0	15.7	22	⎬ 5
8	Veins	4.200	25	100.0		35	⎪
9	Main veins	8.000	4	30.0		6	⎭
Total						200	
Heart							
–	Atria		2			30	⎫ 5
–	Ventricles		2			54	⎭
Total						84	
Total circulation						**1690**	**100**

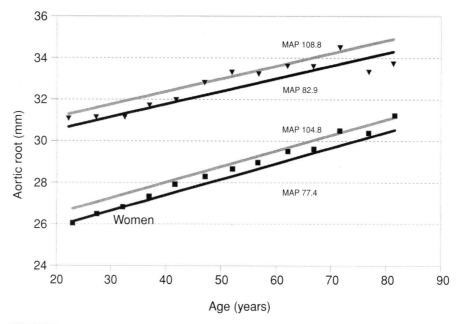

Figure 4.3

Aortic root size as a function of age and mean arterial pressure (MAP) in men and women. The upper two lines represent plots of aortic root size for men in the lower and upper quartiles of MAP, while the lower two lines are comparable plots for women. Triangles and squares represent observed values of aortic root size in men and women, respectively. These values were measured using echocardiography from approximately 4000 ostensibly normal men and women enrolled in the Framingham Heart Study. From Vasan, Larson and Levy [5]. Reproduced with kind permission of Lippincott Williams & Wilkins.

root diameter depends on age, height, weight, sex, and mean arterial pressure, and can vary from about 2.6 to 3.4 cm in healthy individuals [5] (Fig. 4.3).

It is of interest to note how physiological functionality is distributed throughout the vascular system. From an engineering viewpoint, the vasculature must fulfill several functions:

Enable mass transfer between blood and the surrounding tissue. This is accomplished by the capillaries and, to a much lesser extent, by the venules and arterioles. The capillaries are efficient for mass transfer because of their very large wall surface area,[1] and in some locations, because of the very permeable structure of the capillary wall.

Be able to regulate blood distribution, in response to the metabolic needs of perfused tissues. From values given in Fig. 4.2, one can compute that most

---

[1] The area of capillary walls in the adult human is approximately $300\,m^2$, which is approximately 1.5 times the surface area of a single's tennis court.

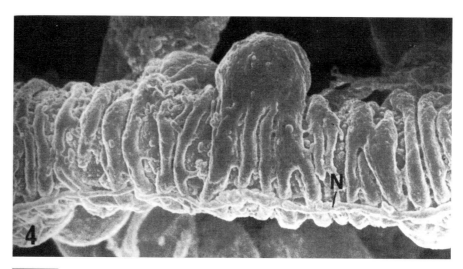

Figure 4.4

Scanning electron micrograph of smooth muscle cells (pericytes) surrounding a precapillary arteriole of diameter approximately 6 μm. When these pericytes relax or contract they alter the caliber of the arteriole and hence regulate flow resistance and blood distribution to the distal capillary beds. A nerve fiber can be seen running along the capillary (N). From Fujiwara and Uehara [6]. Reprinted with permission of John Wiley & Sons, Inc.

flow resistance is generated in the arterioles and capillaries. Moreover, the flow resistance of a given arteriole is regulated through the action of smooth muscle cells, which surround the arteriole and which can contract or relax in response to various stimuli (Fig. 4.4). By controlling vessel caliber in this way, blood flow can be shunted away from tissues where it is not needed and enhanced in tissues with a greater circulatory demand. Thus, a major function of the arterioles is to control flow distribution.

Maintain a reservoir of blood in case of loss.   This role is played by the veins, which contain approximately two thirds of the 5 liters of blood in an adult human.

Figures 4.5 (color plate) and 4.6 show the anatomy of some of the major arteries and veins. The student should know that the major systemic artery leaving the heart is the *aorta*, and that the two major systemic veins returning to the heart are the *superior vena cava* and the *inferior vena cava*.

4.2 The heart

The heart, although a single organ, is functionally two closely coupled positive-displacement pumps in series. Each pump makes up half of the heart, and the pumps share a common wall and common electrical system for stimulation and control.

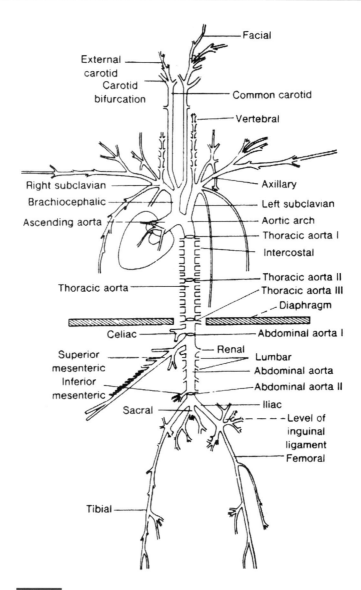

Figure 4.6

An overview of the anatomy of the major arteries in the dog. The drawing is to scale, except that the scale for the arterial diameter is twice that for arterial length. From Nichols and O'Rourke [8]. Reprinted by permission of Edward Arnold.

The right heart accepts oxygen-depleted blood, and pumps this blood through the lungs to deliver it to the left heart. The left heart accepts the oxygenated blood and pumps it to the remainder of the body.

It is of interest to ask why the heart has two sides, rather than being composed of a single large pump. The reason can be understood by comparing the pressure–flow characteristics of the pulmonary and systemic circulatory systems.

The pulmonary capillary beds, consisting of numerous small vessels distributed throughout the lungs, are the site of gaseous exchange between blood and air in the lungs. Efficient blood gas transfer requires that the pulmonary capillary walls be extremely thin. This, in turn, implies that the pulmonary capillaries can be easily ruptured, and so the blood pressure in the pulmonary circulation must be low. As a result, the pulmonary circulation is a low-pressure, low-resistance flow loop, with inlet pressure approximately equal to 15 mmHg and outlet (return) pressure of approximately 5 mmHg. In contrast, the flow resistance of the systemic circulation is much higher than that of the pulmonary circulation, and the pressure drop in the systemic loop is large. This necessitates a large inlet pressure on the systemic side in order to maintain an adequate systemic flow rate. Fortunately, the systemic vessels are much more robust than the pulmonary capillaries and can, therefore, tolerate higher pressures. In short, the systemic circulation is best described as a high-pressure, high-resistance flow loop. The use of two pumps in series is, therefore, an excellent design that satisfies the constraints of low pulmonary pressure and higher systemic pressures.

Of course, the presence of two pumps rather than a single pump introduces a number of complications. Most importantly, the output of the two halves of the heart must be very closely matched over a wide range of conditions (resting to vigorous exercise). If not, blood would accumulate or be depleted from the pulmonary circulation, with serious consequences. Output matching is in part accomplished by the two half-hearts sharing a common "wiring system" used to control heart rate. Additionally, each half of the heart is able to adjust its output, within certain limits, so that it delivers whatever amount of blood it receives. This ability, known as the *Frank–Starling mechanism*, is sufficient to guarantee that fluid will not accumulate in either the pulmonary or the systemic circulations. Fundamentally, the Frank–Starling mechanism relies on the fact that the contractile force generated by the cardiac muscle increases if the cardiac muscle is stretched before it begins to contract, this stretching being greater in a heart that has received more blood during the filling phase [9].

4.2.1 Gross anatomy of the heart

The main mass of the heart is a specialized muscle type known as *cardiac muscle*. Pumping is achieved by rhythmic contraction and relaxation of the cardiac muscle. More specifically, on each side of the heart there is:

- a main pumping chamber (the *ventricle*)
- a holding chamber for each ventricle (the *atrium*)
- two check valves, one each for the ventricle and the atrium.

The chambers and major structures of the heart are shown in Fig. 4.7 (color plate).[2] Most of the mass of the heart is contained in the ventricles, since these are the major pumping chambers.

The valves are made up of thin tissue leaflets that seal together to prevent backflow when closed (Fig. 4.8). The valves between the ventricles and the atria are the *left* and *right atrioventricular (AV) valves*.[3] The AV valves are prevented from being everted ("driven backwards") by thin tissue strands (the *chordae tendineae*) connected to muscles (Fig. 4.7, color plate). The valve between the left ventricle and the aorta is the *aortic valve*, and that between the right ventricle and the pulmonary artery is the *pulmonary valve*. The aortic and pulmonary valves do not have chordae tendineae to prevent eversion. Instead, the aortic and pulmonary valves have three leaflets that, when in the closed position, lean up against one another (called *coaptation*), thereby sealing the valve and preventing the leaflets from being everted.

4.2.2 Qualitative description of cardiac pumping

To describe the pumping sequence of the heart, two special terms are used, *diastole* and *systole*. The technical definitions are given below, but as a rule of thumb, systole corresponds approximately to the period of ventricular contraction, while diastole corresponds approximately to the period of ventricular dilation, or relaxation. Below is the sequence of events for a single beat of the left heart (numbers refer to the lower portion of Fig. 4.9).[4]

1. **Atrial filling (mid to late diastole)**   Blood flows into the atria. The AV valves are open, so there is communication between the atria and the ventricles. The pulmonary and aortic valves are closed, so no blood can get out of the ventricles. In response to an electrical stimulus, both atria begin to contract, forcing blood into the ventricles. The atria then stop contracting.

2. and 3. **Ventricular contraction (systole)**   The ventricles are now stimulated to contract. As they do so, ventricular pressure increases and the AV valves snap shut. This produces a sound that can be detected with a stethoscope, known as the *first heart sound*. The ventricles are now sealed shut, and the ventricular muscle tenses, increasing the ventricular pressure. Since blood is incompressible, ventricular volume is unchanged (hence, *isovolumetric contraction*). Eventually ventricular

---

[2] In discussing the anatomy of the heart, it is important to know that "left" and "right" refer to the owner's viewpoint. However, by convention, the heart is always drawn as seen by the surgeon in the chest. Hence, left and right are transposed in diagrams of the heart.

[3] The right AV valve is also called the *tricuspid* valve, since it consists of three tissue leaflets. The left AV valve is also known as the *mitral* valve.

[4] The sequence for the right heart is very similar, with valve names modified as appropriate.

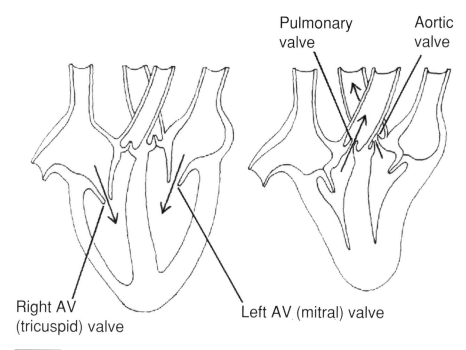

Pulmonary valve

Aortic valve

Right AV (tricuspid) valve

Left AV (mitral) valve

Figure 4.8

Diagram of valve motion during the cardiac cycle. The situation during ventricular filling is shown at left, while the situation during ventricular ejection (systole) is shown at right. The aortic and the pulmonary valves move passively in response to the ambient pressure field and rely on their geometry to prevent leakage/backflow (*regurgitation*). Specifically, their leaflets fit together when closed (*coapt*) so as to be able to resist the retrograde pressure gradient and provide a blood-tight seal. The situation for the tricuspid (right atrioventricular [AV]) and mitral (left atrioventricular) valves is different: they have *chordae tendineae* and *papillary muscles* (Fig. 4.7). The papillary muscles contract during systole to prevent the atrioventricular valves from bulging back into the atria (*prolapsing*). Modified from Vander *et al.* [10]. With kind permission of The McGraw-Hill Companies.

pressure exceeds aortic pressure, and the aortic valve pops open. Blood is ejected from the ventricle and aortic pressure increases. However, the ventricle is already stopping its contraction, so that ventricular pressure reaches a maximum and then starts to decline. Once it falls below aortic pressure, the aortic valve snaps shut (the second heart sound). This ends systole.

4. **Ventricular relaxation (early diastole)**   Once again the ventricle is sealed off. It relaxes at constant volume and ventricular pressure rapidly falls until it is below atrial pressure. At this point the AV valve opens and ventricular filling begins again.

The sequence then repeats . . .

With reference to Fig. 4.9, the following more precise definitions of systole and diastole can be made:

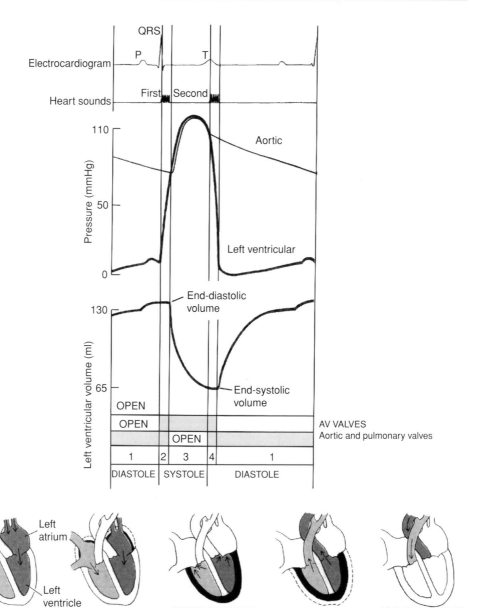

Figure 4.9

Summary of events in the left heart during a cardiac cycle. The electrocardiogram (ECG), heart sounds, aortic and left ventricular pressures, and left ventricular volume are shown in the upper portion of the figure. The lower portion of the figure shows the contracting portions of the heart in black. Readers unfamiliar with the ECG should consult the Appendix. Modified from Vander *et al.* [10]. With kind permission of the McGraw-Hill Companies.

- systole is the period starting at AV valve closure and ending when the aortic valve closes
- diastole is the remainder of the cardiac cycle, beginning when the aortic valve closes and ending when the AV valve closes.

As mentioned above, there are two *heart sounds* per cardiac cycle, occurring at the beginning of systole and at the beginning of diastole. These produce the characteristic "heartbeat," which can be detected with a stethoscope. Normal heart sounds should not be confused with a *heart murmur*, which can indicate cardiac pathology. For example, heart murmurs can be caused by *valvular insufficiency* ("leaky" valves) or a *septal defect* (a small hole between the left and right hearts). Blood leaking through such defects creates turbulence and associated sound, which can be stethoscopically detected.

4.2.3 Cardiac pumping power and ventricular function

A very useful view of cardiac function can be obtained if we plot left ventricular volume against left ventricular pressure to create a *pressure–volume loop*. (We can also make such a plot for the right ventricle, but since the left side of the heart does most of the work, it is usually the left ventricular pressure–volume loop that is more relevant.) Such a loop is shown in Fig. 4.10A, from which we can identify the ventricular filling, isovolumetric contraction, ejection and isovolumetric relaxation phases of the cardiac cycle, as well as times when the various valves open and close.

The shape of the pressure–volume loop for the left ventricle, and specifically the end-diastolic and end-systolic volumes, depend on a number of factors. On the one hand, if the right heart delivers more blood (a condition known as increased *preload*), then the loops are displaced to the right, as shown in Fig. 4.10B. A line drawn along the isovolumetric filling phase of the loops is called the end-diastolic pressure–volume relationship. It depends on the passive elastic properties of the left ventricle wall and the geometry of the ventricle. On the other hand, if mean arterial pressure is increased (a condition known as increased *afterload*), different end-systolic points will be obtained. A curve drawn through these points is called the end-systolic pressure–volume relationship. Empirically, it is found that this curve is approximately linear and independent of ventricular contraction history. A great deal of work has been done to relate the shape of the pressure–volume curve to the elastic and contractile properties of cardiac muscle, see for example the review by McCulloch [11].

It is natural to ask how much energy is transmitted to the blood by the heart with each heart beat. This is known as the *external work* of the heart. The reader will

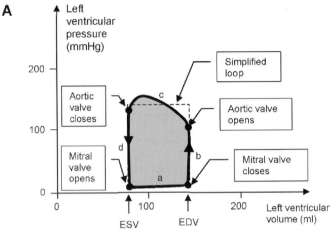

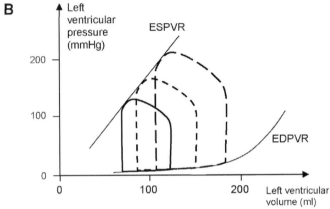

Figure 4.10

Relationship of pressure and volume in the heart. (A) Schematic plot of variation of left ventricular pressure as a function of left ventricular volume during a cardiac cycle. This is known as the work diagram of the left ventricle. a: Passive filling phase, in which work is done by the blood on the left ventricle; b: Isovolumetric contraction, in which no work is done on the blood, but elastic energy is stored in the heart muscle; c: Ejection phase, in which work is done on the blood by the ventricle; d: Isovolumetric relaxation, in which no work is done on the blood but some stored elastic energy is dissipated. The dashed line shows the replacement of the actual pressure–volume loops by a simplified approximate shape (rectangle) having the same enclosed area. ESV and EDV are end-systolic and end-diastolic volume, respectively, and the difference between the two is the stroke volume. (B) A series of pressure–volume loops, showing the end-diastolic pressure–volume relationship (EDPVR) and the end-systolic pressure–volume relationship (ESPVR). These loops are not quantitatively correct; see Fig. 4.11 for actual loops in human left ventricles.

recall that all pumps impart energy to the fluid. It is conventional to represent the useful energy content of an incompressible fluid by its "head," defined as:

$$h = \frac{p}{\rho g} + \frac{V^2}{2g} + z, \tag{4.1}$$

where p is fluid pressure, ρ is fluid density, V is local fluid velocity, and z is elevation above some datum. From the above definition, it is clear that head is made up of pressure head ($p/\rho g$), kinetic head ($V^2/2g$), and potential head (z).

The net head added by a pump, h_p, is given by

$$h_\mathrm{p} = h_2 - h_1, \tag{4.2}$$

where h_2 is the total fluid head at the outlet of the pump and h_1 is the total fluid head at the pump inlet. In the case of the heart, the elevation difference is small and can be neglected. The difference in kinetic head is also small, representing only 2–6% of the total head gain across the heart in humans at rest [4]. That means that the head increase is mainly caused by an increase in pressure across the heart.

Let us return to the pressure–volume curve for the left ventricle (Fig. 4.10). In any process where work is done on the system by pressure, the net work over a cycle, W, is given by the integral

$$W = \oint p\,dV, \tag{4.3}$$

where $\oint$ represents an integral over one cycle and dV is an incremental change in volume. Neglecting the small amount of energy present as kinetic energy in the blood, the external work done on the blood by the left ventricle is proportional to the area enclosed by the heavy loop in Fig. 4.10A. Because this area has a rather complex shape, it is convenient to replace the pressure–volume contour in Fig. 4.10A by a simple rectangular region of width ΔV and height Δp, where ΔV and Δp are chosen so that the area inside the dashed line matches the shaded area: that is, so that $W = \Delta p \Delta V$.

Based on values from the literature appropriate for adult humans at rest (averaged over men and women, Fig. 4.11) we choose ΔV to be 90 ml and Δp to be 110 mmHg. Then, for the left heart we calculate

$$W = (90 \times 10^{-6}\,\mathrm{m}^3)\,(110\,\mathrm{mmHg}) \left(\frac{133\,\mathrm{N/m}^2}{\mathrm{mmHg}} \right) = 1.32\,\mathrm{J}. \tag{4.4}$$

A similar calculation for the right heart using the values $\Delta V = 90$ ml and $\Delta p = 10$ mmHg gives $W = 0.12$ J. Therefore, the total energy imparted to the blood over a single cardiac cycle by the entire heart is 1.44 J. This is easily converted into a *rate* of energy transfer, or external pumping power, by noting that the resting heart rate in the adult human is approximately 70 beats/min; therefore

$$Power = work \times frequency$$
$$= (1.44\,\mathrm{J})\,(70\,\mathrm{beats/min}) \left(\frac{\mathrm{min}}{60\,\mathrm{s}} \right)$$
$$= 1.68\,\mathrm{watt}, \tag{4.5}$$

A

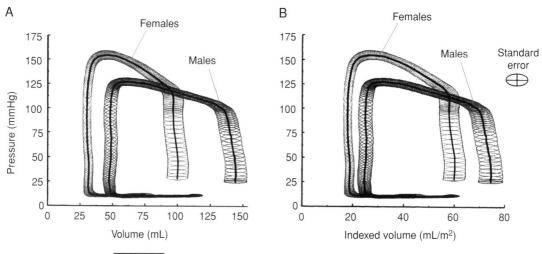

Figure 4.11

Averaged left ventricular pressure–volume loops in males and females showing loops based on volume (A) and on volume divided by the body surface area, or indexed volume (B). The dark line is the average, over which are shown ellipses representing the standard errors in volume and pressure at each point. These data were obtained by making simultaneous left ventricular pressure versus time and left ventricular volume versus time measurements in 14 males and 16 females (average age 60 years) and averaging in time. Because of differences in cardiac period, the averaging procedure truncated each trace to the time of the shortest cardiac period, which explains the "gap" in the average loops. Each individual pressure–volume trace was closed. Reprinted from Hayward *et al.* [12], with permission from the European Society of Cardiology.

which is approximately equal to 0.002 horsepower. This rather crude calculation can be compared with more careful ones reported by Milnor [4], where the external power of the right and left ventricles in humans were 0.24 W and 1.72 W, respectively, for a total external pumping power of 1.96 W. These values pertain to rather large men and so are expected to be a little larger than our averaged values.

It is also interesting to compare the external power produced by the heart with that which would be required to produce the same flow rate against the same circulatory resistance if the flow were steady rather than pulsatile. This latter value is about 85% of the actual external power. What happens to the other 15%? It must be used to create phenomena that are not associated with steady flow. In fact, we will see in Section 4.3 that this remaining 15% generates a pressure pulse wave that propagates throughout much of the vascular system.

Although the *rate* of power transfer to the blood is small, it occurs continuously over an entire lifetime. For fun, let us calculate how much energy is transferred to the blood over an idealized 75 year lifespan.[5] During this period, the heart would

[5] We say that this calculation is for an "idealized" lifespan because we do not account for the effect of exercise on cardiac pumping power, nor for the fact that cardiac pumping power for a child is different than that for an adult.

beat approximately 2.76 billion times, assuming a rate of 70 beats/min. Therefore, the total energy imparted to the blood during this period is nominally

$$\text{Total energy} = 2.76 \times 10^9 \text{ beats } \frac{1.44 \text{ J}}{\text{beat}} = 3.96 \times 10^9 \text{ J}. \quad (4.6)$$

This is sufficient to lift a 50 ton weight from sea level to the top of Mount Everest!

4.3 Arterial pulse propagation

Because of the manner in which the heart pumps, flow in the systemic and pulmonary circulations is pulsatile i.e., unsteady.[6] Since the capillary beds tend to damp pulsatility, flow and pressure pulsations are most pronounced on the arterial side of the circulatory tree and are significantly weaker in the veins.

The effect of pulsatility, in conjunction with the fact that the arteries and other vessels are elastic (distensible), greatly complicates the analysis of blood flow in the circulatory system. For example, if the pressure is elevated locally at some point in an artery, two effects will occur. First, fluid will accelerate so as to move from a region of high pressure to a region of lower pressure, just as would occur in a rigid vessel. Second, the vessel will locally distend, thereby "storing" fluid. In practice, a balance between these two effects causes distension waves to propagate throughout the arteries. Before discussing these effects in greater detail, it is necessary to introduce some terminology describing pressure variations over a cardiac cycle.

4.3.1 Systolic and diastolic pressure

In light of the above discussion, we expect pressure at a given site in an artery to vary during the course of one cardiac cycle. It is conventional to characterize the pressure history at a given location by measuring the maximum (*systolic*) and minimum (*diastolic*) pressures over a single cardiac cycle. They are reported as systolic pressure/diastolic pressure, with units of millimeters of mercury assumed; for example, blood pressure of 120/80 means a systolic pressure of 120 mmHg and a diastolic pressure of 80 mmHg.

Non-invasive blood pressure measurement is accomplished by using the *sphygmomanometer* (from the Latin *sphyg*, meaning pulse). The main element of the sphygmomanometer is an inflatable cuff with a pressure gauge attached. The cuff is placed on the upper arm and inflated until the cuff pressure, p_{cuff}, exceeds the

---

[6] In fact, it is hard to imagine designing a heart from biological material that would produce a steady output.

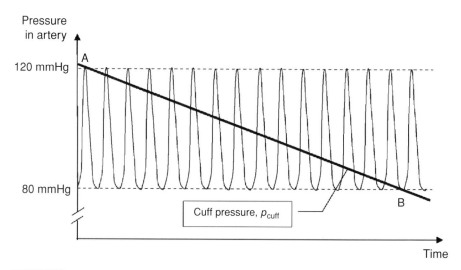

Figure 4.12

Schematic representation of arterial and cuff pressures during sphygmomanometry. The thin solid line is local arterial pressure, while the heavy line represents pressure in the sphygmomanometer cuff. A is the systolic pressure where the artery is just opening. This produces a ticking noise. B is the diastolic pressure, where flow is fully laminar and no noise can be detected. Between A and B, the artery is partially open. "Swooshing" and ticking noises can be heard between A and B because of turbulence and the artery opening and closing.

pressure in the brachial artery,[7] p_{brachial}. The cuff pressure is transmitted through surrounding tissue to the artery, causing the artery to collapse and thereby occlude flow. The pressure in the cuff is then slowly lowered until p_{cuff} equals the systolic pressure, at which point the artery just opens at peak systole. The opening and closing of the artery produces a ticking noise that can be heard with a stethoscope. Continued reduction of p_{cuff} allows the artery to stay open for a greater and greater fraction of the cardiac cycle. While in the partially open state, blood flow in the region of the constriction becomes turbulent, producing characteristic "whooshing" noises, which can be heard with the stethoscope. As p_{cuff} falls further, the artery remains open for a greater fraction of the cardiac cycle, and the "whooshing" noises diminish. Finally, when p_{cuff} equals the diastolic pressure, the artery is fully open and no noises can be heard. This procedure is summarized diagrammatically in Fig. 4.12.

4.3.2 Windkessel model

We now turn to the effects of pulsatility on flow in the vasculature. The simplest model of such effects is the *Windkessel* model, developed by Otto Frank in 1899 [13]

---

[7] The *brachial* artery is the major artery in the upper arm. It runs near the center of the arm.

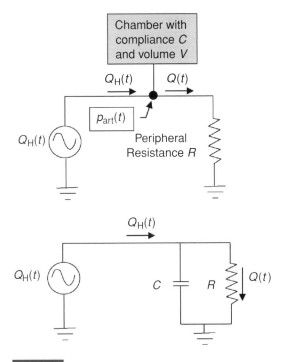

Figure 4.13

The top panel shows the Windkessel model of the circulatory tree. The heart is represented as a variable current source, the distensible arteries as a compliant chamber, and the peripheral vessels as a resistor. The bottom panel shows the equivalent electrical circuit. Note that this is not the only possible "lumped parameter" circuit representation of the circulation; see for example [8]. More sophisticated models include three or four elements; see for example [15] and [16].

(for a translation of the original German paper, see Sagawa *et al.* [14]). Although it is rather crude in a number of ways, and in general does not give very accurate predictions, it is useful for providing qualitative understanding of the physical processes involved. The key feature of the model is that it accounts for the ability of the vessels to store blood by distension. Physically, blood storage occurs primarily (but not exclusively) in the larger central vessels, which in the model are lumped together into a single chamber (Windkessel) that experiences a spatially uniform, but time-varying, arterial pressure, $p_{art}(t)$. The peripheral circulation (arterioles, capillaries, and venules) are represented as a pure flow resistance. The outflow from the heart is represented as a time-varying flow rate $Q_H(t)$, and the flow rate entering the peripheral circulation is $Q(t)$ (see Fig. 4.13).

We assume that an incremental change in volume of the compliant vessels, dV, is proportional to an incremental change in arterial pressure, dp_{art}

$$C = \frac{\mathrm{d}V}{\mathrm{d}p_{art}}. \qquad (4.7)$$

The proportionality coefficient, C, is the effective arterial compliance, which will be assumed to be constant. We will take the flow resistance of the peripheral circulation to be R, which is also assumed to be constant. Therefore, if we neglect the small back pressure in the veins, the arterial pressure at any given instant can be expressed as

$$p_{art}(t) = Q(t)R. \tag{4.8}$$

With reference to Fig. 4.13, the instantaneous difference between the input from the heart (Q_H) and the blood delivery rate to the peripheral circulation (Q) must be from blood that is accumulating in the compliant arteries. Since blood is effectively incompressible, this accumulation/loss of blood must correspond to a change in arterial volume. Conservation of mass then yields

$$\frac{dV}{dt} = Q_H - Q. \tag{4.9}$$

Combining Equations (4.7), (4.8), and (4.9), we obtain a single ordinary differential equation for $Q(t)$

$$RC\frac{dQ}{dt} + Q = Q_H. \tag{4.10}$$

The forcing term, Q_H is determined by the cardiac output. Suppose that we approximate this term by a constant flow output, Q_0, plus a simple sinusoid, i.e., $Q_H(t) = Q_0 + Q_1 \sin(\omega t)$, where Q_0 and Q_1 are constants. Then $Q(t)$ is readily determined to be (see Box 4.1)

$$Q(t) = \frac{Q_1}{\sqrt{1 + (RC\omega)^2}} \sin(\omega t - \phi) + Q_0. \tag{4.11}$$

It is clear that the presence of the chamber produces a phase shift between Q and Q_H. This results from the storage of fluid in the complaint arteries during high rates of inflow from the heart, with subsequent release during low rates of inflow.

Several other aspects of this model deserve comment. First, the cardiac output is not sinusoidal. However, the same physical process occurs with more complex cardiac output waveforms: blood is stored in the arteries during systole and is released during diastole. Mathematically, a more complex cardiac output waveform can be treated by Fourier decomposition.

Second, in practice, the volume flow waveforms predicted by this model are incorrect. This is because of an important fact overlooked by the Windkessel model: different parts of the arterial tree distend at different times during the cardiac cycle. This is partly because the pressure pulse travels at a finite speed down the arteries, so that distal sites distend later than do proximal sites, and partly because of the

Box 4.1 **Solution of Equation (4.10)**

Here we present some of the mathematical details of the solution of Equation (4.10), with $Q_H = Q_1 \sin(\omega t)$. Since the system is linear, and the forcing term Q_H is periodic in ωt, we expect that the solution will also be periodic in ωt, and therefore write

$$Q = Q_1[A_1 \sin(\omega t) + A_2 \cos(\omega t)], \qquad (4.12)$$

with A_1 and A_2 to be determined. Substitution of Equation (4.12) into Equation (4.10) yields two linear equations for A_1 and A_2

$$RC\omega A_1 = -A_2$$
$$-RC\omega A_2 = 1 - A_1. \qquad (4.13)$$

Solution of Equations (4.13) and substitution into Equation (4.12) yields

$$\begin{aligned} Q(t) &= \frac{Q_1}{1 + (RC\omega)^2}[\sin(\omega t) - RC\omega \cos(\omega t)] \\ &= \frac{Q_1}{1 + (RC\omega)^2}\left[\frac{\cos\phi \, \sin(\omega t) - \sin\phi \cos(\omega t)}{\cos\phi}\right] \end{aligned} \qquad (4.14)$$

where $\phi = \tan^{-1}(RC\omega)$. This can be rewritten slightly by using standard trigonometric identities to yield the unsteady term in Equation (4.11).

effects of reflection. It is observed that pressure pulses are reflected and altered at bends and junctions throughout the arterial tree; consequently, the exact form of the arterial distension versus time waveform is different at each site in the arterial tree (Fig. 4.14). To analyze such effects, we need to consider the propagation of elastic waves within the artery wall. To understand this phenomenon, we must first learn a little bit about the elastic properties of the arteries, to which we now turn.

4.3.3 Arterial wall structure and elasticity

The artery wall is a three-layered structure (Figs. 2.57 and 4.15 [color plate]). The innermost (blood-contacting) layer is known as the *tunica intima*, and in a young, healthy artery is only a few micrometers thick. It consists of endothelial cells and their basal lamina, containing type IV collagen, fibronectin, and laminin. The endothelial cells have an important barrier function, acting as the interface between blood components and the remainder of the artery wall. The middle layer is the *tunica media*, which is separated from the intima by a thin elastin-rich

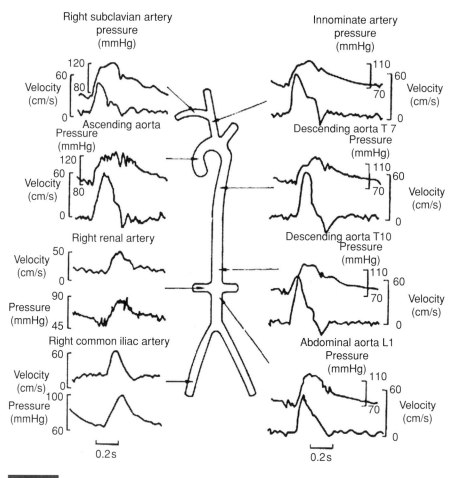

Figure 4.14

Simultaneous pressure and blood velocity waveforms at selected points in the human arterial tree. All waveforms were measured in the same patient except for those from the right renal artery and the right common iliac artery. Changes in the instantaneous pressure are approximate indicators of changes in arterial caliber, and thus of local blood storage. From Caro *et al.* [17], based on data reported in Mills *et al.* [18]. Reproduced with kind permission of Oxford University Press and the European Society of Cardiology.

ring known as the *internal elastic lamina*. The outermost layer is known as the *tunica adventitia*, which is separated from the media by the *outer elastic lamina*. The adventitia is a loose connective tissue that contains type I collagen, nerves, fibroblasts, and some elastin fibers. In some arteries, the adventitia also contains a vascular network called the *vasa vasorum*, which provides nutritional support to the outer regions of the artery wall. Biomechanically, the adventitia helps to tether the artery to the surrounding connective tissue.

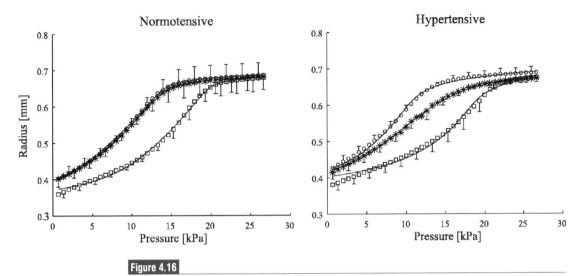

Figure 4.16

Pressure–radius relationship for the carotid arteries of normotensive and hypertensive rats measured under static conditions. The symbols are experimental measurements averaged over six or more rats, the error bars represent standard deviation, and the lines are the fit of a constitutive model to the experimental data. Each graph has three datasets with different vascular smooth muscle (VSM) tone: fully relaxed VSM (○), normal VSM tone (∗), and maximally contracted VSM (□). It can be seen that the state of the vascular smooth muscle has a profound effect on the mechanical properties of the artery wall, and that the contribution of vascular smooth muscle is different in normotensive and hypertensive arteries. Reproduced from Zulliger *et al.* [20] with permission of the American Physiological Society.

The media is the most important layer for determining the biomechanical properties of the artery wall. It contains smooth muscle cells, elastin, types I, III, and V collagen, and proteoglycans. Smooth muscle cells are oriented circumferentially and, as can be seen from Fig. 4.16, have an important influence on arterial stiffness, especially in the smaller arteries. They also provide control of arterial caliber. The collagen is oriented largely circumferentially [21] with a slight helical pattern. The relative proportion of elastin to collagen changes with position in the vascular tree: in the dog, the proportion of elastin in the thoracic aorta is about 60%, but this value decreases significantly near the diaphragm and then gradually falls to about 20% for the peripheral arteries [8]. As elastin content falls, smooth muscle content increases, and arteries are, therefore, classified as being either *elastic* (the large central arteries) or *muscular* (the smaller peripheral arteries).

As we will see in Chapter 9, most biological materials demonstrate highly non-linear stress–strain behavior, and the artery wall is no exception. Figure 4.16 shows pressure–radius data gathered from static inflation tests on excised rat carotid arteries. In these tests, a segment of artery is excised and mounted on a test apparatus that allows the artery lumen to be pressurized while the artery is bathed in a physiological saline solution. The outer arterial diameter is then measured as the lumenal pressure is increased. For a linearly elastic, thin-walled artery undergoing small deformations, a linear pressure–radius relationship is expected. It can be seen from the data in Fig. 4.16 that a linear relationship is not present, with the artery experiencing significant stiffening as the lumenal pressure exceeds approximately 15 kPa (110 mmHg) in the relaxed state. This reflects the strain-stiffening behavior of the collagen and elastin in the artery wall.

In addition to its non-linearity, the artery wall is anisotropic (stiffness is different in different directions) and viscoelastic. Therefore, it is not possible to characterize the stiffness of a given artery wall completely by a single number, such as a Young's modulus. Nonetheless, it is possible to make operational measurements of the stiffness of the artery by observing the change in outer radius, ΔR_0, as the lumenal pressure is varied. This leads to a quantity called the "pressure–strain" modulus, E_p, defined by [22]

$$E_p = R_0 \frac{\Delta p}{\Delta R_0},\tag{4.15}$$

where R_0 is the average outer radius of the artery as the pressure is changed by Δp. Measured values of this quantity for arteries in humans are shown in Table 4.2.

Based on work by Bergel for the distension of an isotropic, thick-walled elastic tube (see summary in Milnor [4]), Gow and Taylor [29] related E_p to the incremental Young's modulus for the artery wall, E_{inc}, by the expression

$$E_{inc} = E_p \frac{2(1 - \nu^2)\left(1 - \dfrac{t}{R_0}\right)^2}{1 - \left(1 - \dfrac{t}{R_0}\right)^2}\tag{4.16}$$

where t is artery wall thickness and ν is Poisson's ratio. This should be interpreted as a circumferential elastic modulus: $E_{\theta\theta}$. The value of the ratio t/R_0 depends on age and increases with distance from the heart, but a typical value for humans aged 36–52 years is approximately 0.15 [8]. Using this value, as well as a Poisson's ratio of 0.5, we compute from Equation (4.16) that $E_{inc} \approx 4E_p$. Using this, as well as the values in Table 4.2, we see that incremental Young's modulus values range from about 3×10^6 to 20×10^6 dynes/cm^2 for the systemic arteries. A more careful analysis, accounting for variations in wall thickness with position, gives elastic modulus values between 8×10^6 and 25×10^6 dynes/cm^2 [8]. Numerous constitutive relationships for vessel walls have been developed and have been summarized by Vito and Dixon [30].

4.3.4 Elastic waves

We now turn to the important question of how pressure pulses are propagated throughout the arterial tree. As will be seen, they travel as transverse elastic waves

Table 4.2. Values of the "pressure–strain modulus", E_p, and other arterial parameters in humans. From more complete listings in Milnor [4] and Nichols and O' Rourke [8], except the values for aortic root.

Artery	No.[a]	R_0 (cm)[b]	Pressure (mmHg) (mmHg)	Radial pulsation (%)[c]	E_p (dyn/cm^2)[d]	Source
Aortic root[e]	1	1.6	–	±4.7	–	Jin et al. [23]
Ascending aorta	10	1.42	79–111	±2.9	0.76×10^6	Patel and Fry [24]
Thoracic aorta	12	1.17	98–174	±2.6	1.26×10^6	Luchsinger et al. [25]
Femoral	6	0.31	85–113	±0.6	4.33×10^6	Patel et al. [24]
Carotid	11	0.44	126–138	±0.5	6.08×10^6	Patel et al. [24]
Carotid	16	0.40	96	±7.4	0.49×10^6	Arndt [26]
Carotid	109	–	–	–	0.63×10^6	Riley et al. [27]
Pulmonary (main)	8	1.35	16	±5.6	0.16×10^6	Greenfield and Griggs [28]
Pulmonary (left)	5	1.07	25	±6.2	0.17×10^6	Luchsinger et al. [25]
Pulmonary (right)	13	1.13	27	±5.8	0.16×10^6	Luchsinger et al. [25]
Pulmonary (main)	8	1.43	18–22	±5.4	0.16×10^6	Patel et al. [24]

[a] Number of arteries studied.
[b] Mean outer radius.
[c] Pulsation about the mean radius from normal pulse pressures (i.e., 100 × one-half the total radial excursion in each cardiac cycle (systolic–diastolic) divided by the average radius).
[d] Calculated from Equation (4.15) using total excursion of pressure and radius during natural pulsations; therefore represents a dynamic modulus.
[e] Measured using MRI.

within the walls of the artery. The physics is similar to the water hammer phenomenon, except that essentially all the elastic energy is stored in the artery walls rather than in compression of the blood.

To analyze elastic wave propagation in greater detail, we consider the simplest possible case of a long straight tube with constant radius and a linearly elastic wall material. We will suppose that the tube is filled with an incompressible inviscid fluid. For the initial stages of the discussion, we will also suppose that the fluid in the tube is (on average) at rest. In other words, there is no net flow in the tube, although, as we will see, the fluid can oscillate back and forth. We also assume that the pressure at one end of the tube is varied periodically, corresponding to the heart pumping at the root of the aorta. As we show below, this periodic variation

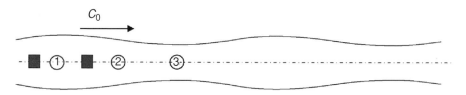

Schematic showing fluid movement in a distensible artery in response to a locally elevated pressure. Fluid elements are represented by squares. See text for definitions of locations ①, ②, and ③. Note that pressure is maximum at ①. c_0 is the elastic wave speed. Note that this figure is not to scale: wavelengths are much longer in the arterial system, and distensions are less severe than shown here.

in the pressure produces an elastic distension wave within the tube wall. We will denote the velocity of this wave by c.

Why is there a distension wave? Suppose that at some location (e.g., site 1, in Fig. 4.17) the pressure is locally elevated so that the wall is locally distended. Then fluid will be displaced from regions of high pressure to those of low pressure (e.g., from just to the right of site 1 to site 2 in Fig. 4.17). If the tube were rigid, this would cause fluid to be displaced all the way down the tube, which would be resisted by the inertia of the fluid filling the tube. In the case of a distensible tube, inertia still resists axial fluid displacement, but an additional physical mechanism exists: fluid can move to site 2 by locally distending the tube at this location. In so doing, the pressure is reduced at site 1 and elevated at site 2, and the distension wave shifts to the right. The same process is then repeated from site 2 to site 3 and so on, with the net result that a pressure wave is propagated down the artery.

Some notes on this phenomenon are in order:

1. We re-emphasize that fluid is not moving along the tube. Fluid just to the right of site 1 moves slightly to the right in the above scenario, but only enough to cause fluid displacement at site 2. When the next bulge travels down the tube to cause distension at point 1, the displaced fluid packet moves back towards site 1. Thus, fluid particles execute back-and-forth oscillatory motions but never go anywhere in the sense that their mean (cycle-averaged) velocity is zero.

2. A fluid packet just to the left of site 1 moves to the left in the above discussion and so contributes to the next "bulge" traveling down the tube.

3. The entire process depends critically on a balance between fluid inertia and wall distension. If fluid inertia is large, then relatively more wall distension will occur. On the other hand, if the wall is very stiff, then less distension

will occur and fluid particles will oscillate back and forth over a greater distance.

4. The same process will occur if an oscillatory flow (with zero mean) is imposed at one end of the tube, rather than an oscillatory pressure.

We wish to estimate the wave speed c. Based on the above discussion, we expect that this will depend on the elastic properties of the tube, as well as on the inertia of the fluid. We characterize the tube's elastic properties by its distensibility β

$$\beta = \frac{1}{A}\frac{dA}{dp} = \frac{2}{D}\frac{dD}{dp} \approx \frac{2}{D}\frac{\Delta D}{\Delta p}, \tag{4.17}$$

where A is the tube cross-sectional area, D is the tube diameter, p is the transmural pressure, and we have assumed that the tube cross-section remains circular. Equation (4.17) simply indicates that β is proportional to the fractional change in tube diameter induced by a unit change in transmural pressure.

It is simplest to estimate c by dimensional analysis. In addition to β, we expect that the wave speed will depend on the fluid inertia, which is characterized by the fluid density, ρ. The relevant parameters have dimensions of[8]

$$[c] \sim LT^{-1} \tag{4.18}$$

$$[\beta] \sim M^{-1}LT^2 \tag{4.19}$$

$$[\rho] \sim ML^{-3}. \tag{4.20}$$

From these three parameters, one Π-group can be formed, namely $\Pi = \rho\beta c^2$. Since there is only one dimensionless group, Π must be a constant, and thus

$$c^2 = \frac{constant}{\rho\beta}. \tag{4.21}$$

A more detailed analysis (presented in Box 4.3, below) shows that the constant in Equation (4.21) is equal to 1, so we may write

$$c^2 = \frac{1}{\rho\beta}. \tag{4.22}$$

Equation (4.22) describes how the pulse wave speed c depends on artery distensibility β and fluid density. For practical application of this result, it is convenient to relate arterial distensibility β to the material and geometric properties of the

---

[8] The syntax $[\cdot] \sim$ means "has dimensions of", where M represents mass, L represents length, and T represents time.

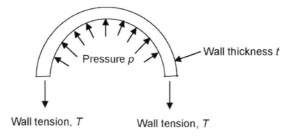

Figure 4.18

Schematic showing forces acting on a half-section of artery wall. The internal diameter of the artery is D and p is the transmural pressure. The wall tension T is per unit depth into the page.

artery wall, as follows. We consider a section of artery of internal diameter D and wall thickness t and assume that the thickness is small, i.e., that $t \ll D$. The artery is pressurized, with a transmural (internal minus external) pressure p, as shown in Fig. 4.18. This transmural pressure is supported by stresses in the artery wall, which, by virtue of the assumption $t \ll D$, are hoop stresses.

Denoting the hoop stresses by σ, we may express the tension within the wall per unit depth into the page, T, as $T = \sigma t$. A static force balance in the vertical direction yields $pD = 2T$, so the hoop stress may be written as

$$\sigma = p\frac{D}{2t}. \tag{4.23}$$

Assuming that the wall material acts as a Hookean solid with modulus E, we may write the circumferential strain in the arterial wall, ε, as

$$\varepsilon = p\left[\frac{D}{2Et}\right]. \tag{4.24}$$

If we now consider only small deformations, then the diameter and wall thickness will not change substantially, and to first order, we may treat the bracketed term in Equation (4.24) as a constant. The increase in strain $\Delta\varepsilon$ resulting from an increment in transmural pressure Δp is, therefore, approximated by

$$\Delta\varepsilon = \Delta p\left[\frac{D}{2Et}\right]. \tag{4.25}$$

However, the change in circumferential strain is equal to $\Delta D/D$, so we may combine Equations (4.17) and (4.25) to write

$$\beta = \frac{2\Delta\varepsilon}{\Delta p} = \frac{D}{Et}. \tag{4.26}$$

From this, the wave speed can be written as:

$$c_0 = \sqrt{\frac{Et}{\rho D}}. \tag{4.27}$$

The wave speed given in Equation (4.27) is known as the *Korteweg–Moens* wave speed, and is related to the water hammer wave speed (see Box 4.3). Note that a subscript 0 has been added to c in this expression, since the Korteweg–Moens wave speed represents an idealized wave speed that is only strictly true under a number of simplifying assumptions:

- The artery can be treated as thin walled. In practice this is not the case, since arterial wall thicknesses are usually 4 to 10% of arterial diameter [5]. Problem 4.4 at the end of this chapter considers the effects of finite wall thickness on wave speed.
- The deformations of the artery wall are small. In vivo, this is a reasonable assumption, as most measurements indicate that arterial diameter varies by approximately $\pm 4\%$ in the systemic arteries (Table 4.2).
- The artery wall can be treated as a Hookean solid. This is definitely not true, since artery walls exhibit both elastic non-linearity and viscoelastic behavior. If the constant E in Equation (4.27) is interpreted as an incremental Young's modulus (tangent modulus), then non-linear effects are not too important in view of the relatively small deformations typically experienced in vivo.

Despite the limitations of the Korteweg–Moens wave speed formula, it is approximately correct and is useful for first-order treatment of problems involving elastic wave propagation in artery walls.

It is of interest to calculate c_0 for a normal artery. We will assume typical property values shown below, based on measured data.[9]

$$\rho = 1.06 \text{ g/cm}^3 \quad \text{(blood)}$$
$$t/D = 0.07 \quad \text{(typical value)}$$
$$E = 8 \times 10^5 \text{ to } 20 \times 10^5 \text{ Pa} \quad \text{(typical value)}$$

For purposes of calculation, we will assume a Young's modulus of $E = 10 \times 10^5 \text{ Pa} = 10 \times 10^6 \text{ dynes/cm}^2$. The Korteweg–Moens wave speed is

---

[9] As discussed above, values of t/D and E depend on the artery, age and health of the subject.

Box 4.2 A more complete derivation of Equation (4.22)

We will model the artery as a long, straight cylindrical tube of radius $R(x, t)$ filled with an incompressible, inviscid fluid (Fig. 4.19). We seek to derive an equation describing the propagation of a transverse elastic wave along the artery, that is, an equation describing the dependence of R on axial position x and time t. To accomplish this, we will apply conservation of fluid mass, a fluid momentum balance (unsteady Bernoulli equation), and a constitutive relationship for the tube wall. Assumptions made at each step in the derivation will be pointed out.

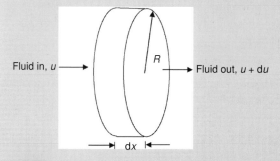

Fluid in, u R   Fluid out, $u + du$

dx

Figure 4.19

Definition sketch showing an arterial segment. The radius of the artery is R, which depends on axial position x and time t. Average fluid velocity entering the segment is u, exit velocity is $u + du$, and length of segment is dx.

First, we apply conservation of mass (fluid continuity). We then consider a small pillbox-shaped control volume of radius R and length dx (Fig. 4.19). Next, observe that the pressure pulse travels as a wave whose wavelength is much longer than the arterial radius. This implies that on the length scale of the control volume, the dependence of R on axial position can be neglected. Denoting the cross-sectionally averaged axial velocity by u, we see that fluid accumulation in the control volume is balanced by net fluid influx. For an incompressible fluid, this can be expressed as

$$\frac{\partial}{\partial t}[\pi R^2 \, dx] + \pi R^2 \, du = 0. \tag{4.28}$$

In the limit as dx approaches 0, this can be rewritten as

$$\frac{\partial u}{\partial x} + \frac{2}{R}\frac{\partial R}{\partial t} = 0. \tag{4.29}$$

Next, we apply a momentum balance, embodied in the unsteady Bernoulli equation [31]. Because the fluid is inviscid, the loss terms can be neglected. Applying this equation between two points 1 and 2, located a distance dx apart, and neglecting elevation differences between points 1 and 2, we obtain

$$\frac{u_1^2}{2g} + \frac{p_1}{\rho g} = \frac{u_2^2}{2g} + \frac{p_2}{\rho g} + \frac{1}{g} \int_1^2 \frac{\partial u}{\partial t} \, ds. \tag{4.30}$$

Note that in writing Bernoulli's equation in this way, we have assumed that the velocity profile is uniform across the tube, which is acceptable for an inviscid fluid when wall deformations are small. We now use the fact that points 1 and 2 are very close, and thus in view of the long wavelength of the pressure pulse, we can assume that $\partial u / \partial t$ is constant over dx, so that the above integral equals $\partial u / \partial t$ dx. Similarly, for small wall deformations and a long pressure pulse, $u_1^2 - u_2^2$ can be neglected. Finally, in the limit as dx becomes small, $(p_2 - p_1)/dx$ equals the axial pressure gradient, dp/dx. Thus, the momentum equation simply reduces to a balance between fluid inertia and the axial pressure gradient,

$$\rho \frac{\partial u}{\partial t} = -\frac{\partial p}{\partial x}. \tag{4.31}$$

Finally, we write the constitutive equation for the tube wall, which is simply Equation (4.17) modified to account for infinitesimal changes in pressure and tube radius:

$$\beta = \frac{2}{R} \frac{dR}{dp}. \tag{4.32}$$

If the inertia of the wall and surrounding tissue can be neglected, and if the wall is purely elastic, then variations in pressure will be in phase with variations in radius. In this case β will be a real number, otherwise it will be complex.

If we combine Equations (4.31) and (4.32), and define $c^2 = 1/(\rho\beta)$, we obtain

$$R \frac{\partial u}{\partial t} + 2c^2 \frac{\partial R}{\partial x} = 0. \tag{4.33}$$

Differentiating this equation with respect to x, and Equation (4.29) with respect to t, combining the results, and assuming small deformations (i.e., neglecting first-order variations in R), yields

$$\frac{\partial^2 R}{\partial t^2} - c^2 \frac{\partial^2 R}{\partial x^2} = 0. \tag{4.34}$$

This is a hyperbolic equation for $R(x, t)$, which admits traveling wave solutions of the form $R(x, t) = f(x \pm ct)$, where f is an arbitrary function. This proves that, under the stated assumptions, transverse elastic waves travel at speed $1/\sqrt{\rho\beta}$.

Box 4.3 Derivation of the Korteweg–Moens wave speed

The Korteweg–Moens wave speed can also be derived from the formula for water hammer wave speed

$$c = \sqrt{\frac{K/\rho}{1 + \dfrac{KD}{Et}}} \qquad (4.35)$$

where K is the bulk modulus of compressibility of the fluid in the tube. In the case of an artery filled with blood $Et \ll KD$, and in this limit Equation (4.35) reduces to Equation (4.27).

---

then given by Equation (4.27) as

$$c_0 = \sqrt{\frac{(10 \times 10^6)(0.07)}{1.06} \frac{\text{cm}^3}{\text{g}} \frac{\text{g}}{\text{cm s}^2}} = 812 \text{ cm/s}. \qquad (4.36)$$

In an adult human, the distance from the heart to the wrist is approximately 90 cm. Therefore, it takes about $(90 \text{ cm})/(812 \text{ cm/s}) = 0.11 \text{ s}$ for the pressure wave to reach the wrist from the heart. (When we palpate this pressure wave we feel "the pulse.") More generally, it is evident that arteries at different distances from the heart will distend at different times during the cardiac cycle. This important fact invalidates the Windkessel model.

Based on the expression for the Korteweg–Moens wave speed, it can be seen that c_0 will increase if either E or t increases. This occurs naturally in aging: the arteries of elderly people are stiffer and have thicker walls than those of young people. Hence, we expect wave speed (c_0) to increase with age, which is confirmed by data from Caro et al. [17], who quote values of c_0 of 520 and 860 cm/s at ages 5 and 84 years, respectively.

4.3.5 Pressure–flow relationships: purely oscillatory flow

In the previous section it was pointed out that the Windkessel model is invalid because different parts of the arterial tree distend at different times. A second problem with the Windkessel model is that even if the peak arterial pressure occurred simultaneously for all arteries throughout the body, the detailed shape of the pressure–time waveform is different at different arterial locations, as shown by Fig. 4.14. This is because of the presence of *bifurcations* (branches), bends, and arterial taper, all of which act to modify the shape of the pressure pulse. In order to quantify these effects, it is necessary to develop relationships between pressure and flow in oscillatory systems, to which we now turn.

To recap, the combination of distensible artery walls and a time-varying (periodic) pressure input from the heart causes elastic waves (pressure pulses) to travel along the arteries. Mathematically, the pressure, therefore, depends on time t as well as on axial location in the arterial tree, x. If we assume the simplest possible case, in which the pressure varies harmonically in time with zero mean at a given location, we can express the pressure as[10]

$$p(x, t) = p_0 \cos\left[\frac{2\pi}{\lambda}(ct - x)\right]$$

$$= p_0 \cos\left[\omega t - \frac{2\pi x}{\lambda}\right]. \tag{4.37}$$

Here p_0 and λ are the amplitude and wavelength of the pressure pulse, respectively, while ω is its circular frequency,

$$\omega = \frac{2\pi c}{\lambda}. \tag{4.38}$$

We can differentiate Equation (4.37) to obtain the pressure gradient

$$\frac{\partial p}{\partial x} = \frac{2\pi}{\lambda} p_0 \sin\left[\omega t - \frac{2\pi x}{\lambda}\right], \tag{4.39}$$

which is responsible for forcing the oscillatory fluid motion.

The pressure gradient $\partial p/\partial x$ can be related to blood velocity by using Newton's second law. It is assumed that the blood is inviscid (no shear stresses are present) and that fluid motion is purely axial, i.e., that vessel wall displacement is small. If we then consider a thin fluid element with cross-sectional area A and volume V (Fig. 4.20) we may write

$$\xrightarrow{+} \sum F_x = \rho V \frac{\partial u}{\partial t}$$

$$-\mathrm{d}p\, A = \rho\, \mathrm{d}x\, A \frac{\partial u}{\partial t}, \tag{4.40}$$

which yields the equation

$$\rho \frac{\partial u}{\partial t} = -\frac{\partial p}{\partial x}. \tag{4.41}$$

---

[10] To convince yourself that this is the correct form of $p(x, t)$, note that p is constant if $ct - x =$ constant, i.e., if $x = ct +$ constant. This, therefore, represents a wave propagating in the positive x-direction at velocity c.

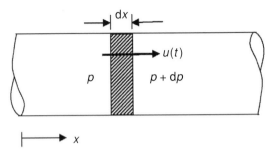

Force balance on a fluid element of length dx and cross-sectional area A in unsteady flow. The fluid is assumed to be inviscid.

Using Equation (4.39) for the pressure gradient, we can integrate Equation (4.41) to obtain

$$u(x, t) = \frac{2\pi}{\rho \omega \lambda} p_0 \cos \left[\omega t - \frac{2\pi x}{\lambda} \right] = \frac{2\pi}{\rho \omega \lambda} p(x, t). \qquad (4.42)$$

Thus, using Equation (4.38), the axial fluid velocity can be related to the pressure by

$$u(x, t) = \frac{p(x, t)}{\rho c}. \qquad (4.43)$$

Note that Equation (4.42) implies that the magnitude of u will be small if ρ or ω is large. The former case corresponds to fluid with a great deal of inertia, which resists motion driven by the oscillatory pressure gradient $\partial p / \partial x$. The latter case corresponds to frequencies so high that fluid elements only just get started moving in one direction before the pressure gradient changes sign; as a result, the net motion and velocity are never allowed to become large.

In an inviscid flow, the velocity profile is uniform across the tube. Consequently, the instantaneous flow rate $Q(x, t)$ is simply the tube cross-sectional area A multiplied by $u(x, t)$, i.e.,

$$Q(x, t) = \frac{A}{\rho c} p(x, t). \qquad (4.44)$$

We have previously used the analogy between fluid flow and current flow to define a flow resistance for *steady flow* (ratio of pressure drop to flow rate; see Box 4.4). By similar analogy, we define the *characteristic impedance*, Z_0, as the ratio of pressure to flow in an *oscillatory* system

$$Z_0 = \frac{p}{Q}. \qquad (4.45)$$

> ## Box 4.4 The electrical/fluid flow analogy
>
> The electrical/fluid flow analogy can be made more precise by considering which physical features contribute to characteristic impedance, as tabulated below.
>
> | Physical effect | Name | Electrical analogue |
> | --- | --- | --- |
> | Viscous flow resistance | Resistance | Resistance |
> | Elastic vessel walls | Compliance | Capacitance |
> | Fluid inertia | Inertance | Inductance |
>
> The analogy between electrical capacitance and fluid compliance is a little bit tricky; an electrical capacitor has an inlet and an outlet, and the current in must equal the current out. A fluid compliance chamber often has only one inlet, and flow rate in does not have to equal flow rate out. This can be handled by suitable placement of the elements in the electrical and fluid circuits (e.g., Fig. 4.13).

From the above discussion, we can write

$$Z_0 = \frac{\rho c}{A} \tag{4.46}$$

for an inviscid fluid. Note that Z_0 is a real number (but see Section 4.3.7).

It is important to understand the meanings of p and Q in the above derivation: p is the oscillatory pressure (measured about some mean value), while Q is the corresponding oscillatory flow. For flow with zero mean, as considered above, that means that Q can be thought of as a "sloshing" flow, in which fluid oscillates back and forth but (on average) never goes anywhere.

4.3.6 Pressure–flow relationships: mean flow effects

To this point, we have considered only the case of a purely oscillatory pressure gradient, which leads to a purely oscillatory flow (i.e., one in which the flow averaged over a single cardiac cycle is zero). Evidently this is not the situation in vivo, since the cycle-averaged flow rate in the cardiovascular system is non-zero. This can be accounted for by adding a steady mean flow (and corresponding pressure gradient) to the above description. In this case, the motion of a fluid particle consists of a steady translation with a sinusoidal oscillation superimposed on it. The steady component of the flow does not affect the pulse propagation.

4.3.7 Pressure–flow relationships: deviations from ideality

The model that we have developed is quite idealized, corresponding to an infinitely long straight tube filled with an inviscid fluid. This situation does not occur physiologically, and we must, therefore, examine what effect relaxing each of the above assumptions has on pulse wave propagation and flow dynamics. The most important effects are described below.

Viscous losses

In the analysis above, no account has been taken of viscous losses, which will arise from the action of fluid viscosity and arterial wall viscoelasticity. Generally speaking, viscosity dissipates energy and, therefore, diminishes the amplitude of the pressure pulse. It also decreases c_0, since it tends to impede fluid motion.

More specifically, inclusion of fluid viscosity affects the above analysis in two ways. First, the amplitude of $Q(x, t)$ is reduced compared with that predicted by Equation (4.44). Second, the flow rate waveform lags the pressure waveform. These effects can be summarized by [8]

$$Q(x, t) = \frac{A M'_{10}}{\rho c} p(x, t - \phi) \tag{4.47}$$

where $\phi = \varepsilon_{10}/\omega$ (compare Equation [4.44]). The coefficients M'_{10} and ε_{10} have been computed as a function of flow parameters [8]; M'_{10} is <1. Because of phase differences between p and Q in this case, Z_0 becomes complex.

Variation of arterial properties

Moving distally (i.e., away from the heart), arteries typically become stiffer and smaller in caliber. Examination of Equation (4.27) implies that the pulse wave speed c_0 will, therefore, increase distally. This causes the amplitude of the pressure pulse to increase distally, and the amplitude of the flow waveform to decrease distally, as the following qualitative argument shows.

Consider fluid moving from point 1 to point 2 in Fig. 4.17, where motion from left to right is away from the heart. If the artery is smaller and stiffer at point 2, a given fluid displacement will cause a proportionally larger pressure increase at point 2 than would occur if arterial properties were uniform. In other words, if the amplitude of the flow waveform is kept constant, the amplitude of the pressure waveform will increase distally. If, however, the pressure wave amplitude is kept

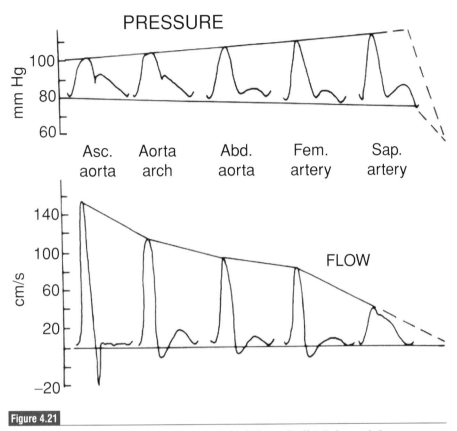

Figure 4.21

Matched records of pressure and blood velocity at different sites in dog arteries. Note the increase in the pressure wave amplitude and the decrease in the flow wave amplitude with distance from the heart. Asc., ascending; Abd., abdominal; Fem., femoral; Sap., saphenous. From Nichols and O'Rourke [8]. Reprinted by permission of Edward Arnold.

constant, then the amount of arterial "stretch" (and hence the oscillatory component of the flow) will be smaller at point 2 than at point 1 (i.e., the amplitude of the flow waveform will decrease distally).

In practice, neither the pressure nor the flow waveforms have constant amplitude. Instead, the amplitude of the flow waveform decreases somewhat, while the amplitude of the pressure pulse increases somewhat, moving distally in the arterial tree (Fig. 4.21). This represents a compromise between the two scenarios outlined in the previous paragraph.

Effects of branching and taper

The arteries are not infinitely long and straight, but rather are tapered, curved, and typically delimited by branches (*bifurcations*) both upstream and downstream. These effects, particularly branching, have a major effect on pulse wave

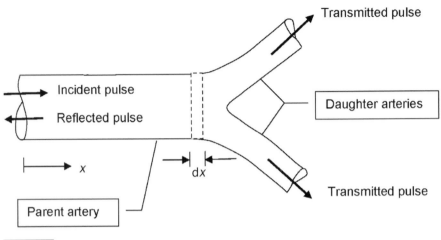

Figure 4.22

Sketch of an arterial bifurcation, showing the parent artery and two daughter arteries. The incident pressure pulse travels from left to right; the reflected and transmitted pressure waves are also shown. The control volume used in the analysis is shown by the dashed lines.

propagation. Here we analyze in some detail the behavior of the pressure wave at a single bifurcation.

Consider an incident pressure pulse arriving at the bifurcation shown in Fig. 4.22. The characteristic impedance of the daughter arteries will in general be different from that of the parent artery, since c and A will differ from daughter to parent. This means that the ratio of the pressure pulse amplitude to the flow pulse amplitude must change as the pulse wave passes through the junction. In other words, the impedance mismatch across the junction will force the amplitude of the pressure pulse passing through the junction to be different from the amplitude of the incident pressure pulse. Since the pressure pulse carries energy with it, and there are no losses in the system, some energy must be reflected back into the parent artery in the form of a reflected pressure wave. Hence, the junction gives rise to both a transmitted and a reflected pressure pulse. We seek to characterize the amplitude of these pulses in terms of arterial properties.

To develop such a characterization, the arteries and junction are treated as a one-dimensional system in which pressure is uniform across the artery at each axial station, while pressure and flow rate vary axially. The daughter tubes are assumed to be identical. Consider a very thin control volume (thickness dx) positioned at the junction (Fig. 4.22). Since the control volume is very thin, the mass within it can be considered negligible. Consequently, the forces acting on fluid within this control volume must sum to zero at every instant; if not, a finite force would act on an infinitesimal mass, thereby producing an infinite acceleration. Since the

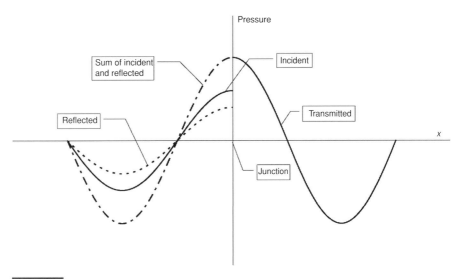

Graph showing pressure as a function of axial distance x around the junction of Fig. 4.22 at a given instant. Note that the dependence of pressure on x is harmonic, and that the sum of pressures upstream of the junction equals the pressure downstream of the junction.

cross-sectional area of the control volume is identical on the upstream and downstream faces, and since pressure forces are the only forces acting, this implies that the pressure on the upstream face of the control volume must balance the pressure on the downstream face at every instant. The pressure on the upstream face has two contributions, namely the incident and reflected pulse waves. Therefore, we may state: *the sum of the incident and reflected pressure pulses must equal the transmitted pressure pulse* (Fig. 4.23). Referring to Equation (4.37), we can write the pressures at the junction as

$$p_i(x_{\text{junction}}, t) = p_{0,i} \cos(\omega t + \theta), \tag{4.48}$$

$$p_r(x_{\text{junction}}, t) = p_{0,r} \cos(\omega t + \theta), \tag{4.49}$$

$$p_t(x_{\text{junction}}, t) = p_{0,t} \cos(\omega t + \theta), \tag{4.50}$$

where the subscripts i, r, and t refer to the incident, reflected, and transmitted waves, respectively, and p_0 refers to the amplitude of a pressure pulse. The force balance can then be written as

$$[p_{0,i} + p_{0,r}] \cos(\omega t + \theta) = p_{0,t} \cos(\omega t + \theta) \tag{4.51}$$

which implies that the pressure pulse amplitudes must satisfy

$$p_{0,i} + p_{0,r} = p_{0,t}. \tag{4.52}$$

In Equations (4.48) to (4.50), θ is a constant that depends on the location of the junction, x_{junction}. The important fact is that ω and θ are the same for all three pressure waves. This implies that the three pressure pulses are in phase, which Equation (4.51) shows must be the case if the pressures are to balance at every instant.[11]

The next step is to consider the volume flow rates entering the junction. Before doing so, we must make a brief detour to discuss sign conventions. Although we did not explicitly say so previously, our convention is that positive oscillatory flow corresponds to the direction of propagation of the pressure wave.[12] This means that the reflected wave is producing a left-going "slosh" at the junction at the same time as the incident and transmitted waves are producing a right-going "slosh." This will be important when we balance fluid mass, a task to which we now return.

As dx approaches zero, no fluid can be stored in the control volume, and thus the volume flow rate entering the control volume must balance the volume flow rate leaving at every instant. Adopting notation similar to that describing the pressure waves, we write

$$Q_{\text{i}}(x_{\text{junction}}, t) = Q_{0,\text{i}} \cos(\omega t + \theta) \tag{4.53}$$

$$Q_{\text{r}}(x_{\text{junction}}, t) = -Q_{0,\text{r}} \cos(\omega t + \theta) \tag{4.54}$$

$$Q_{\text{t}}(x_{\text{junction}}, t) = Q_{0,\text{t}} \cos(\omega t + \theta) \tag{4.55}$$

Note the sign conventions in these equations, where for positive $Q_{0,\text{r}}$ we must add a negative sign on the right-hand side of equation (4.54) since it represents a left-going "slosh."

After accounting for the fact that there are *two* daughter vessels and for the sign of Q_{r}, setting the net flow into the control volume equal to zero yields

$$Q_{0,\text{i}} - Q_{0,\text{r}} = 2Q_{0,\text{t}} \tag{4.56}$$

We now use the definitions of characteristic impedance in the parent (subscript p) and daughter (subscript d) tubes

$$Z_{0,\text{p}} = \frac{p_{0,\text{i}}}{Q_{0,\text{i}}} = \frac{p_{0,\text{r}}}{Q_{0,\text{r}}} \tag{4.57}$$

$$Z_{0,\text{d}} = \frac{p_{0,\text{t}}}{Q_{0,\text{t}}} \tag{4.58}$$

---

[11] The reader will note that there is another possibility: two of the pressure waves could be 180 degrees out of phase. But this is the same as them being in phase but with one wave having a negative amplitude, so we need not consider this case any further.

[12] To convince yourself of this fact, examine Equation (4.44) and the subsequent derivation. It shows that for a right-propagating wave, the oscillatory flow was taken as positive in the x-direction, i.e., rightwards. The situation is different for a left-propagating wave: Equation (4.37) must be modified so that the minus sign in the bracketed term becomes a plus sign. To maintain a positive impedance in this case, oscillatory flow must be taken as positive when it is propagating leftward.

to rewrite Equation (4.52) as

$$\frac{Z_{0,\text{p}}}{Z_{0,\text{d}}}[Q_{0,\text{i}} + Q_{0,\text{r}}] = Q_{0,\text{t}} \tag{4.59}$$

A quantitative measure of the fraction of the incident pressure pulse transmitted through the junction is the transmission ratio T, defined by

$$T = \frac{p_{0,\text{t}}}{p_{0,\text{i}}} = \frac{Z_{0,\text{d}}}{Z_{0,\text{p}}}\frac{Q_{0,\text{t}}}{Q_{0,\text{i}}} \tag{4.60}$$

Eliminating $Q_{0,\text{r}}$ between Equations (4.56) and (4.59) allows us to write T as

$$T = \frac{2Z_{0,\text{d}}}{2Z_{0,\text{p}} + Z_{0,\text{d}}} \tag{4.61}$$

Similarly, the reflected pulse ratio R is

$$R = \frac{p_{0,\text{r}}}{p_{0,\text{i}}} = \frac{Z_{0,\text{d}} - 2Z_{0,\text{p}}}{Z_{0,\text{d}} + 2Z_{0,\text{p}}} \tag{4.62}$$

Several notes on these results are in order.

1. T is always positive, as Equation (4.61) shows. However, R can be positive, negative, or zero. If R is zero, or nearly zero, the junction is said to be *matched*, since all, or nearly all, of the pulse energy passes through the junction.
2. The maximum positive value of R is 1, which occurs when $Z_{0,\text{d}} \gg Z_{0,\text{p}}$. This corresponds to the daughter tubes having a very high impedance, which can occur, for example, if $A_\text{d} \ll A_\text{p}$. This situation is therefore called a *closed end reflection*, analogous to what would occur if the end of the artery was completely blocked. In this situation, $Q_{0,\text{t}}$ is close to zero, and equation (4.56) shows that $Q_{0,\text{i}} \approx Q_{0,\text{r}}$. The net pressure at the junction is therefore

$$p_{\text{net}}(x_{\text{junction}}, t) = 2p_{0,\text{i}}\cos(\omega t + \theta) \tag{4.63}$$

since $p_{0,\text{i}} \approx p_{0,\text{r}}$ in this case. An entry to a capillary bed approximates this condition. At such sites, the incident and reflected pressure waves are in phase, while the incident and reflected flow waves are 180° out of phase. The minimum value of R is -1, which occurs when $Z_{0,\text{d}} \ll Z_{0,\text{p}}$. This corresponds to replacing the daughter tube with a reservoir and is called an open end reflection. In such a system, the pressure is effectively constant at the end of the tube, and Equation (4.52) shows that $p_{0,\text{i}} = -p_{0,\text{r}}$. Thus, the incident and reflected pressure waves are 180° out of phase, while the incident and reflected flow waves are in phase. This case does not normally occur in vivo.

3. Wave reflection occurs primarily at distal capillary beds in humans. Most other junctions are fairly well matched.

4. Since the arterial tree is actually a complex branching network, with each artery having its own reflection characteristics, the relationship between flow and pressure at any given location is very complex, depending on the multiple reflection sites throughout the artery. This in large part explains why the shape of the flow and pressure waveforms changes throughout the arterial tree. At any location, it is useful to characterize the net impedance of all downstream elements by a lumped quantity called the *input impedance*. Most measurements in humans have been done for the input impedance at the root of the aorta, that being the impedance seen by the heart (or by an artificial heart). This is particularly important since to obtain maximum effectiveness and output from an artificial heart, the heart's output should be tuned to the input impedance of the aorta.

5. Real pulse waveforms are not simple sinusoids but have more complex shapes. Therefore, it is necessary to decompose the waveforms into individual Fourier modes, and to know the impedance for each mode. This leads to plots such as that shown in Fig. 4.24.

From the above discussion, it should be clear that some reflection will occur whenever the characteristic impedance changes with location in an artery. Therefore, arterial taper, stiffening, and bifurcations can all be considered as special cases of a general change in Z_0 with location.

4.4 The capillaries

The capillaries are the "business end" of the circulatory system where mass is exchanged between the blood and surrounding tissue. More precisely, fluid and dissolved species are exchanged between the blood and the extracellular fluid that bathes the cells in the tissue surrounding the capillary. There is then an exchange of dissolved species between the cells and the extracellular fluid. This can be envisioned as mass exchange between three fluid compartments: the blood, the extracellular fluid, and the intracellular fluid. From the above discussion, it is clear that a necessary step in blood–cellular fluid exchange is fluid transport across the capillary wall. In this section we examine this fluid transport.

The structure of the endothelium controls the ease with which fluid and dissolved species can cross the capillary wall. Three types of endothelium are recognized:

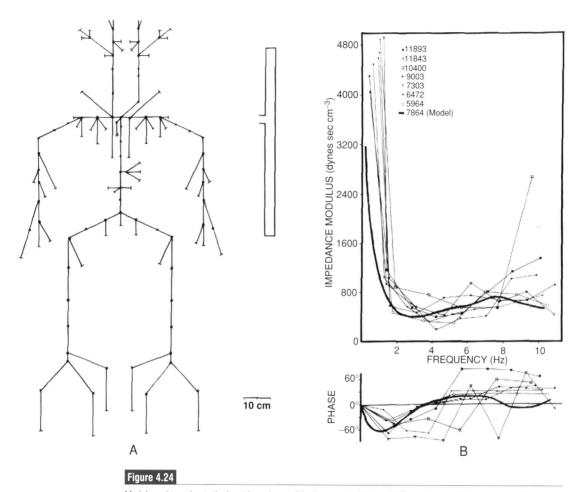

Figure 4.24

Model used to estimate the input impedance of the human arterial tree. (A) The major arterial segments used in the model. (B) The measured and calculated values of the ascending aortic input impedance for different harmonics. (Because of viscous effects, the impedance is a complex number, and, therefore, phase and magnitude must be displayed to characterize the impedance.) Experimental data are gathered from seven human subjects (symbols), and the calculated values (thick line) are obtained from the model shown in (A). From Nichols and O'Rourke [8]. Reprinted by permission of Edward Arnold.

Continuous endothelium. Adjacent endothelial cells are tightly joined, and little fluid passes across the endothelium. In such capillaries, solute movement across the capillary wall occurs by active cellular transport in a controlled fashion. This type of capillary is typical in tissues where the composition of the extravascular environment is rigidly controlled (e.g., the brain).

Fenestrated endothelium. Similar to continuous endothelium, except that there are isolated locations where the endothelium becomes extremely thin enabling opposite surfaces of the membrane to join to form *fenestrae* (Fig. 4.25). This type of endothelium is more permeable than continuous endothelium.

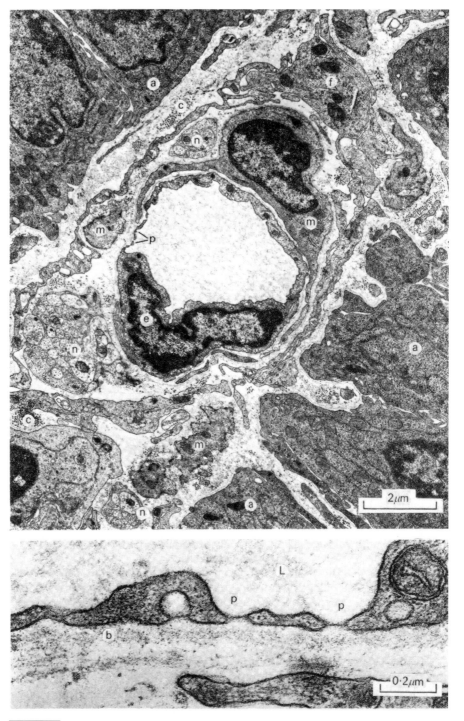

Figure 4.25

Transmission electron micrograph of a fenestrated capillary within the mucosa of the guinea-pig ileum (third section of intestine). The fenestrae are shown at higher magnification in the lower panel. a, epithelial absorbtive cells; b, basement membrane; c, collagen fibrils; e, endothelial cell; f, fibroblast; L, lumen; m, smooth muscle cell; n, nerve; p, fenestrations. From Caro *et al.* [17]. Reproduced with kind permission of Oxford University Press and Dr. G. Gabella, University College, London.

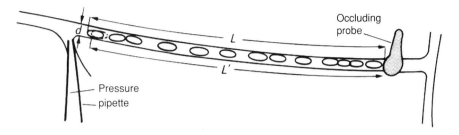

Figure 4.26

Diagram showing Landis' experiment, in which a capillary is occluded at its upstream end, and red cell motion is observed. After occlusion, the red cells pack slowly towards the probe, moving a distance $L - L'$ in a time t. The vessel internal diameter is d. Modified from Caro *et al.* [17]. Reproduced with kind permission of Oxford University Press.

Discontinuous endothelium. There are frank intercellular gaps, allowing relatively free passage of fluid and solutes across the endothelial layer.

4.4.1 Capillary filtration: the experiments of Landis

Now that the nature of the capillary endothelium is better understood, we proceed to a qualitative description of fluid flow patterns across the capillary walls. A crucial set of experiments in this area was conducted by Landis in 1927 [32]. Using in vivo microscopy, he was able to examine the motion of red cells in a capillary in a living animal, while manipulating this capillary. More specifically, he first occluded a capillary at the arteriolar (upstream) end so that the capillary was connected to the rest of the vascular tree through its venular (downstream) end. This caused the pressure in the capillary to be equal to venular pressure and prevented net flow of blood along the capillary (Fig. 4.26). Landis noted that the red cells within the occluded capillary oscillated back and forth, as might be expected since the venular pressure was changing somewhat during each cardiac cycle. More interestingly, however, the red cells slowly moved apart from one another. When the experiment was repeated with the venular end of the capillary occluded, so that capillary pressure equaled local arteriolar pressure, the finding was reversed: the red cells in the occluded capillary packed closer together.

In order to interpret these data, it must be realized that under normal circumstances red cells cannot cross the capillary walls, but (for the capillaries that Landis used) fluid is quite highly permeable across the capillary wall. Therefore, a gradual spreading apart of red cells implies that fluid is flowing into the capillary from the surrounding tissue. Conversely, a packing together of red cells means that fluid is leaving the capillary and flowing into the surrounding tissue. Landis, therefore, concluded that:

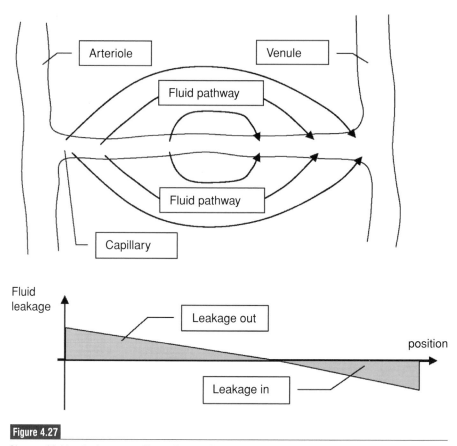

Figure 4.27

Diagram of fluid motion between capillary and surrounding tissue, showing fluid leaving the blood at the arteriolar end and re-entering the blood at the venular end of the capillary. There is also some fluid resorption from tissue into the postcapillary venule (not shown).

- when capillary pressure equals arteriolar pressure, there is net fluid leakage out of the capillary
- when capillary pressure equals venular pressure, there is net fluid leakage into the capillary.

Since pressure normally varies from arteriolar pressure to venular pressure over the length of a capillary, in the normal course of events, fluid must leave the blood for the surrounding tissues at the arteriolar end and re-enter the blood at the venular end (Fig. 4.27). Some thought will indicate that this is, in fact, a very efficient mechanism for solute exchange between blood and surrounding tissue. These experiments confirmed a postulate first advanced by the British physiologist Starling, and so this pattern of filtration and resorption, or more specifically, the forces that drive this pattern (see below), is called *Starling's hypothesis.*

All seems to make sense, until measurements are made of the pressures within the capillary and in the surrounding tissue. These show that capillary pressure at the venular end exceeds surrounding tissue pressure, which indicates that fluid re-entering the capillary is flowing from a region of low pressure to one of high pressure. For slow flows such as are present in tissue, this appears to violate the laws of motion. This paradox is resolved when it is realized that in addition to a fluid pressure difference across the capillary wall, there is a second force driving the fluid, namely the *osmotic pressure*. To understand this effect, it is necessary to make a slight diversion into the world of osmotic pressure.

4.4.2 Osmotic pressure

Osmotic pressure is a thermodynamically generated net force on *solvent* molecules that requires three things:

* solvent
* one (or more) solutes
* a *semipermeable membrane*.[13]

Consider a two-chambered vessel divided by a semipermeable membrane. We fill both chambers with solvent and dissolve different amounts of solute in each chamber, so that the concentrations of the resulting two solutions are different (Fig. 4.28). There is then a net effective force causing solvent molecules to cross the membrane from the *low* concentration solution to the *higher* concentration solution. This driving force can be interpreted as an equivalent driving pressure, known as the *osmotic pressure* (symbol: π).

As a specific example, consider a vessel with two chambers separated by a membrane that is permeable to water but not to sodium (Na^+) or chloride (Cl^-) ions. On one side of the barrier is pure water, while the other side contains a NaCl solution. The pressure and osmotic pressure on side 1 are p_1 and π_1, respectively; on side 2 they are p_2 and π_2.

It can be shown both experimentally and theoretically that the driving force for solvent flow across the membrane is not the fluid pressure p, but instead is the fluid pressure minus the osmotic pressure, $p - \pi$. Thus, the flux of solvent from chamber 1 to chamber 2, Q_{12}, is written as

$$Q_{12} = constant \ [(p_1 - \pi_1) - (p_2 - \pi_2)]. \tag{4.64}$$

---

[13] A semipermeable membrane or barrier is a device that allows solvent molecules to pass freely, while preventing or impeding the passage of solute molecules. In the context of the present discussion, the solutes are plasma proteins, while the barrier is the capillary wall. Physiologists sometimes call an osmotic pressure generated by proteins *oncotic pressure*.

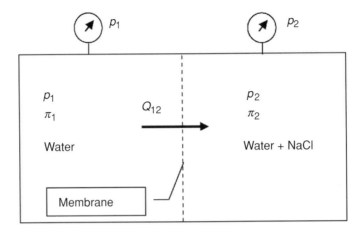

A semipermeable membrane separating two chambers, one of which contains pure water and the other of which contains a NaCl solution. The presence of salt generates an osmotic pressure in chamber 2, leading to a net solvent flow Q_{12} when $p_1 = p_2$.

In other words, the system in Fig. 4.28 is at equilibrium (no fluid crossing the membrane) when $p_1 = p_2 + \pi_1 - \pi_2$. Thus, if $\pi_1 \neq \pi_2$, the pressures on the two sides of the membrane will not be equal at equilibrium.

We are used to the conventional pressure p. What about π? We cannot measure it directly; however, empirically and theoretically, π is a function of the molar concentration of solute, c, and the temperature of the solution. For an *ideal* solution, π is given by *van't Hoff's law*

$$\pi = RT \sum_j c_j \tag{4.65}$$

where R is the universal gas constant (8.314 J/mol K), T is the *absolute* temperature of the solution, and c_j is the *molar* concentration of species j in solution. This equation holds if the solution is sufficiently dilute that interactions between solvent molecules can be neglected.[14] As an example, consider the system shown in Fig. 4.28, taking the NaCl concentration as 1 molar (1 M or 1 mol/l) and the temperature as 300 K. What is π?

To answer this question, we note that there is no salt on side 1, so π_1 equals zero. One side 2, NaCl dissociates into two species, Na^+ and Cl^-. This dissociation is nearly complete, so that to a very good approximation, each of Na^+ and Cl^- has a concentration in solution of 1 molar. Assuming ideal behavior, we, therefore,

---

[14] In real life, solutions are not ideal, and van't Hoff's law must be modified by adding an empirical activity coefficient γ_j, such that $\pi = RT \sum \gamma_j c_j$.

calculate that

$$\pi_2 = \left(8.314\frac{\text{J}}{\text{mol K}}\right)(300\,\text{K})\left(2\,\frac{\text{mol}}{\text{liter}}\right)\left(\frac{\text{liter}}{10^{-3}\,\text{m}^3}\right) = 4.99 \times 10^6\,\text{Pa} \quad (4.66)$$

This is a pressure of 49.2 atmospheres, or 724 psi, equivalent to a column of water 497 metres high! Thus, at equilibrium in the above system, the transmembrane pressure difference $p_2 - p_1$ is 724 psi.[15] This shows that very substantial osmotic pressures can be generated by modestly concentrated solutions.

When dealing with mixtures of ions, each ionic species contributes to π. The total ionic concentration is, therefore, of interest and is expressed in units of *osmolars* (abbreviation: Osm). The osmolarity of a solution is simply the sum of the molar concentrations of each ionic species. In the above example, we, therefore, have a 2 osmolar solution. Nearly all physiological fluids have an osmolarity of approximately 285 milliosmolar (mOsm).

4.4.3 Quantitative analysis of capillary leakage

With this background, it is possible to return to the capillary filtration problem and to estimate the net potential that drives fluid across the capillary wall. Treating the capillary as chamber 2 and the surrounding tissue as chamber 1, the driving potential for fluid to leave the capillary can be written as

$$\text{driving potential} \sim (p - \pi)_{\text{capillary}} - (p - \pi)_{\text{tissue}} = \Delta p - \Delta \pi \quad (4.67)$$

where $\Delta p = p_{\text{capillary}} - p_{\text{tissue}}$ and $\Delta \pi = \pi_{\text{capillary}} - \pi_{\text{tissue}}$. For dogs and humans, the relevant numbers are:

$(p - \pi)_{\text{tissue}} = -5\,\text{cmH}_2\text{O}$
$\pi_{\text{capillary}} = 25\,\text{cmH}_2\text{O}$
$p_{\text{capillary}} = 40\,\text{to}\,10\,\text{cmH}_2\text{O}$ (arteriolar to venular).

It is difficult to measure p or π alone for the surrounding tissue, but a reasonable estimate is $\pi_{\text{tissue}} \approx 10\,\text{cmH}_2\text{O}$ and $p_{\text{tissue}} \approx 5\,\text{cmH}_2\text{O}$. We then obtain

$\Delta p \approx 35\,\text{to}\,10\,\text{cmH}_2\text{O}$ (arteriolar to venular)
$\Delta \pi = 15\,\text{cmH}_2\text{O}$.

In other words, the fluid pressure difference is always such as to force fluid out of the capillary, but on the venular end, this difference is overcome by the osmotic pressure difference, which draws fluid from the tissue into the blood (Fig. 4.29).

---

[15] In order to remember the direction in which the osmotic pressure acts, it is convenient to think that the solvent "wants" to dilute the solution by flowing from chamber 1 to chamber 2. In order to counteract this tendency and prevent solvent flow, a higher pressure in chamber 2 is needed.

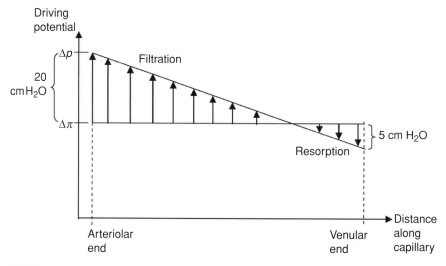

Filtration and resorption across a capillary wall, using data given in the text. Note that for purposes of this diagram, the fluid pressure is assumed to vary linearly within the capillary, while tissue pressure, osmotic pressures, and capillary wall permeability are assumed to be constant.

Although the above calculation is only approximate, it suggests that there is more filtration out of the capillaries than there is fluid resorption into the capillaries. This is partly offset by two effects:

- capillary wall permeabilities are higher at the venular end than at the arteriolar end, which means that resorption is faster than filtration, for a given driving potential
- resorption takes place to some extent in the venules as well as in the capillaries.

Even after accounting for these effects, there is still an excess of filtration over resorption. Somewhere between 20 and 30 liters of fluid are filtered out of the non-renal capillaries per day. Of this, 85–90% is resorbed, leaving 2 to 4 liters of fluid per day "behind." This fluid is taken up by the lymphatic system, principally by the *lymphatic capillaries*. The lymphatic collecting system is equipped with a series of one-way valves oriented so that collected fluid drains towards the *thoracic duct*, eventually returning to the bloodstream via the right heart. Thus, there is a net flow of fluid out of the capillaries through the surrounding tissue.

4.5 The veins

Because most of the systemic pressure drop occurs across the arterioles and capillaries, and because much of the flow pulsatility is damped out in the capillaries, flow in the veins is at a much lower pressure and is less pulsatile than flow in the arteries. Also, because the volume of blood in the veins is greater than that in the arteries (implying larger lumen cross-sections), flow velocities in the veins are smaller than those in the arteries. Veins are much less prone to the formation of mural disease, such as atherosclerosis, than are arteries.

Because pressure in the veins is less than in the arteries, vein walls are significantly thinner than artery walls. However, the composition of the vein wall is fairly similar to that of the arterial wall, and the vein wall shows strain stiffening as does the arterial wall. Typical values for the elastic modulus of the vein wall in humans are of order 1×10^6 to 10×10^6 dynes/cm^2. For example, the human saphenous vein has an incremental elastic modulus in the circumferential direction of 0.27×10^6 to 25.1×10^6 dynes/cm^2 as the lumenal pressure increases from 10 to 150 mmHg [33].

One interesting feature of the veins, particularly those in the leg, is their system of one-way valves. These valves consist of passively moving tissue leaflets, similar in operation to those in the heart (Fig. 4.30). They are oriented so as to direct blood towards the heart. Venous return is dependent on the action of these valves, in concert with the rhythmic pumping of the veins from the contraction and relaxation of muscles in the surrounding tissue. This is particularly true in the case of the veins in the lower legs, where the pumping is accomplished by contraction and relaxation of the calf muscles. For this reason, people who are sedentary for extended periods experience blood pooling in the veins in the calf. This promotes thrombus formation, which can be dangerous if the thrombi break loose and travel into the cerebral or pulmonary circulations. For this reason, it is inadvisable to keep the legs motionless for extended periods of time. (Travellers on long-haul flights, for example, are advised to occasionally walk about or at least flex the calf muscles from time to time to avoid "economy class syndrome.")

4.6 Scaling of hemodynamic variables

It has long been recognized that many physiological variables depend on body size. This is true both within a species (children have a more rapid heart rate than adults) and between species (the heart rate in mice is greater than that in elephants). The standard way of expressing such a relationship is through a scaling law:

$$y = aM^b \tag{4.68}$$

where y is any measurable quantity of interest, M is body mass, and a and b are coefficients. Although the notation in the literature is somewhat inconsistent, this is sometimes referred to as the *allometric equation*, and if the variable y obeys Equation (4.68) with constant exponent $b \neq 1$, then the scaling is said to be *allometric*, with allometric power b. If $b = 1$, the scaling is said to be *isometric*.

Proximal

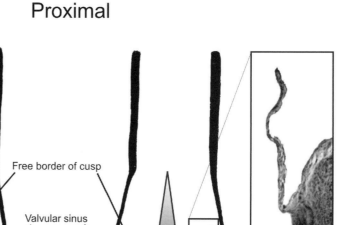

Free border of cusp

Valvular sinus

Valvular agger

Distal

Figure 4.30

Cross-sectional view through a venous valve in the closed (left) and open (right) positions. Blood flow, shown by the large arrows, is from bottom to top when the valve is open. The inset at right shows a histological cross-section through a valve in the open position. After Gottlob and May [34].

An example of an important physiological quantity that is proportional to body mass is *basal metabolic rate* (BMR), which represents the rate of energy expenditure in an adult organism at rest. This energy is expended for all of the "housekeeping" tasks that are essential for homeostasis: pumping ions across cell membranes, synthesizing proteins, circulating the blood, etc. The BMR can be expressed in terms of O_2 consumption per unit time. There has been considerable controversy in the literature as to whether the allometric exponent for BMR in mammals is 3/4 or 2/3, and the generally accepted value has been 3/4. However, a recent meta-analysis [35] suggested that if suitable corrections are made for body temperature, the exponent should actually be 2/3. Figure 4.31 shows how the available data fall onto a single line on a log–log plot; the correlation is quite remarkable when it

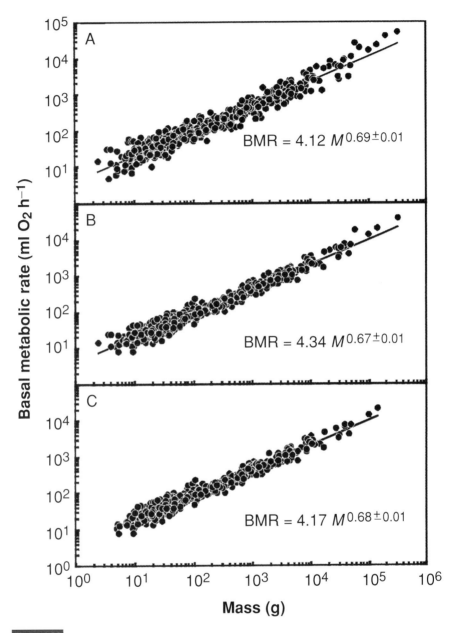

Figure 4.31

Relationship between basal metabolic rate (BMR) and body mass in mammals. (A) All data ($n = 619, r^2 = 0.94$) is shown; (B) the same data corrected to a common body temperature of $36.2\,^\circ$C ($n = 507, r^2 = 0.96$); (C) data from (B) excluding several species whose unusual digestive processes might affect BMR ($n = 469, r^2 = 0.96$). Exponents are shown with 95% confidence intervals. From White and Seymour [35]. Copyright 2003 National Academy of Sciences, U.S.A.

is noted that body mass spans almost five orders of magnitude in this graph. This suggests that there must be some fundamental principle that governs metabolic rate across all mammals; for example, it has been suggested that BMR is governed by the need to reject heat through the skin to the surroundings. This raises the tantalizing possibility of being able to discover fundamental physiological design constraints from looking at how physiological quantities of interest vary with body mass. Here we will look at some scaling relationships that pertain specifically to the cardiovascular system.

A physiological variable that would be expected to correlate very closely with body mass is body surface area. Readers familiar with the concept of geometrical similarity will recognize that body volume is directly proportional to body mass for organisms with similar densities, and body surface area is proportional to (body volume)$^{2/3}$ for similarly shaped organisms. Such animals should obey an allometric relationship of the form:

$$\text{body surface area [m}^2] = a \, M \, [\,\text{kg}]^{2/3}, \tag{4.69}$$

which is indeed what is found, with coefficient $a \approx 0.1$ for a wide variety of species. For humans, $a = 0.11$; for species-specific values of the coefficient a for other mammals, the reader is referred to Guyton *et al.* [36]. For humans specifically, a number of relationships have been proposed that are somewhat more accurate than Equation (4.69). The best known of these is the formula due to DuBois, dating from 1916, but a more accurate form [37] is that of Mosteller [38]

$$\text{body surface area [m}^2] = \frac{1}{60} \, M[\text{kg}]^{1/2} \, H[\text{cm}]^{1/2} \tag{4.70}$$

where H is body height. Since H is expected to be proportional to $M^{1/3}$, the form of Equation (4.70) is consistent with Equation (4.69).

The relationship between body surface area and body mass is presented here because, for historical reasons, the literature contains many instances where cardiovascular parameters are correlated with surface area rather than body mass. Further, the surface area–body mass relationship is important in deciding dosage for some drugs and patient groups, e.g. children. For example, cardiac output is often divided by body surface area to give a quantity called the *cardiac index*. Stroke volume is also sometimes normalized in this way (e.g., Fig. 4.11B). In humans, the mean cardiac index is 3.59 l/min per m^2 (standard deviation, 0.55 l/min per m^2), corresponding to a mean cardiac output of 6.5 l/min [39]. (The surface area of an "average" human is approximately 1.8 m^2.)

When multiple species are considered, Milnor suggests correlating cardiac output against body mass, rather than using cardiac index [4]. He obtained an exponent of 0.98 when considering 12 species spanning three orders of magnitude in body

Table 4.3. Allometric scaling relationships for selected cardiovascular parameters. *a* and *b* are coefficients in the equation $y = aM^b$, where M is body mass in kg; *LV*, left ventricle; *RV*, right ventricle; ECG, electrocardiogram.

Parameter	a	b	Source
Heart rate (s^{-1})	3.60	−0.27	Adolph [41]
	4.02	−0.25	Stahl [42]
	2.36	−0.25	Holt *et al.* [43]
Stroke volume (LV) (ml)	0.74	1.03	Juznic and Klensch [44]
	0.66	1.05	Holt *et al.* [43]
Cardiac output (ml/min)	108	0.98	Milnor [4]
	100	0.97	Patterson *et al.* [45]
	187	0.81	Stahl [42]
	225	0.79	Lindstedt and Schaeffer [40]
	166	0.79	Holt *et al.* [43]
	200	0.75	Juznic and Klensch [44]
Arterial pressure (mmHg)	87.8	0.033	Gunther and Guerra [46]
Pulse velocity (cm/s)	446.0	0.0	Li [47]
Diameter of aorta (cm)	0.41	0.36	Holt *et al.* [48]
Length of aorta (cm)	17.5	0.31	Li [47]
	16.4	0.32	Holt *et al.* [48]
Aortic area (cm^2)	0.00179	0.67	Stated in Li [49][a]
	0.074	0.94	Milnor [4]
Blood density (g/cm^3)	1.05	0	Stated in Li [49][a]
Duration ECG (P) (s)	0.0139	0.168	Stated in Li [49][a]
Duration ECG (QRS) (s)	0.0111	0.165	Stated in Li [49][a]
End systolic volume (LV) (ml)	0.59	0.99	Holt *et al.* [43]
End systolic volume (RV) (ml)	0.62	0.99	Holt *et al.* [43]
End diastolic volume (LV) (ml)	1.76	1.02	Holt *et al.* [43]
End diastolic volume (RV) (ml)	2.02	1.02	Holt *et al.* [43]
Total peripheral resistance (dynes/s per cm^5)	3.35×10^6	−0.68	Li [49]
Ventricular weight (LV) (g)	1.65	1.11	Holt *et al.* [43]
Ventricular weight (RV) (g)	0.74	1.06	Stated in Li [49][a]
Blood viscosity (Poise)	0.03	0	Stated in Li [49][a]

[a] "Stated in Li" means that a result is given in Li [49] that is unreferenced; see caveats in text about use of these values.

mass. Others have obtained exponents ranging from 0.74 to 0.79 (see reviews by Milnor [4] and Lindstedt and Schaeffer [40] and Table 4.3). Cardiac output is a hard thing to measure, which may explain some of the scatter in the data and the resulting range of allometric exponents.

There have been many studies of scaling in the cardiovascular system, and results from some of these studies are shown in Table 4.3. It can be seen that heart rate scales approximately as body mass to the $-1/4$ power, a finding that seems to be consistent over several studies. Furthermore, left ventricular stroke volume seems to be approximately proportional to body mass, a result that is intuitively appealing if we believe that stroke volume is a constant fraction of left ventricle volume and that left ventricle volume is proportional to body volume. Since cardiac output is the product of heart rate and stroke volume, we therefore expect cardiac output to scale with body mass to the $3/4$ power. As can be seen from Table 4.3, a fairly wide range of scaling exponents for cardiac output have been reported, some of which are quite a bit different from the expected $3/4$. This highlights an important caveat: if the input data span only a small range of body masses, or if only a modest number of data points are included in the dataset, there can be significant uncertainty in the allometric exponent b. It is best if the authors specify a confidence interval for b, as was done in Fig. 4.31. In light of this comment, it is suggested that readers refer to the primary sources to judge the reliability of the data before using values in Table 4.3.

Bearing this caveat in mind, it is interesting that there are two variables in Table 4.3 that are essentially independent of body mass, namely systemic pressure and pulse velocity. The constancy of these quantities over mammals of differing sizes has long been noted by others (e.g., discussion in Milnor [4]) and suggests that some fundamental physiological constraints control these parameters. In the case of blood pressure, it has been suggested that a similar mean arterial pressure is obtained in mammals because a minimum pressure is required to ensure brain perfusion, renal filtration, and perfusion of other capillary beds, while similar diastolic pressures are needed to provide sufficient subendocardial coronary perfusion [50]. It is also interesting that, in addition to similarity in pressure magnitudes, cardiac flow waveform and pressure waveform shapes are comparable across different sizes of mammals. Perhaps this is a consequence of constraints on heart design, constraints on arterial elasticity, or (most likely) both.

4.7 Problems

4.1 In Section 4.2.3 we concluded that the pumping power of the human heart is approximately 2 W (assuming a normal cardiac output of 5 liters/min).

Like any pump, however, the heart is not 100% efficient, and, therefore, the power that is supplied to the heart muscle will actually exceed 2 W. In this question we will estimate the heart's pumping efficiency η by calculating how much energy is supplied to the heart muscle from the blood. (Note that the heart muscle has its own vasculature, called the coronary circulation.)

(a) When we calculated the pumping power of the heart, we only considered the head gain across the heart from the pressure increase. In general, there could also be changes in elevation (very small) and in kinetic energy. Estimate the ratio of kinetic energy head to pressure head at peak systole, when pressure is 120 mmHg and blood velocity is 100 cm/s. Can we safely neglect kinetic energy gains in calculating pumping power?

(b) At rest, the coronary blood flow is 225 ml/min, and 65% of the O_2 is removed from the blood as it passes through the coronary vasculature. The oxygen capacity of blood is 19.4 ml O_2/100 ml blood, and in a normal diet 4.83 kcal of food energy is released for every liter of O_2 consumed. From this data, estimate η for the heart. State assumptions.

(c) The basal metabolic rate of a normal individual is 72 kcal/h. What fraction of this is consumed by the heart?

4.2 In the Windkessel model (see Equations [4.10] or [4.11]), the product RC has units of time and can be thought of as a characteristic response time for the cardiovascular system. The purpose of this question is to estimate the magnitude of RC.

(a) Starting from the definition of C given in Equation (4.7) show that $C = VD/Et$, where V is the volume of the compliant arteries, D is arterial diameter, t is arterial wall thickness, and E is the Young's modulus of the artery wall. To derive this formula, it is useful to first relate C to the arterial distensibility, β, as defined by Equation (4.17).

(b) Taking the volume of the compliant arteries as 700 ml and cardiac output as 5 l/min, estimate RC. You can use numerical values given in the text, supplemented by any assumptions you feel are necessary. For full credit, you must clearly state each assumption that you make.

4.3 Consider *steady* flow of blood in a tube whose thin walls are made of a linearly elastic Hookean material, so that changes in hoop stress are directly proportional to changes in hoop strain. You may treat the blood as Newtonian, with viscosity μ.

(a) When blood is flowing, it is observed that tube diameter decreases with axial position along the tube, x (Fig. 4.32). Why is this?

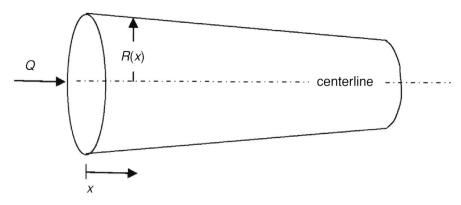

Figure 4.32

For Problem 4.3.

(b) Show that a change in internal pressure dp causes a change in radius dR given by:

$$dR = \frac{R^2}{Eh}dp \qquad (4.71)$$

where h is the tube wall thickness.

(c) If the tapering of the tube is not too severe, then the flow is "nearly" Poiseuille and can be described by:

$$Q = -\frac{\pi R^4}{8\mu}\frac{dp}{dx}. \qquad (4.72)$$

Using this expression, as well as the results of part (b) above, derive an expression for R in terms of the inlet radius (R_o), μ, the flow rate (Q), E, h, and the axial position within the tube, x.

(d) Blood (effective Newtonian viscosity 3.5 cP) flows in such a tube with $E = 100\,\text{dynes/cm}^2$ and $h = 1$ mm. If Q is 100 ml/min and the radius at the inlet is 1 cm, determine the tube radius at 20 cm from the inlet.

4.4 In deriving the Korteweg–Moens equation for elastic wave speed we assumed that there was no longitudinal tension in the vessel and that the vessels were thin walled. In fact, these assumptions are not true, and the wave speed predicted by the Korteweg–Moens equation is slightly incorrect. From the theory of elasticity it can be shown [8] that the incremental Young's modulus, E, of an isotropic thick-walled cylinder with closed ends and constant length is

$$E = \frac{2\Delta p\, R_i^2(1 - \nu^2)}{R_o^2 - R_i^2}\frac{R_o}{\Delta R_o} \qquad (4.73)$$

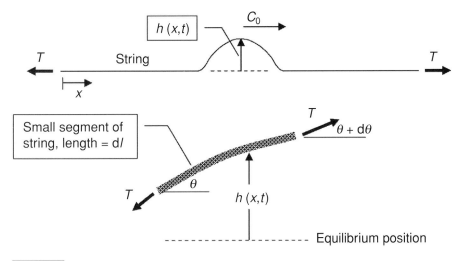

Figure 4.33

For Problem 4.5.

where R_o and R_i are the outer and inner radii of the cylinder, ΔR_o is the change in external radius due to a change Δp in internal pressure, and v is Poisson's ratio for the cylinder wall material.

(a) Using this equation, extend the Korteweg–Moens equation to the case of a thick-walled vessel with closed ends and constant length. In your derivation, you may assume that the diameter in Equation (4.17) refers to the outer diameter.

(b) What limiting form is obtained if the vessel wall thickness (t) is small? In particular, for a Poisson ratio of 0.5, what is the percentage difference in the wave speed predicted by the wave speed from part (a) versus that predicted by the Korteweg–Moens equation?

4.5 The propagation of elastic pulse waves in arteries is analogous to a wave travelling down an elastic string that has been "flicked." Consider a string under tension T that has mass per unit length M (Fig. 4.33).

(a) Call the speed of an elastic wave c_0. Show dimensionally that $c_0 = constant \sqrt{T/M}$.

(b) Consider a small piece of the string (of length dl) that has been displaced a distance $h(x, t)$ from equilibrium. From a force balance show that

$$-T \sin \theta + T \sin(\theta + d\theta) = M \, dl \frac{\partial^2 h}{\partial t^2}. \qquad (4.74)$$

(c) Using the approximations (valid for small deflections):

$$\sin(\theta + d\theta) \doteq \sin\theta + d\theta$$

$$dl \doteq dx$$

$$\theta \doteq \frac{\partial h}{\partial x}, \tag{4.75}$$

show that:

$$\frac{\partial^2 h}{\partial x^2} - \frac{M}{T}\frac{\partial^2 h}{\partial t^2} = 0. \tag{4.76}$$

(d) Prove that the above equation has solutions of the form $h(x, t) = f(x \pm c_0 t)$, where f is any function. Hint: use the chain rule. Note that $f(x - c_0 t)$ describes a wave propagating to the right while $f(x + c_0 t)$ describes a wave going to the left.

4.6 In this problem, we will look more carefully at the propagation of an elastic wave inside the arterial tree. We will consider an infinitely long artery of radius $R(x, t)$, filled with an incompressible, inviscid fluid.

(a) Consider a cylindrical shaped control volume of length dx and radius $R(t)$, i.e., one which fills up the vessel's cross-section and extends a distance dx along the vessel. If the blood's axial velocity is $u(x, t)$, show that from conservation of mass we must have:

$$R\frac{\partial u}{\partial x} + 2\frac{\partial R}{\partial t} = 0. \tag{4.77}$$

(b) The momentum balance for an inviscid fluid can be written as:

$$\rho\frac{\partial u}{\partial t} = -\frac{\partial p}{\partial x}. \tag{4.78}$$

This assumes that the blood only flows axially, which is only approximately correct. Combine this with the constitutive relation for the tube wall:

$$\beta\, dp = \frac{2}{R}dR \tag{4.79}$$

to show that

$$R\frac{\partial u}{\partial t} + 2c_0^2\frac{\partial R}{\partial x} = 0. \tag{4.80}$$

(c) Combine the results of (a) and (b) to get the equation:

$$\frac{\partial^2 R}{\partial t^2} - c_0^2\frac{\partial^2 R}{\partial x^2} = 0 \tag{4.81}$$

You will need to assume that $\partial R/\partial t$ and $\partial R/\partial x$ are small compared with second derivatives of R.

(d) Prove that the above equation has solutions of the form $R(x, t) = f(x \pm c_0 t)$, where f is any function. Hint: use the chain rule. Note that $f(x - c_0 t)$ describes a wave propagating to the right while $f(x + c_0 t)$ describes a wave going to the left.

4.7 Expanding the artery wall requires energy. Therefore, when a pressure wave propagates in an artery, it carries energy with it. This question explores this effect. *Note: in answering this question you do not have to re-derive any formulae shown in the text.*

(a) Starting from the expression $\Psi = \sigma^2 V/(2E)$, show that the energy stored in the artery wall per unit wall volume is $p^2 D^2/(8Et^2)$. Here Ψ is the energy stored in an elastic solid of modulus E and volume V owing to a stress σ; t is the thickness of the artery wall; D is the arterial diameter, and p is the local pressure in the artery.

(b) Show that the energy stored in the artery wall per unit wavelength from a traveling pressure wave is $\pi \lambda p_0^2 D^3/(16Et)$, where p_0 is the amplitude of the pressure wave and λ is its wavelength. You may assume that the distention of the artery from the pressure wave is very small. Hint: $\int_0^\lambda \cos^2(2\pi x/\lambda + \phi)\,\mathrm{d}x = \lambda/2$.

(c) Show that the rate at which energy is transported along the artery with the pressure wave is $p_0^2/(4Z_0)$.

(d) Based on conservation of energy, as well as the results of part (c), shown that the reflection and transmission coefficients at a junction such as discussed in Section 4.3.7 must satisfy the expression

$$R^2 + 2\frac{Z_{0,\mathrm{p}}}{Z_{0,\mathrm{d}}}T^2 = 1.$$

In deriving this result you may not use Equations (4.61) and (4.62).

4.8 A patient who has normal blood pressure (systolic and diastolic) also has hardening of the arteries, including the brachial artery in the upper arm. Will this patient's measured blood pressure be larger than the actual value, and why or why not?

4.9 In Section 4.3.7, we derived an expression for the ratio of pressure pulses at a junction when the two daughter tubes were identical. Perform the same calculation assuming that the two daughter tubes have different characteristic impedances. Specifically, what are the ratios R and T? To simplify the

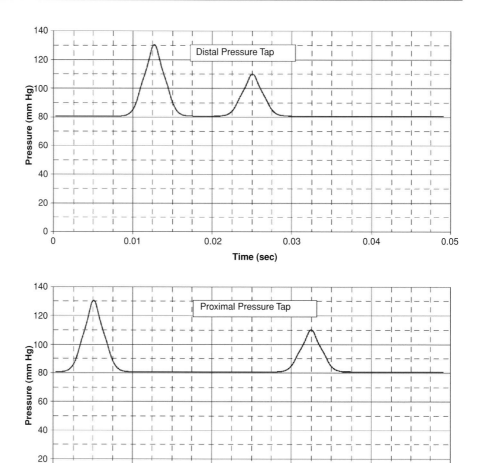

Figure 4.34

For Problem 4.10.

notation, call Z_0 the impedance of the parent tube, and Z_1, Z_2 the impedances of the two daughter tubes.

4.10 A dual-port catheter can simultaneously measure pressure in an artery at two different locations, or "taps." Idealized measurements from one such catheter are shown in Fig. 4.34. It is known that there is a bifurcation distal to the measurement locations.

(a) If the two ports are 4 cm apart, estimate the wave speed in the artery containing the catheter. Explain your logic.

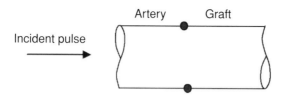

Artery   Graft

Incident pulse

Figure 4.35

For Problem 4.11.

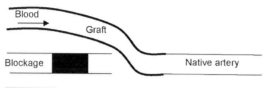

Blood

Graft

Blockage   Native artery

Figure 4.36

For Problem 4.12.

Table 4.4. For Problem 4.12. Data from Tai *et al.* [51].

Measured quantity	Artery	ePTFE[a]
Average diameter (mm)	6·8	5·4
Average compliance (mmHg^{-1})	8·0 × 10^{-4}	1·2 × 10^{-4}

[a] Teflon, a graft material.

(b)   If the ratio of parent:daughter artery cross-sections at the distal bifurcation is 3:1, estimate the wave speed in the daughter arteries. Hint: first estimate the reflection coefficient at the bifurcation.

4.11   A graft (impedance Z_g) is joined to an artery (impedance Z_a) in the end-to-end anastomosis shown in Fig. 4.35. What is the transmission coefficient T? In three sentences (or less), describe the physics of why T is not equal to one.

4.12   A common surgical procedure is bypass graft implantation, in which a natural or synthetic arterial substitute ("the graft") is used to carry blood around an obstructed portion of an artery (Fig. 4.36). Data suggest that these grafts perform best if their compliance matches that of the native artery that they are attached to. Compliance is a measure of the elasticity of the vessel walls; here it is defined as $C = (D_s/D_d - 1)/(P_s - P_d)$ where D_s and D_d are systolic and diastolic diameter, and P_s and P_d are systolic and diastolic pressure, respectively.

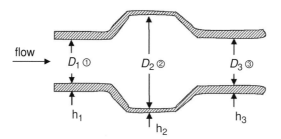

flow

D_1 ① D_2 ② D_3 ③

h_1 h_2 h_3

Figure 4.37

For Problem 4.13.

(a) One material used for synthetic grafts is ePTFE (Teflon), which is very stiff compared with native arteries. Knowing that Young's modulus for ePTFE is 40 times larger than that for an artery, and given the properties in Table 4.4 measured from actual grafts, estimate the ratio of graft wall thickness to artery wall thickness.

(b) One design approach is to make the wall of the graft thin enough so that the graft compliance matches that of the artery. Discuss the "pros and cons" of doing this from a biomedical engineering design viewpoint.

(c) Why might compliance mismatch promote redevelopment of disease? It will be useful to think about what endothelial cells can "sense" and to know that there is a postoperative healing response that includes the growth of endothelial cells onto the new graft surface.

4.13 An aneurysm is a dilatation (or ballooning out) of the artery wall, causing inner arterial diameter to increase from D_1 to D_2 (Fig. 4.37). If artery wall elastic modulus (E) and artery wall cross-sectional area do not change in the aneurysm, what is the net pulse transmission coefficient between points ① and ③? Do not consider multiple reflections within the aneurysm. You may use the fact that the pulse transmission coefficient at a single junction is given by:

$$T = \frac{2}{1 + Z_{0,u}/Z_{0,d}} \tag{4.82}$$

where $Z_{0,u}$ and $Z_{0,d}$ are the characteristic impedances upstream and downstream of the junction, respectively. Dimensions are $D_1 = D_3 = 2\,\text{cm}$ and $D_2 = 4\,\text{cm}$. The wall thickness at points ① and ③ is $h = 2\,\text{mm}$. Take blood density as $1.05\,\text{g/cm}^3$.

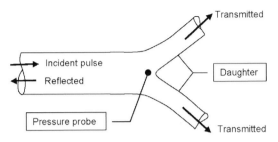

Figure 4.38

For Problem 4.14.

Table 4.5. For Problem 4.14

i	C_i (Pascals)	ω_i (rad/s)
1	45	6.2
2	16	12.4
3	9	18.6

4.14   An incident pressure waveform approaching a bifurcation was measured
by a pressure probe (see Fig. 4.38). Analysis showed that the unsteady
portion of the incident waveform, $p(t)$, could be written as the sum of three
harmonics:

$$p(t) = \sum_{i=1}^{3} C_i \cos(\omega_i t) \tag{4.83}$$

where amplitudes C_i and frequencies ω_i are shown in Table 4.5.

At time $t = 0.6$ s, compute the magnitude of the reflected and transmitted
pressure waves. You may assume that the wave speed is the same in the parent
and daughter arteries, and that each daughter artery has a cross-sectional
area equal to 40% of the parent cross-sectional area. Be careful to use radians
(not degrees).

4.15   Consider pressure pulse propagation in an artery that gradually tapers, such
that the diameter decreases linearly with x according to the relationship
$D(x) = D_1 - \alpha x$, where α is some positive number and D_1 is the diameter
at $x = 0$. The wall thickness and Young's modulus of the wall material do
not change with position.
  (a)   Show that this implies that the characteristic impedance changes
according to

$$\frac{dZ}{dx} = \frac{5\alpha Z}{2D}. \tag{4.84}$$

(b) The transmission coefficient for a junction with a single parent and a single daughter is $T = 2Z_d/(Z_p + Z_d)$. Consider each segment of length dx of the tapering artery to be such a junction, with the "parent" having impedance $Z(x)$ and the "daughter" having impedance $Z(x) + dZ$. Show that in this case, the transmission coefficient for this "junction" is

$$T(x) = 1 + \frac{1}{2}\frac{dZ}{Z(x)} \tag{4.85}$$

(c) At location x, suppose the incident pressure waveform amplitude is $p_0(x)$ and the transmitted pressure waveform amplitude is $p_0(x) + dp_0$. Using results from parts (a) and (b), show that the pressure waveform amplitude must be proportional to $(D_1 - \alpha x)^{-5/4}$.

4.16 Consider a water wave of wavelength λ propagating in otherwise still water of depth h. If $h \ll \lambda$, it is known that such a wave propagates at speed $c = \sqrt{gh}$, where g is the acceleration due to gravity. In this case, the elevation of the free surface, η, can be described by [52]:

$$\eta = \eta_{max} \sin[2\pi(x - ct)/\lambda] \tag{4.86}$$

while the volume flow rate passing any cross-section can be written as

$$Q = c\eta_{max} \sin[2\pi(x - ct)/\lambda] \tag{4.87}$$

(a) Develop an analogy between the propagation of pressure waves in the arteries and the propagation of water waves. In particular, identify the characteristic impedance of a channel of depth h to the propagation of water waves.

(b) What is the amplitude of the transmitted wave if an incident wave of maximum height 10 cm passes over the step shown in the right panel of Fig. 4.39? In answering this question, you will find it useful to examine Equation (4.82).

4.17 In 1926, C. D. Murray published a paper [53] where he showed that there is an optimal parent:daughter diameter ratio in branching arterial networks. This optimal ratio is based on minimizing the rate of energy expenditure of the body, which consists of two parts:

- the power (rate of energy expenditure) required to synthesize blood and its formed elements; this is assumed to be proportional to the volume of blood in the arterial network

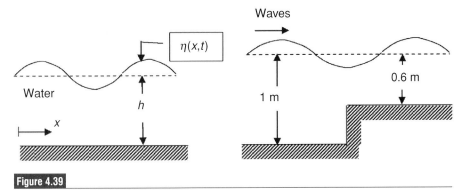

Figure 4.39

For Problem 4.16.

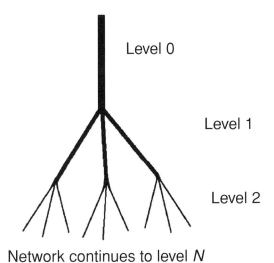

Level 0

Level 1

Level 2

Network continues to level *N*

Figure 4.40

For Problem 4.17.

- the power required to overcome viscous losses while pumping blood around the vascular tree.

Consider a branching network with $N + 1$ levels, where level 0 is the largest artery, level 1 is the next largest, and so on. Suppose that the radius of the arteries in level k is R_k, the length is L_k, and the number of branches is n_k. (See Fig. 4.40, where $n_0 = 1, n_1 = 3, n_2 = 9$, etc.)

(a) Show that the power required to pump the blood through this network, under the assumption of steady, Newtonian flow, is

$$Q^2 \sum_{k=0}^{N} \frac{8\mu L_k}{\pi n_k R_k^4},$$ (4.88)

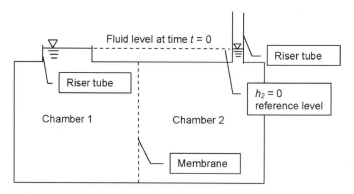

Figure 4.41

For Problem 4.18.

where Q is the flow rate in level 0, assumed constant. Hint: the power to pump fluid through a tube in Poiseuille flow is $q\Delta p$, where q is the flow in the tube and Δp is the total pressure drop along the tube.

(b) Show that minimizing the rate of energy expenditure with respect to the radius of a given level, R_j, leads to the conclusion that $n_j R_j^3$ must be a constant. This is known as Murray's law. Mathematical hint:

$$\frac{\mathrm{d}}{\mathrm{d}R_j} \sum_{k=0}^{N} f_k\,(R_k) = \frac{\mathrm{d}f_j}{\mathrm{d}R_j}, \qquad (4.89)$$

where f_k is any function that depends only on level k quantities.

(c) Show that this same conclusion could also have been reached by requiring the shear stress on the artery walls to be constant throughout the entire branching network.

(d) What is the reflection coefficient for a bifurcation that obeys Murray's law? Assume that the ratio of wall thickness to diameter and the Young's modulus are constants throughout the network. Hint: Murray's law holds for every level j in the network.

4.18 A membrane of hydraulic permeability L_p and area A_m separates two chambers, each of volume V. At time $t = 0$, m_1 moles of solute are added to chamber 1 and m_2 moles are added to chamber 2, where $m_2 > m_1$. You may assume that the solute mixes and dissolves in each chamber very rapidly, that the membrane is impermeable to the solute, and that the densities of the resulting two solutions are approximately equal to ρ. Attached to each chamber is a riser tube (see Fig. 4.41) of cross-sectional area A_1 (chamber 1) or A_2 (chamber 2), where $A_2 \ll A_1$. At $t = 0$, the fluid levels in the two tubes are equal, as shown.

What is the height of solution in tube 2 as a function of time, i.e., what is $h_2(t)$? Take the $h_2 = 0$ reference point as shown in Fig. 4.41. State relevant assumptions. Hint: $J_H = L_p A_m (\Delta p - \Delta \pi)$.

4.19   When considering the effects of osmotic pressure, we assumed that the membrane was completely impermeable to the solute. In real life, membranes can be *partially* permeable to a solute. This situation can be described by a slightly modified form of Equation (4.64)

$$Q_{12} = AL_p(\Delta p - \sigma \Delta \pi) \tag{4.90}$$

where Q_{12} is the fluid flow rate across the membrane, A is the membrane area, L_p is the membrane permeability, Δp is the pressure difference across the membrane, σ is the *osmotic reflection coefficient* for the solute, and $\Delta \pi$ is the osmotic pressure difference across the membrane.[16] The new feature here is the reflection coefficient, σ, which depends on the molecular weight (size) of the solute. *Note that $\sigma = 1$ for a perfectly rejected solute and $\sigma = 0$ for a freely permeable solute.*

(a)   Suppose that the solute is spherical. Show by a simple proportionality argument that the effective radius of the solute, r_s, should vary as $r_s \sim (MW)^{1/3}$, where MW is the molecular weight of the solute.

(b)   Consider L_p, A, and Δp to be constants, and suppose a fixed mass of solute is added to the water on the right side of the membrane in Fig. 4.42. Argue that there must be a certain solute size, r_s, that maximizes Q_{12}. You do not need any mathematics for this part of the question; a written argument is sufficient. Hint: consider two limiting cases: a very small solute particle and a very large one. Think about what happens to the molar concentration as the MW gets large for a fixed mass of solute.

(c)   When the solute radius is close to the membrane pore radius, r_p, Ferry [56] showed that the reflection coefficient varies as $\sigma = 1 - 2(1 - \eta)^2 - (1 - \eta)^4$, where $\eta = r_s/r_p$, for $\eta \leqslant 1$. Show that in this case, the maximum Q_{12} occurs for a solute radius $r_s/r_p = 2 - (5/2)^{1/2}$. You may assume van't Hoff's law holds for the solute and neglect the fourth-order term in the expression for σ.

[16] Equation (4.90) is one half of a set of equations known as the *Kedem–Katchalsky equations*, which describe the coupled transport of solvent and a solute through a membrane [54,55]. Equation (4.90) is for solvent transport; the solute transport equation is $J_s = Q_{12}(1 - \sigma_f)\overline{C} + PA\Delta c$, where J_s is the rate of solute transport across the membrane, σ_f is the filtration rejection coefficient, $\overline{C}$ is the mean solute concentration in the membrane, P is the membrane permeability coefficient, and Δc is the solute concentration difference across the membrane.

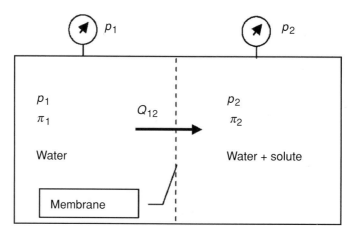

Figure 4.42

For Problem 4.19.

4.20 Red blood cells are normally in osmotic equilibrium with the surrounding plasma. However, if they are placed in a hypotonic solution (one having lower than physiological osmotic pressure), they will swell, since the cell membrane is effectively impermeable to ions but allows the water to pass through. Consider a cell of initial internal volume V_i having initial internal total ionic concentration c_i, which at time zero is placed in a very large hypotonic bathing solution having total ionic concentration $c_o \neq c_i$ (o, outside). The red cell membrane allows no ions to pass and has a water permeability L_p (flow rate per unit membrane area per unit transmembrane pressure difference). Assume van't Hoff's law holds, $\pi = \sum_j RT c_j$.

   (a) Derive a relationship between the cell's volume V and elapsed time t. State relevant assumptions that you have made.

   (b) If $V_i = 98\ \mu m^3$, $c_i = 300$ mM, and the red cell surface is $130\ \mu m^2$, predict the lowest value of c_o that the cell can tolerate without bursting.

4.21 The graph shown in Fig. 4.43 is based on data from a classic 1927 paper [32] in which Landis proved the existence of Starling's hypothesis by occluding capillaries, as discussed in the text. The ordinate is the rate at which fluid leaks out (or re-enters) the capillary per unit capillary wall area, J_H. (These experiments were done in frogs, so the numbers for osmotic pressure etc. do not exactly match those given in the text for humans.)

   (a) Assuming that $p - \pi$ for the interstitium is $-5\,\mathrm{cmH_2O}$, estimate plasma oncotic pressure from Fig. 4.43.

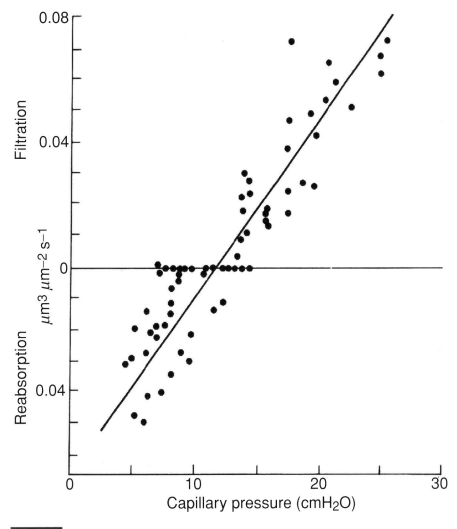

Figure 4.43

For Problem 4.21. The graph shows transmural flow rate as a function of capillary pressure. From [17], redrawn from the original data of Landis [32]. Reproduced with kind permission of Oxford University Press and the American Physiological Society.

(b) Estimate the filtration coefficient L_p for this capillary, defined as:

$$L_p = \frac{J_H}{\Delta p - \Delta \pi} \tag{4.91}$$

(c) Consider a capillary 0.05 cm long of diameter 8 μm, for which arteriolar and venular luminal pressure are 25 and 5 cmH$_2$O, respectively. Assume that L_p and π are constant and that the pressure drop varies linearly along the capillary. What is the net rate of fluid loss (gain) from the capillary?

4.22   In analyzing capillary leakage in the previous question, we assumed that capillary plasma protein osmotic pressure did not change with axial position within the capillary, x. Strictly speaking, this is not true, since leakage of water from the capillary increases the effective protein concentration in the capillary, and hence increases osmotic pressure.

(a)   Using the data given in the previous question, find the balance point for the capillary, i.e., the location where the leakage rate $J_H = 0$. Note that answering parts (a) and (b) of this problem requires you to first solve the previous problem.

(b)   How much fluid is lost per unit time from the capillary between the inlet and the balance point, assuming $\pi_{cap} = constant$?

(c)   The total flow rate entering the capillary is 4.2×10^4 μm^3/s, consisting of 42% red cells, 4% plasma proteins, and 54% water (by volume). The protein concentration at any location x is proportional to

$$\frac{\text{protein volume flow rate at } x}{(\text{water} + \text{protein}) \text{ volume flow rate at } x}. \qquad (4.92)$$

What is the percentage change in protein osmotic pressure at the balance point (referenced to protein osmotic pressure at the inlet)? Assume van't Hoff's law holds, and no protein or red cell leakage through the capillary wall. Is our assumption of approximately constant π_{cap} a valid one?

4.23   When blood flows in a capillary of radius R, some plasma leaks out into the surrounding tissue. This causes the volumetric flow rate Q and the pressure gradient dp/dx within the capillary to be a function of x. The leakage rate J_H (volume of fluid leaked per unit area per unit time) through the wall is given by

$$J_H = constant (p - p_{ex} - \Delta\pi) \qquad (4.93)$$

where p_{ex} is the external pressure and $\Delta\pi$ is the osmotic pressure difference across the wall. Both p_{ex} and $\Delta\pi$ may be assumed constant. Assuming that J_H is small, derive an expression for the pressure in the capillary. It will involve two constants, which you do not need to determine. Hint: assume the flow is everywhere locally Poiseuille.

4.24   The composition of major components in intracellular fluid is listed in Table 4.6, column 2. Assume that a neutrophil having this intracellular fluid composition is placed in a solution having the composition shown in column 3.

Table 4.6. For Problem 4.24. From Vander *et al.* [10]. Reproduced with kind permission of The McGraw-Hill Companies.

Component	Intracellular fluid concentration (mM)	Extracellular fluid concentration (mM)
Na^+	15	90
K^+	150	4
Ca^{2+}	1.5	1
Mg^{2+}	12	1.5
Cl^-	10	110
HCO_3^-	10	30
P_i	40	2
Amino acids	8	2
Glucose	1	5.6
ATP	4	0
Protein	4.0	0.2

(a) Estimate the osmotic pressure difference across the cell membrane arising from the concentration differences listed in Table 4.6. Assume the cell membrane is impermeable to all of the listed components, and that the cell is at body temperature (37 °C).

(b) Knowing that the pressure in the cell equals the pressure outside the cell, and that an osmotic pressure creates a stress across the membrane in the same way that pressure does, compute the stress acting in the cell membrane from osmotic effects. Assume that the thickness of the membrane is 0.1 μm, that the cell is a perfect sphere of radius 12 μm, and that all transmembrane stresses are carried by the cell membrane (i.e., ignore the cytoskeleton).

(c) If Young's modulus for the neutrophil's membrane is 300 pN/μm², compute the resulting strain in the membrane. Note that $1\,pN = 10^{-12}\,N$.

(d) Does this strain seem reasonable? Can you think of factors that we have neglected that will change this calculation?

References

1. J. B. d. C. M. Saunders and C. D. O'Malley. *The Illustrations from the Works of Andreas Vesalius of Brussels* (Cleveland: World Publishing, 1950).

2. A. Vesalius. *Fabric of the Human Body.* Books III & IV: *Veins, Arteries, Nerves.* (San Francisco, CA: Norman, 2002).

3. A. Despopoulos and S. Silbernagl. *Color Atlas of Physiology* (New York: Georg Thieme Verlag, 1986).

4. W. R. Milnor. *Hemodynamics*, 2nd edn (Baltimore, MD: Williams & Wilkins, 1989).

5. R. S. Vasan, M. G. Larson and D. Levy. Determinants of echocardiographic aortic root size. The Framingham Heart Study. *Circulation*, **91** (1995), 734–740.

6. T. Fujiwara and Y. Uehara. The cytoarchitecture of the wall and the innervation pattern of the microvessels in the rat mammary gland: a scanning electron microscopic observation. *American Journal of Anatomy*, **170** (1984), 39–54.

7. J. W. Rohen, C. Yokochi and E. Lütjen-Drecoll. *Color Atlas of Anatomy: A Photographic Study of the Human Body*, 5th edn (Philadelphia, PA: Lippincott Williams & Wilkins, 2002).

8. W. W. Nichols and M. F. O'Rourke. *McDonald's Blood Flow in Arteries* (Philadelphia, PA: Lea & Febiger, 1990).

9. K. Sagawa, W. L. Maughan, K. Sunagawa and H. Suga. *Cardiac Contraction and the Pressure–Volume Relationship* (New York: Oxford University Press, 1988).

10. A. J. Vander, J. H. Sherman and D. S. Luciano. *Human Physiology: The Mechanisms of Body Function*, 4th edn (New York: McGraw-Hill, 1985).

11. A. D. McCulloch. Cardiac biomechanics. In *Biomechanics: Principles and Applications*, ed. D. J. Schneck and J. D. Bronzino. (Boca Raton, FL: CRC Press, 2003), pp. 163–188.

12. C. S. Hayward, W. V. Kalnins and R. P. Kelly. Gender-related differences in left ventricular chamber function. *Cardiovascular Research*, **49** (2001), 340–350.

13. O. Frank. Die Grundform Des Arteriellen Pulses Erste Abhandlung: Mathematische Analyse. *Zeitschrift für Biologie*, **37** (1899), 483–526.

14. K. Sagawa, R. K. Lie and J. Schaefer. Translation of Otto Frank's Paper "Die Grundform Des Arteriellen Pulses" *Zeitschrift für Biologie*, **37**: 483–526 (1899). *Journal of Molecular and Cell Cardiology*, **22** (1990), 253–277.

15. N. Westerhof, G. Elzinga and P. Sipkema. An artificial arterial system for pumping hearts. *Journal of Applied Physiology*, **31** (1971), 776–781.

16. M. Yoshigi and B. B. Keller. Characterization of embryonic aortic impedance with lumped parameter models. *American Journal of Physiology: Heart and Circulatory Physiology*, **273** (1997), H19–H27.

17. C. G. Caro, T. J. Pedley, R. C. Schroter and W. A. Seed. *The Mechanics of the Circulation* (Oxford: Oxford University Press, 1978).

18. C. J. Mills, I. T. Gabe, J. H. Gault, D. T. Mason, J. Ross, Jr. *et al.* Pressure–flow relationships and vascular impedance in man. *Cardiovascular Research*, **4** (1970), 405–417.

19. D. W. Vaughan. *A Learning System in Histology* (Oxford: Oxford University Press, 2002).

20. M. A. Zulliger, A. Rachev and N. Stergiopulos. A constitutive formulation of arterial mechanics including vascular smooth muscle tone. *American Journal of Physiology: Heart and Circulatory Physiology*, **287** (2004), H1335–H1343.

21. P. B. Canham, E. A. Talman, H. M. Finlay and J. G. Dixon. Medial collagen organization in human arteries of the heart and brain by polarized light microscopy. *Connective Tissue Research*, **26** (1991), 121–134.

22. L. H. Peterson, R. E. Jensen and J. Parnell. Mechanical properties of arteries in vivo. *Circulation Research*, **8** (1960), 622–639.

23. S. Jin, J. Oshinski and D. P. Giddens. Effects of wall motion and compliance on flow patterns in the ascending aorta. *Journal of Biomechanical Engineering*, **125** (2003), 347–354.

24. D. J. Patel, Jr. and D. L. Fry. In vivo pressure–length–radius relationship of certain blood vessels in man and dog. In *Pulsatile Blood Flow*, ed. E. O. Attinger. (New York: McGraw-Hill, 1964), pp. 293–302.

25. P. C. Luchsinger, M. Sachs and D. J. Patel. Pressure–radius relationship in large blood vessels of man. *Circulation Research*, **11** (1962), 885–888.

26. J. O. Arndt. [On the mechanism of the intact human common carotid artery under various circulatory conditions.] *Archiv für Kreislaufforschung*, **59** (1969), 153–197.

27. W. A. Riley, D. S. Freedman, N. A. Higgs, R. W. Barnes, S. A. Zinkgraf *et al.* Decreased arterial elasticity associated with cardiovascular disease risk factors in the young. Bogalusa Heart Study. *Arteriosclerosis, Thrombosis, and Vascular Biology*, **6** (1986), 378–386.

28. J. C. Greenfield and D. M. Griggs. Relation between pressure and diameter in main pulmonary artery of man. *Journal of Applied Physiology*, **18** (1963), 557–559.

29. B. S. Gow and M. G. Taylor. Measurement of viscoelastic properties of arteries in the living dog. *Circulation Research*, **23** (1968), 111–122.

30. R. P. Vito and S. A. Dixon. Blood vessel constitutive models 1995–2002. *Annual Review of Biomedical Engineering*, **5** (2003), 413–439.

31. R. L. Daugherty, J. B. Franzini and E. J. Finnemore. *Fluid Mechanics With Engineering Applications*, 8th edn (New York: McGraw-Hill, 1985).

32. E. M. Landis. Micro-injection studies of capillary permeability: II. The relation between capillary pressure and the rate at which fluid passes through the walls of single capillaries. *American Journal of Physiology*, **82** (1927), 217–238.

33. R. L. Wesly, R. N. Vaishnav, J. C. Fuchs, D. J. Patel and J. C. Greenfield, Jr. Static linear and nonlinear elastic properties of normal and arterialized venous tissue in dog and man. *Circulation Research*, **37** (1975), 509–520.

34. R. Gottlob and R. May. *Venous Valves* (New York: Springer Verlag, 1986).

35. C. R. White and R. S. Seymour. Mammalian basal metabolic rate is proportional to body mass$^{2/3}$. *Proceedings of the National Academy of Sciences USA*, **100** (2003), 4046–4049.

36. A. C. Guyton, C. E. Jones and T. G. Coleman. *Circulatory Physiology: Cardiac Output and Its Regulation* (Philadelphia, PA: W. B. Saunders, 1973).

37. T. K. Lam and D. T. Leung. More on simplified calculation of body-surface area. *New England Journal of Medicine*, **318** (1988), 1130.

38. R. D. Mosteller. Simplified calculation of body-surface area. *New England Journal of Medicine*, **317** (1987), 1098.

39. O. L. Wade and J. M. Bishop. *Cardiac Output and Regional Blood Flow* (Oxford: Blackwell, 1962).

40. L. Lindstedt and P. J. Schaeffer. Use of allometry in predicting anatomical and physiological parameters of mammals. *Laboratory Animals*, **36** (2002), 1–19.

41. E. F. Adolph. Quantitative relations in the physiological constitutions of mammals. *Science*, **109** (1949), 579–585.

42. W. R. Stahl. Scaling of respiratory variables in mammals. *Journal of Applied Physiology*, **22** (1967), 453–460.

43. J. P. Holt, E. A. Rhode and H. Kines. Ventricular volumes and body weight in mammals. *American Journal of Physiology*, **215** (1968), 704–715.

44. G. Juznic and H. Klensch. [Comparative physiological studies on the behavior of the indices for energy consumption and performance of the heart.] [In German]. *Pflugers Archiv für Die Gesamte Physiologie Des Menschen und Der Tiere*, **280** (1964), 38–45.

45. J. L. Patterson, Jr., R. H. Goetz, J. T. Doyle, J. V. Warren, O. H. Gauer *et al.* Cardiorespiratory dynamics in the ox and giraffe, with comparative observations on man and other mammals. *Annals of the New York Academy of Sciences*, **127** (1965), 393–413.

46. B. Gunther and E. Guerra. Biological similarities. *Acta Physiologica Latino Americana*, **5** (1955), 169–186.

47. J. K.-J. Li. *Arterial System Dynamics* (New York: New York University Press, 1987).

48. J. P. Holt, E. A. Rhode, W. W. Holt and H. Kines. Geometric similarity of aorta, venae cavae, and certain of their branches in mammals. *American Journal of Physiology: Regulatory, Integrative and Comparative Physiology*, **241** (1981), R100–R104.

49. J. K.-J. Li. *Comparative Cardiovascular Dynamics of Mammals* (Boca Raton, FL: CRC Press, 1996).

50. N. Westerhof, N. Stergiopulos and M. I. M. Noble. *Snapshots of Hemodynamics: An Aid for Clinical Research and Graduate Education* (New York: Springer, 2005).

51. N. R. Tai, H. J. Salacinski, A. Edwards, G. Hamilton and A. M. Seifalian. Compliance properties of conduits used in vascular reconstruction. *British Journal of Surgery*, **87** (2000), 1516–1524.

52. B. S. Massey. *Mechanics of Fluids*, 6th edn (London: Van Nostrand Reinhold, 1989).

53. C. D. Murray. The physiological principle of minimum work. I. The vascular system and the cost of blood volume. *Proceedings of the National Academy of Sciences USA*, **12** (1926), 207–214.

54. A. Katchalsky and P. F. Curran. *Nonequilibrium Thermodynamics in Biophysics* (Cambridge, MA: Harvard University Press, 1965).

55. O. Kedem and A. Katchalsky. Thermodynamic analysis of the permeability of biological membranes to non-electrolytes. *Biochimica et Biophysica Acta*, **27** (1958), 229–246.

56. J. D. Ferry. Statistical evaluation of sieve constants in ultrafiltration. *Journal of General Physiology*, **20** (1936), 95–104.

5 The interstitium

The interstitium surrounds the capillaries. It consists of cells that are embedded in a matrix of extracellular biomacromolecules known as the extracellular matrix (ECM[1]; see Section 2.4). Depending on the tissue, the cells can account for a large fraction of interstitial volume or can be relatively few in number. An important task of the resident cells, or at least of some subpopulation of the resident cells, is to synthesize and secrete the extracellular biopolymers that make up the ECM.

5.1 Interstitial fluid flow

We have seen in Section 4.4 that there is a continuous transport of fluid from capillary blood into the lymphatic system, via the interstitium. The net flow rate from capillaries to lymphatics is determined by the driving potential (the difference in $p - \pi$ between the capillaries and the lymphatics), as well as the hydrodynamic resistance of the intervening interstitium. Interstitial flow resistance is therefore very important in fluid homeostasis, or maintaining proper tissue hydration.

Intuition suggests that interstitial flow resistance should depend on tissue structure, or more specifically on:

- the local concentrations of ECM components
- the local number of cells per unit tissue volume
- the manner in which cells and ECM components are packed together.

Cells and ECM components are packed together in a random manner. Because of this random character, it is challenging to describe detailed flow patterns on a cellular or subcellular length scale. However, it is possible to characterize and quantify tissue composition if a statistical approach is used, in which larger tissue regions containing many cells are considered. In this case, the average flow rate through the interstitium can be related to average interstitial composition by using

---

[1] The interstitium is but one member of a class of tissues called connective tissues having similar properties and structure. Other members of this class include cartilage, tendons, sclera, cornea, and dermis. We will consider some of these connective tissues in Ch. 9.

porous media theory. Since it is usually the average flow through a tissue area that is of practical interest, this approach is entirely adequate. These flow patterns are important in fluid clearance but can also have important biological effects. For example, flow-induced gradients of vascular endothelial growth factor, a potent angiogenic factor, have been shown to drive capillary formation [1].

5.1.1 Darcy's law

At this point, it is necessary to take a small diversion to describe the essentials of porous media theory. A porous material consists of a solid matrix permeated by a network of pores, usually having a very complex topology. Porous media theory is typically used to describe flow through the material when it is too complex or difficult to describe the flow through each pore on an individual basis.

The basic law describing average flow through the tissue is due to Darcy [2][2]. In a series of experiments, Darcy forced fluid through porous bodies having cross-sectional area A and flow-wise length L. The pressure difference driving flow was Δp, the working fluid had viscosity μ, and the resulting flow rate Q was measured (Fig. 5.1). Empirically, Darcy found that his data were correlated by the relationship

$$\frac{Q}{A} = \frac{K}{\mu} \frac{\Delta p}{L}. \tag{5.1}$$

In Equation (5.1), K is a material property called the *permeability* (or *hydraulic permeability*), having dimensions of length squared. The permeability is a purely local property of the porous material: that is, it does not depend on the overall size of the sample being considered. It characterizes the ease with which fluid can pass through the porous material.

We now know that Darcy's law is valid only for slow (low Reynolds number) single-phase flow of a Newtonian fluid through a porous matrix. However, that is precisely the situation that occurs when fluid flows through the interstitium, and we therefore take Equation (5.1) as the starting point of a quantitative description of flow in tissue. Noting that Q/A is the average (or superficial) velocity of the fluid in the porous material, and that $\Delta p/L$ is the pressure gradient in the flow-wise (x) direction, Equation (5.1) can be written as

$$u = -\frac{K}{\mu} \frac{\mathrm{d}p}{\mathrm{d}x} \tag{5.2}$$

---

[2] An English translation of Darcy's original work is available [3].

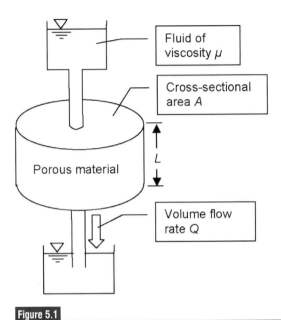

Apparatus used to deduce Darcy's law. Porous material fills the cylinder of length L and cross-sectional area A. The driving head is equivalent to a pressure Δp, and results in a flow rate Q.

where u is the *superficial velocity* in the x direction.[3] Equation (5.2) can be generalized to three-dimensional flows by replacing (scalar) u and dp/dx with the superficial velocity vector $\mathbf{u}$ and the pressure gradient ∇p, respectively, to obtain

$$\mathbf{u} = -\frac{K}{\mu}\nabla p \qquad (5.3)$$

Equation (5.3) states that the mean fluid transport rate ($\mathbf{u}$) is proportional to the driving potential ∇p, with proportionality constant K/μ. This is Darcy's law.

5.1.2   Clearance of edema

We will apply Darcy's law to estimate the clearance time of edematous (swollen) tissue, such as a bruise. Everyone is familiar with the concept of bruising: after sustaining blunt trauma, localized swelling and discoloration results, which is cleared over the course of several days. Physiologically, the trauma causes a series of events leading to a loss of capillary endothelial integrity. This allows fluid, plasma proteins, and formed elements to enter the interstitium. The excess fluid produces

---

[3] The superficial velocity needs to be distinguished from the *interstitial velocity*, the fluid velocity in the pores of the porous media. Because the superficial velocity is determined by averaging over the total cross-sectional area of the porous body (including solid portions), it is less than the interstitial velocity.

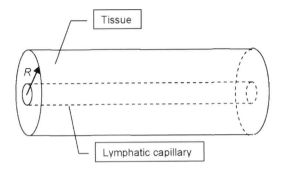

A single lymphatic capillary (inner dashed cylinder) surrounded by the interstitium that it is assumed to drain. The drainage region is a long cylinder of radius R.

local swelling (edema). Once the capillary endothelium heals, the surrounding tissue is left in an overhydrated state. Because significant amounts of plasma protein are present in the edematous tissue, the oncotic pressure in the interstitium is close to that in the capillary, so little fluid drainage occurs into the capillaries. Rather, fluid leaves the edematous tissue by draining into the lymphatic capillaries. We seek to estimate how long it will take the tissue to drain as a function of the properties of the interstitium.

Evidently a given tissue region will contain many lymphatic capillaries, and, all other things being equal, fluid will drain into the closest lymphatic capillary. Assuming that all capillaries are similar, it is therefore sufficient to consider a single lymphatic capillary, which we assume is responsible for draining fluid from a cylindrically shaped region (or domain) of radius R, as shown in Fig. 5.2. The entire tissue can be conceptually broken up into such regions, each one draining into its central capillary. The radius of each region will be a function of the number of capillaries per unit volume: more capillaries mean that each capillary must drain a smaller region (smaller R) and *mutatis mutandis*.

To analyze this problem, it is necessary to understand the driving force that causes fluid to leave the tissue. All tissues have an equilibrium or homeostatic level of hydration. In the case of edema, the tissue is overhydrated, which causes the ECM components (particularly the proteoglycans) to swell beyond their equilibrium value. Thermodynamically, it is favorable for the proteoglycans to contract and return to their equilibrium configuration; in so doing, fluid must be driven out of the tissue. This thermodynamic effect is manifested as an excess tissue pressure (*tissue swelling pressure*), which drives fluid out of the tissue whenever the tissue is overhydrated.[4] The closest analogy is a sponge that has had water injected into it

---

[4] The converse also occurs: if tissue is under-hydrated and exposed to water, there is a negative effective pressure within the tissue, which tends to draw water in.

under pressure, causing it to swell excessively. In this case the elastic fibers making up the solid matrix of the sponge have been stretched beyond their equilibrium lengths. These fibers try to contract elastically, which is manifested as a tendency of the sponge to expel water, and which can be quantified through the concept of an excess pressure within the sponge.

In order to characterize this effect, we denote the tissue swelling pressure by p_{swell}, and note that this quantity will depend on how much water is in the tissue. The hydration of the tissue can be characterized by the tissue porosity, ε, defined as

$$\varepsilon = \frac{\text{fluid volume in tissue}}{\text{total tissue volume}}. \tag{5.4}$$

It is expected that the swelling pressure will increase with increasing tissue porosity, and we denote this rate of increase by the parameter γ (the effective tissue elasticity)

$$\gamma = \frac{\mathrm{d}p_{swell}}{\mathrm{d}\varepsilon} \tag{5.5}$$

For present purposes, we will assume that γ is a constant, which depends on tissue composition but which can be taken as known.

Returning to the edema clearance problem, the physics of the process can now be clearly understood. The problem is fundamentally unsteady, with the rate of fluid clearance from the tissue being proportional to the changing tissue hydration. Fluid drainage is driven by the swelling pressure within the tissue, which forces the water out of the interstitium and into the capillary. However, fluid drainage is opposed by the flow resistance of the interstitium, such that the net rate at which the tissue drains results from a balance between tissue swelling and interstitial hydrodynamic resistance.

Although it is possible to analyze this problem in a detailed quantitative manner (see below), it is simpler to use dimensional analysis to obtain the time scale for edema clearance, τ. The value of τ is expected to depend on the size of the region to be drained (characterized by domain radius R), the effective tissue elasticity γ, and the flow resistance of the interstitium, characterized by μ/K.[5] Its value will also depend on the radius of the capillary, δ, but we assume that this dependence is not a strong one. Recalling the notation from Section 4.3.4, the quantities τ, R, γ, and μ/K have the following dimensions:

$$[\tau] \sim \mathrm{T}$$
$$[R] \sim \mathrm{L}$$
$$[\gamma] \sim \mathrm{ML}^{-1}\mathrm{T}^{-2}$$
$$[\mu/K] \sim \mathrm{ML}^{-3}\mathrm{T}^{-1}$$

---

[5] The ratio μ/K appears, rather than μ and K independently, since these two quantities appear only in Darcy's law, and appear there as a ratio: μ/K represents an effective tissue flow resistance (pressure gradient per unit fluid velocity).

From these parameters one Π-group can be formed, namely

$$\Pi_1 = \tau \frac{K}{\mu} \frac{\gamma}{R^2} \tag{5.6}$$

Since there is only one Π-group, it must be constant, which indicates that

$$\tau \sim \frac{\mu R^2}{\gamma K} \tag{5.7}$$

This makes physical sense: the clearance time increases if a larger region must be drained or if fluid viscosity increases, and the clearance time decreases if tissue permeability increases or if tissue elasticity is large.

It is of interest to estimate the value of τ based on this analysis. Making the assumption that the proportionality constant implied by Equation (5.7) is one, and using the following values

$$\mu = 0.007 \text{ g/(cm s)}$$
$$R = 5.5 \times 10^{-2} \text{ cm}$$
$$K = 8 \times 10^{-14} \text{ cm}^2$$
$$\gamma = 4540 \text{ dynes/cm}^2$$

we obtain a value for τ of 5.83×10^4 s, or approximately 16 h. This accords with practical experience: if we raise a bruise it persists for a day or so.

Detailed parameter derivation

Here we justify some of the parameter values selected in the above calculation, as well as deriving the form of the equation that governs tissue swelling. We begin with the governing equation.

Consider a small tissue-containing region of volume V. The void volume within the region is $\int_V \varepsilon \, dV$, and the fluid mass within the region is $\int_V \rho \varepsilon \, dV$, where ρ is fluid density. Recalling that $\mathbf{u}$ is the superficial velocity, the mass flux of fluid out of the cube is $\int_S \rho \mathbf{u} \cdot \hat{\mathbf{n}} \, dS$, where $\hat{\mathbf{n}}$ is the outward normal to the region's surface S. Conservation of mass written in integral form then states

$$\int_V \rho \varepsilon \, dV + \int_S \rho \mathbf{u} \cdot \hat{\mathbf{n}} \, dS = 0. \tag{5.8}$$

Assuming that ρ is constant, using Gauss' theorem, and taking the volume V as arbitrary gives an unsteady differential equation representing conservation of mass

$$\frac{\partial \varepsilon}{\partial t} + \nabla \cdot \mathbf{u} = 0. \tag{5.9}$$

Combining Equations (5.9) and (5.3) yields

$$\frac{\partial \varepsilon}{\partial t} + \frac{K}{\mu} \nabla^2 p_{\text{tissue}} = 0. \tag{5.10}$$

Finally, we use the definition of the effective tissue elasticity γ to write $d\varepsilon = dp_{\text{tissue}}/\gamma$, which when combined with Equation (5.10), yields

$$\frac{\partial p_{\text{tissue}}}{\partial t} + \frac{K\gamma}{\mu} \nabla^2 p_{\text{tissue}} = 0. \tag{5.11}$$

It is convenient to non-dimensionalize the above equation by defining a non-dimensional pressure, time, and radial position by

$$\hat{p}_{\text{tissue}} = \frac{p_{\text{tissue}} - p_0}{p_i - p_0} \tag{5.12}$$

$$\hat{t} = \frac{t}{\tau} \tag{5.13}$$

$$\hat{r} = \frac{r}{R} \tag{5.14}$$

where p_0 is the equilibrium (homeostatic) pressure in the tissue, p_i is the initial tissue pressure (when the tissue is in the swollen state and fluid begins to drain), and τ is a time scale given by

$$\tau = \frac{\mu R^2}{\gamma K}. \tag{5.15}$$

With this set of non-dimensional parameters, the governing equation for the tissue pressure becomes

$$\frac{\partial \hat{p}_{\text{tissue}}}{\partial \hat{t}} = \hat{\nabla}^2 \hat{p}_{\text{tissue}}. \tag{5.16}$$

If we make the (admittedly simplistic) assumption that tissue pressure is uniform within the circular domain at $t = 0$, then the initial condition is $\hat{p}_{\text{tissue}} = 1$ at $t = 0$ for all $\delta/R < \hat{r} < 1$. By virtue of the fact that we divided the entire tissue into cylindrical domains, with one capillary per domain, there is little fluid drainage across the outer surface of a cylindrical domain. This can be expressed by a no-flux boundary condition

$$\frac{\partial \hat{p}_{\text{tissue}}}{\partial \hat{r}} = 0 \qquad \text{at } \hat{r} = 1 \tag{5.17}$$

A second boundary condition is required, which is obtained from the reasonable assumption that the lymphatic capillary is freely draining (i.e., that the capillary walls have little flow resistance), in which case the tissue pressure immediately adjacent to the capillary is homeostatic. Mathematically, this is expressed as $\hat{p}_{\text{tissue}} = 0$ at $\hat{r} = \delta/R$.

Equation (5.16) has a separable solution in terms of Bessel functions of order zero and decaying exponentials, with time scale τ. This confirms that $\tau = \mu R^2/(K\gamma)$ is the relevant time scale for tissue drainage.

We now turn attention to estimating the radius of the cylindrical region R and the effective tissue elasticity γ. Estimation of R is somewhat difficult, since there are few quantitative data on lymphatic capillary distribution. Casely-Smith [4] estimated that there is 0.4 ml of lymph in the heart, which has a mass of approximately 300 g. Therefore, the volume fraction of lymphatics in the heart is roughly $0.4/300 = 0.0013$, assuming that the density of heart tissue is approximately 1 g/cm^3. Taking a typical lymphatic capillary diameter of 20 μm, this gives $R = 20\,\mu\text{m}/\sqrt{0.0013}$, or $R = 550\,\mu\text{m}$. It is unclear as to whether data for the heart can be applied to the interstitium, but since there is little information on capillary distribution in the latter tissue, we will assume that it can.

Estimation of γ is somewhat more complex and requires further examination of the physics behind tissue elasticity. Recall that p_{tissue} arises because the interstitium has become "overhydrated" and the proteoglycans have become overstretched.[6] We ignore the contribution to γ from physical extension of collagen and elastin and assume that the proteoglycans are entangled with other tissue components. This implies that proteoglycans become hydrated at the same rate as the entire tissue, which seems reasonable.

By analogy with the theory of rubber elasticity, the quantity $p_{\text{tissue}} - p_{\text{o}}$ is equal to $\pi_{\text{o}} - \pi_{\text{tissue}}$, where π_{tissue} refers to the osmotic pressure generated by tissue proteoglycans and π_{o} is the equilibrium (homeostatic) value of π_{tissue}. Defining the solid tissue volume fraction $\phi_{\text{tot}} = 1 - \varepsilon$ allows us to write

$$\gamma = \frac{\mathrm{d}p_{\text{tissue}}}{\mathrm{d}\varepsilon} = \frac{\mathrm{d}\pi_{\text{tissue}}}{\mathrm{d}\phi_{\text{tot}}}. \tag{5.18}$$

It seems reasonable to assume that the ratio of proteoglycans to total solid tissue, $\phi_{\text{PG}}/\phi_{\text{tot}}$, is constant. Therefore, we can write

$$\gamma = \frac{\phi_{\text{PG}}}{\phi_{\text{tot}}}\frac{\mathrm{d}\pi_{\text{tissue}}}{\mathrm{d}\phi_{\text{PG}}} = \frac{c_{\text{PG}}}{\phi_{\text{tot}}}\frac{\mathrm{d}\pi_{\text{tissue}}}{\mathrm{d}c_{\text{PG}}} \tag{5.19}$$

where c_{PG} is the concentration of proteoglycans in the tissue. The major contribution to proteoglycan osmotic pressure comes from their glycosaminoglycan components. Johnson [6] stated that the osmotic pressure of the most common glycosaminoglycan, hyaluronan, is given by

$$\pi = A_2 c^2 \tag{5.20}$$

$$A_2 = 7M^{-0.1}S^{-0.5} \tag{5.21}$$

---

[6] This is the rubber elasticity phenomenon described in classical texts on polymer physics [5].

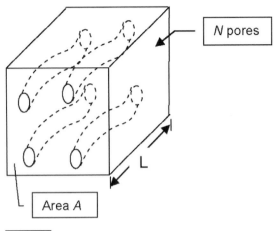

Figure 5.3

For Problem 5.1.

where A_2 is expressed in mmHg/(g/100 ml). We assume a molecular weight M of 500 kDa and physiologic salinity S of 150 mM, to compute γ as 4500 dynes/cm^2.

5.2 Problems

5.1 When analyzing porous materials, it is handy to have a way of estimating the permeability of a material without actually having to force fluid through it. Intuition tells us that the material's permeability should somehow be related to the number and size of the pores within the material. A simple relationship that allows us to estimate permeability from these variables is the Carman–Kozeny equation.

(a) Consider a block of cross-sectional area A and flow-wise length L, which is permeated by N pores, each of length l ($\geqslant L$) and radius R (Fig. 5.3). (Here we allow l to exceed L to account for the *tortuosity* of the pores; that is, the fact that they are not perfectly straight.) Assuming that Poiseuille's law holds in each pore, show that the total flow rate passing through the block, Q, is given by:

$$Q = \frac{N\pi R^4}{8\mu}\frac{\Delta p}{l} \qquad (5.22)$$

where Δp is the pressure drop from front to back.

(b) Compare this result with Darcy's law to show that the permeability, K, can be given by:

$$K = \frac{\varepsilon^3}{2\tau^2 S^2} \qquad (5.23)$$

where τ, ε, and S are the tortuosity, porosity, and specific surface, respectively, defined by

$$\tau = \frac{l}{L}, \qquad (5.24)$$

$$\varepsilon = \frac{\text{total pore volume}}{\text{total block volume}}, \qquad (5.25)$$

$$S = \frac{\text{total pore wall area}}{\text{total block volume}}. \qquad (5.26)$$

Equation [5.23] is the Carman–Kozeny equation. It works very well for most porous materials, even though it makes a lot of simplifying assumptions. To generalize this equation it is conventional to replace the factor $2\tau^2$ in the denominator of Equation (5.23) with a parameter k, called the Kozeny constant. Empirical data shows that for real porous materials with complex pore shapes, the Carman–Kozeny equation works well if the Kozeny constant is assigned a value of 4 or 4.5, so long as the porosity of the porous material is less than approximately 60%.

References

1. C. A. Helm, M. E. Fleury, A. H. Zisch, F. Boschetti and M. A. Swartz. Synergy between interstitial flow and VEGF directs capillary morphogenesis in vitro through a gradient amplification mechanism. *Proceedings of the National Academy of Sciences USA*, **102** (2005), 15779–15784.
2. H. P. G. Darcy. *Les Fontaines Publiques De La Ville De Dijon* (Paris: Dalmont, 1856).
3. P. Bobeck (translator). *The Public Fountains of the City of Dijon by Henry Darcy (1856)*. (Dubuque, IA: Kendall/Hunt, 2004).
4. J. R. Casely-Smith. Lymph and lymphatics. In *Microcirculation*, ed. G. Kaley and B. M. Altura. (Baltimore, MD: University Park Press, 1977), pp. 423–500.
5. P. J. Flory. *Principles of Polymer Chemistry* (Ithaca, NY: Cornell University Press, 1971).
6. M. Johnson. Transport through the aqueous outflow system of the eye. Ph.D. thesis, Massachusetts Institute of Technology (1987).

6 Ocular biomechanics

At first, it may seem like biomechanics play little or no role in the eye, but nothing could be further from the truth. In fact, the eye is a pressurized, thick-walled shell with dedicated fluid production and drainage tissues, whose shape is controlled by biomechanical factors. It has internal and external musculature, a remarkably complex internal vascular system, and a variety of specialized fluid and solute transport systems. Biomechanics play a central role in accommodation (focussing near and far), as well as in common disorders such as *glaucoma*, *macular degeneration*, *myopia* (near-sightedness), and *presbyopia* (inability to focus on nearby objects). To appreciate the role of biomechanics in these processes we must first briefly review ocular anatomy.

6.1 Ocular anatomy

The eye is a remarkable organ. It functions like a camera, with an adjustable compound lens, an adjustable aperture (the *pupil*), and a light-sensitive medium (the *retina*) that converts photons into electrochemical signals (Fig. 6.1, color plate). The eye automatically adjusts pupil size and lens shape so that images are clear under a wide variety of lighting conditions and over a wide range of distances from the observer.

The outer coat of the eye is formed by the *cornea* and *sclera*, two tough connective tissues that together make up the *corneoscleral shell*. This shell is pierced at the back of the eye by the *scleral canal* and at other discrete locations by small vessels and nerves. The "output" from the retina is carried to the visual cortex in the brain by retinal ganglion cell axons,[1] which group together to form the *optic nerve*. The clinically visible portion of the optic nerve is known as the *optic nerve head* and consists of the unmyelinated nerve fibers that converge, turn, and exit the eye. The optic nerve proper, which is 3.5 to 4 mm in diameter external to the eye [2], leaves the eye through the scleral canal. Six extraocular muscles attach to

---

[1] As will be discussed in greater detail in Section 8.1, axons are elongated processes of neural cells along which information is transmitted by traveling electrochemical waves known as *action potentials*.

the outer surface of the sclera and work together with a clever system of active pulleys to rotate the eye [3].

Light enters the eye by passing through the cornea, after which it traverses the *anterior chamber*, pupil, *lens*, and *vitreous body* before striking the retina. The lens is suspended by ligaments (the *zonules*) that attach to the inner fibers of the *ciliary muscle*; alterations in tone of these muscle fibers cause the zonules to tug on the lens so that the lens changes shape to accommodate (i.e., to alter the effective focal length of the eye).

To ensure good optical performance, it is essential that the distance from the retina to the lens be stable. This implies the need for rigidity in the overall structure of the eye. In most other organs, rigidity is conferred by bones. However, no bones are present in the mammalian eye; instead, the eye is maintained at a small positive pressure (the *intraocular pressure* [IOP]) with respect to its surroundings. This causes the eye to be inflated and hence somewhat rigid, just as an inflated soccer ball is fairly rigid.

A closely related issue is how to supply nutrients to the lens and cornea. Both tissues contain cells, and thus have metabolic needs, but must also remain optically transparent. The latter requirement implies that these tissues cannot be perfused with blood, which is the usual way of delivering metabolites and removing catabolites. To solve this problem, the eye secretes a clear, colorless fluid (the *aqueous humor*), which carries nutrients, etc. This fluid is produced by a highly folded and vascularized inner layer of the ciliary body known as the *ciliary processes* [4]. The secreted aqueous humor flows radially inward, bathing the lens, then flows anteriorly through the pupil to fill the anterior chamber, before draining out of the eye through specialized tissues in the angle formed by the iris and cornea, known as the *angle of the anterior chamber*. These outflow pathways have appreciable flow resistance, which together with active production of the aqueous humor by the ciliary processes creates the IOP. In other words, the flow of aqueous humor serves both to nourish the lens and cornea and to "inflate" the eye. The normal mean value of IOP is 15.5 mmHg (standard deviation, 2.6 mmHg) [5]. The vitreous body is a relatively inert and quite porous connective tissue, so it transmits the pressure from the anterior chamber throughout the interior of the eye.

6.2   Biomechanics of glaucoma

Unfortunately, not all eyes are normal. For example, it is estimated that about 65 million people worldwide [6] suffer from some form of glaucoma, making it the second most common cause of blindness in western countries. In the vast majority of such cases, the flow resistance of the outflow tissues is elevated, causing the IOP to be higher than normal. By a mechanism that is not well understood, chronic elevation of IOP causes progressive irreversible damage to the optic nerve, with blindness as the end result. A major aspect of glaucoma research is therefore to try to understand how and why the flow resistance of the outflow tissues is elevated in the glaucomatous eye. We will discuss this in greater detail in Section 6.2.2.

First we concern ourselves with a related clinical problem: how does the practicing ophthalmologist determine if a patient has glaucoma? This is a particularly important question, since people with glaucoma are often not aware of their condition until very significant irreversible optic nerve damage has occurred. Early detection of the disease is therefore crucial. The definitive diagnosis relies on the experience of the ophthalmologist, who judges the appearance of the optic nerve head and measures the patient's *visual field*, a map of how well each part of the retina is working and transmitting its signals to the brain. Unfortunately, such tests are expensive and time consuming, and therefore are not suitable for routine screening. Since the majority of patients with glaucoma have elevated IOP, a measurement of IOP is a valuable screening tool.[2]

What is required, therefore, is a quick, safe, and reliable method of measuring IOP. The most accurate method is to cannulate the eye (introduce a needle into the eye) and attach a tubing line to a pressure sensor. This is obviously not practical for routine screening, and so a more indirect method must be used. This leads to a discussion of tonometry.

6.2.1 Tonometry

Tonometry exploits the fact that the eye is essentially an elastic-walled inflated spherical shell. Consequently, by pushing on the surface of the eye it is possible to effect a deformation (indentation) of the eye. The higher the IOP, the more difficult it is to accomplish this maneuver. In practice, there are several possible approaches [5].

Fixed area applanation tonometry.   The force required to flatten a known region of the cornea is measured. The most common type of fixed applanation tonometry is Goldmann tonometry, which we will discuss below.

Fixed force applanation tonometry.   The amount of corneal flattening caused by application of a known force is measured. This is the basis of pneumotonometry (air-puff tonometry), where a small puff of air is used to deform the cornea, avoiding

---

[2] More generally, it seems that each person has a level of IOP that they can tolerate without damage to their optic nerve [7]. This level varies from person to person: in some people, a normal level of IOP causes damage, in which case the disease is known as low-tension glaucoma. Others are lucky: they can tolerate high pressure with no optic nerve damage. Therefore, measurement of IOP is not 100% specific as a screening tool, but it is still valuable.

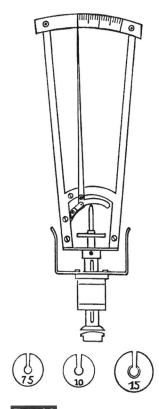

Figure 6.2

A Schiøtz tonometer. The plunger at the lower portion of the tonometer is placed on the cornea, and weights (bottom of figure) are put onto the device. The resulting corneal deflection is measured on the upper scale, and can be related to intraocular pressure. Reprinted from Gloster [8], with kind permission from Elsevier.

mechanical contact with the eye. Unfortunately, pneumotonometry is less accurate than Goldmann tonometry.

Schiøtz (indentation) tonometry.   The indentation of the cornea caused by a predetermined force applied to the eye is measured (Fig. 6.2).

We will restrict attention to Goldmann tonometry, since it is in common clinical use and is considered to be the "gold standard" for tonometry. The cornea is lightly anaesthetized, and the applanation head is placed on the cornea. The flattened corneal area is denoted by A, and the force applied to the applanation head is denoted as W (Fig. 6.3). Then, in the ideal situation, the IOP is given by the *Imbert–Fick law*

$$IOP = \frac{W}{A} \tag{6.1}$$

However, there are two important non-idealities.

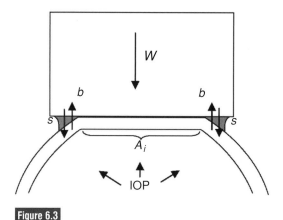

Figure 6.3

The forces acting during applanation tonometry, as seen in a cross-sectional view through the cornea and applanation head. See text for definition of symbols, except for A_i, which represents applanation area. IOP, intraocular pressure. The shaded area represents the tear film. Redrawn, based on Gloster [8], with kind permission from Elsevier.

The cornea is not a thin-walled shell. Therefore, there are bending stresses in the cornea that resist corneal deformation. We denote the net force acting on the applanation head due to these stresses as b.

A tear film is present on the corneal surface. When the applanation device contacts the cornea, a meniscus forms along the contact line. The effect of surface tension forces is to pull the tonometer head towards the cornea. We denote the net force acting on the applanation head due to these stresses as s.

A more accurate force balance on the applanation head is therefore (Fig. 6.3):

$$W + s = A \text{ IOP} + b \tag{6.2}$$

In general, determination of s and b would be very difficult. Luckily, experimental evidence indicates that if A is chosen correctly, s and b are equal in magnitude and thus cancel each other, so the Imbert–Fick law becomes valid. For normal human corneas, this cancellation occurs when the contact area is $7.35 \, \text{mm}^2$ (Fig. 6.4). It is therefore sufficient to ensure that $A = 7.35 \, \text{mm}^2$ when measuring IOP. How is this done in practice? A drop of fluorescein dye is placed on the cornea, which mixes with the tear film and causes the tear film to appear green when viewed under blue light. Tear fluid accumulates in the meniscus that forms along the edge of the contact surface between the cornea and tonometer head. To an observer, this line appears as a green circle, and to ensure $A = 7.35 \, \text{mm}^2$, the circle's diameter D must equal

$$D = \sqrt{\frac{4A}{\pi}} = 3.06 \, \text{mm} \tag{6.3}$$

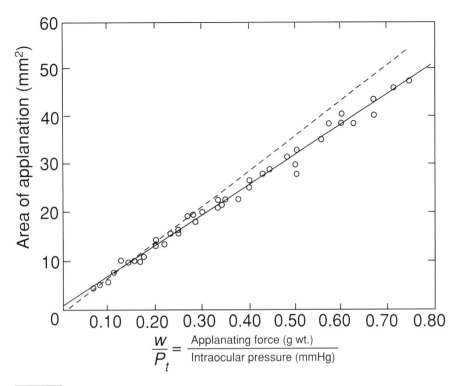

$$\frac{W}{P_t} = \frac{\text{Applanating force (g wt.)}}{\text{Intraocular pressure (mmHg)}}$$

Figure 6.4

Relationship between applied force, applanation area and intraocular pressure (IOP) for human corneas of thickness 550 μm. The dashed line shows the relationship expected from the Imbert–Fick law (Equation [6.1]); the solid line is the regression to the data, $A = 0.67 + 63.1$ (W/IOP), where A is in square millimeters, W is the applanation force in grams weight, and IOP is in millimeters of mercury. Note 1 gram weight is the weight on earth of a 1 gram mass, i.e., 981 dynes. Modified from Gloster [8], with kind permission from Elsevier.

The tonometer head is transparent and has built-in optics such that the top half of the image seen through the head is laterally displaced by 3.06 mm (Fig. 6.5). As seen by the ophthalmologist through the applanation head, the upper half of the fluorescein circle is displaced, as shown in Fig. 6.6. It is therefore sufficient to adjust the tonometer head position until points A and B in Fig. 6.6 touch, in which case the applanation contact area is 7.35 mm². The force required to cause this applanation area is recorded, and by the Imbert–Fick law, is directly proportional to IOP.

Of course, all of the above does not work if the cornea is stiffer or more compliant than normal, which occurs if corneal thickness is abnormal. Luckily, corneal thickness can be easily measured using a device called a *corneal pachymeter*, which employs ultrasound to find the positions of the anterior and posterior boundaries of the cornea. The normal mean thickness of the central cornea as measured by

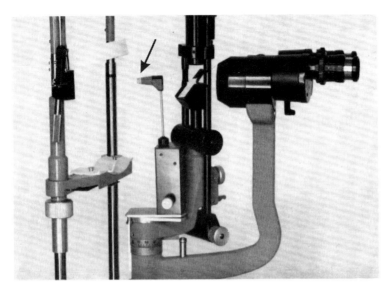

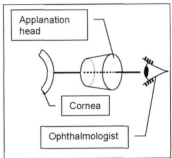

Figure 6.5

Goldmann applanation tonometer (arrow in left panel), mounted on a slit-lamp. The patient places his/her chin and forehead on the rests at left, and the tonometer head (center of figure) is advanced by the ophthalmologist using a joystick until the correct applanation contact area is obtained. The right panel shows the light path as the ophthalmologist looks through the tonometer head at the contact surface between cornea and applanation head. Modified from Gloster [8], with kind permission from Elsevier.

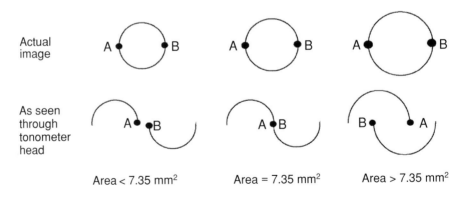

Figure 6.6

Fluorescein contact line as seen through the head of a Goldmann applanation tonometer. The top row shows the contact line as seen without optics. The bottom row shows the contact line as seen through the tonometer head's optics, which displace the upper half of the image laterally by 3.06 mm. When points A and B touch (middle), the actual image is a circle of diameter 3.06 mm (Area $= 7.35$ mm^2).

this technique is 544 μm (standard deviation, 34 μm) [9], and an IOP correction of 2.5–3.5 mmHg needs to be made for each 50 μm deviation of thickness away from this normal value [9,10]. A second problem occurs if the cornea contains significant scar tissue, for example from prior trauma. In such cases, other types of tonometry must be used to get a truer reading of IOP [5,11].

6.2.2   Drainage of aqueous humor in normal and glaucomatous eyes

In Section 6.1 we learned that the eye is pressurized by the continual production and drainage of aqueous humor. The flow of aqueous humor is slow indeed: it is produced at only 2.4 ± 0.6 μl/min (mean $\pm$ standard deviation; daytime measurements in adults aged 20–83 years) [12]. This corresponds to a turnover rate of about 1% of the anterior chamber volume per minute.[3] However, this slow flow is enough to keep the avascular tissues at the front of the eye alive and maintain a positive IOP of approximately 15 mmHg. This leads to several questions. How can such a slow flow generate so much pressure? What controls the pressure in healthy eyes so that it stays in a fairly narrow range? And most importantly, what goes wrong in glaucoma so that the pressure is elevated? The last question is a very important one: we know that elevated IOP is the main risk factor for glaucoma, and that lowering IOP helps to preserve vision [13]. In the vast majority of glaucomas, the elevation in IOP results from too much aqueous humor drainage resistance, typically owing to changes in the conventional drainage tissues. Let us therefore look more closely at the biomechanics of aqueous humor drainage from the eye.

Aqueous humor can exit the eye by two routes: the so-called conventional route and the uveo-scleral (or unconventional) route (Fig. 6.7, color plate). Uveo-scleral outflow normally carries only about 10% of total aqueous outflow [14] and is not thought to be the primary site of flow resistance in glaucoma, although it can act as a "safety valve" under the right conditions [15]. We will not consider it further.

Most aqueous humor drains through the conventional route, consisting of specialized tissues situated in the angle of the anterior chamber, which is located at the conjunction of the iris, cornea, and sclera (Fig. 6.8). Beginning at the anterior chamber and moving exteriorly, these tissues are the *trabecular meshwork*, a porous connective tissue; *Schlemm's canal*, a collecting duct lined by a vascular-like endothelium; and the *collector channels/aqueous veins*. After leaving the aqueous veins, the aqueous humor mixes with blood in the *episcleral veins*, eventually draining back to the right heart.

---

[3] To put this into context, the volume of a typical drop of water is approximately 20 μl, so it takes approximately 8–9 minutes to form one drop of aqueous humor.

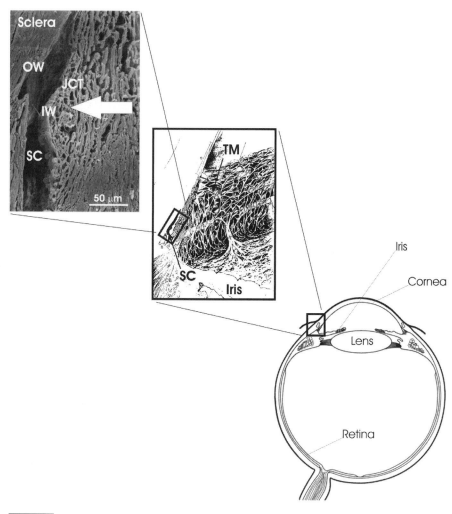

Figure 6.8

Overview of ocular anatomy, showing the position of the conventional aqueous humor drainage route (including Schlemm's canal) in the human eye. Most aqueous humor drains from the eye by passing through the trabecular meshwork. Not shown in this figure are collector channels, which emanate from Schlemm's canal and carry the aqueous humor to mix with blood in the scleral venous circulation. TM, trabecular meshwork: JCT, juxtacanalicular tissue; SC, Schlemm's canal, IW, inner wall, OW, outer wall; large arrow: direction of aqueous humor flow. Middle panel modified from Hogan *et al.* [16]; top panel is a scanning electron micrograph of a human sample cut in cross-section. Reprinted from Ethier *et al.* [17], with permission from Elsevier.

Direct pressure measurements [18,19] and circumstantial evidence [20] indicate that most of the flow resistance is in the trabecular meshwork or the endothelial lining of Schlemm's canal. Since the episcleral venous pressure is approximately 8–10 mmHg [21,22], the resistance of the conventional aqueous drainage tissues is approximately 3–4 mmHg/μl/min in normal eyes, and can reach triple this value

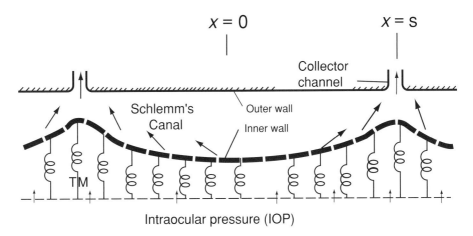

Figure 6.9

Simplified model of Schlemm's canal as a compliant channel. The trabecular meshwork (TM) is treated as a porous elastic body, as shown by the springs. Aqueous humor drains through the trabecular meshwork and the porous inner wall of Schlemm's canal, flows circumferentially in Schlemm's canal, and then leaves the canal by collector channels, here located at $x = \pm s$. From Johnson and Kamm [23]. This work is protected by copyright and is being used with the permission of *Access Copyright*. Any alteration of its content or further copying in any form whatsoever is strictly prohibited.

(or more) in glaucoma. Unfortunately, we do not understand how this resistance is generated, either in normal eyes or in glaucoma.

An early hypothesis about generation of resistance in glaucoma was that Schlemm's canal could collapse, choking off outflow. This possibility was considered in detail by Johnson and Kamm [23], who modeled Schlemm's canal as a compliant channel with a porous, elastic wall. Because the canal is highly elongated in cross-section, they treated the channel as two dimensional, that is as being formed by two parallel plates (Fig. 6.9). The upper plate, representing the side of Schlemm's canal that is bounded by the relatively rigid sclera, or *outer wall*, is immovable. The lower plate, representing the side of Schlemm's canal adjacent to the trabecular meshwork, or *inner wall*, is permeable and can deform as the trabecular meshwork stretches. This means that the local "height" of the canal is a function of position, x, and IOP.

The question is whether Schlemm's canal can collapse enough to create significant flow resistance. The answer to this question depends on a balance between two effects. The stiffness (elasticity) of the trabecular meshwork tends to keep the canal open, while the pressure drop across the trabecular meshwork and inner wall of Schlemm's canal tends to force the canal to close. Any analysis of this process must take these two effects into account. The first step in the analysis is to conserve mass. In the present case, that means that any fluid that enters the canal by crossing the inner wall must increase the local flow rate in the canal, $Q(x)$. The amount of fluid entering the canal depends on the IOP, the local pressure within the canal, $p(x)$, and the flow resistance of the trabecular meshwork and inner wall. More specifically:

$$\frac{dQ}{dx} = \frac{IOP - p(x)}{R_{iw}}$$ (6.4)

where IOP is intraocular pressure (assumed constant) and $1/R_{iw}$ is the hydraulic conductivity of the trabecular meshwork and inner wall, per unit length of inner wall, assumed to be constant. The right-hand side of Equation (6.4) represents the rate at which aqueous humor enters Schlemm's canal per unit length of the canal. For convenience, we take $x = 0$ as the midway point between two collector channels, and $x = \pm s$ as the locations of the nearest collector channel ostia (Fig. 6.9).

The next step is to relate the pressure in Schlemm's canal to the flow in the canal. Because the Reynolds number for flow in Schlemm's canal is $\ll 1$, we can assume the flow is everywhere unidirectional, so the pressure gradient in the canal can be obtained from the solution for laminar flow between parallel plates [24]:

$$\frac{dp}{dx} = 12 \frac{\mu Q(x)}{w h^3(x)}$$ (6.5)

where μ is aqueous humor viscosity, w is the depth of the canal into the plane of the page of Fig. 6.9, and $h(x)$ is the local "height" of the canal (i.e., the spacing between the inner and outer walls of the canal). The last step is to account for the elasticity of the trabecular meshwork. Johnson and Kamm [23] assumed that the trabecular meshwork acted like a linear spring, where local deformation was proportional to the pressure drop across the trabecular meshwork, $IOP - p(x)$. More specifically:

$$\frac{h_0 - h(x)}{h_0} = \frac{IOP - p(x)}{E}$$ (6.6)

where E is the spring stiffness and h_0 is the undeformed canal height, corresponding to the case where IOP equals the pressure in the collector channels.

Equations (6.4), (6.5), and (6.6) represent three equations for the unknowns $h(x)$, $p(x)$, and $Q(x)$. They can be combined to give a second-order non-linear differential equation (see Problem 6.4). We have to specify two pieces of boundary data to close the problem. They are that $Q(x) = 0$ at $x = 0$ and that $p(x) = p_{cc}$ at $x = \pm s$, where p_{cc} is the pressure in a collector channel. The resulting system is non-linear and a closed form solution is not known; therefore, Johnson and Kamm solved it numerically [23]. The result is that for reasonable values of the input parameters, it is predicted that there would be negligible flow resistance within the canal itself, except at extreme pressures ($> 50\,\text{mmHg}$) when the canal collapses. Furthermore, the model predicts that if Schlemm's canal were nearly collapsed, the resistance of the outflow system would be a very non-linear function of IOP, which is not observed experimentally [25]. The conclusion is that Schlemm's canal collapse does not seem to be important in the normal eye. Even when Schlemm's canal is largely collapsed at $50\,\text{mmHg}$, the outflow resistance is less than that seen in glaucomatous eyes, suggesting the glaucoma cannot be explained by collapse of Schlemm's canal.

Let us turn now to a more fundamental question. How does Schlemm's canal "know" how big to be? We learned in Section 2.9.1 that large arteries adjust their caliber in response to the amount of blood flowing in them or, more specifically, to the shear stress acting on their lining endothelial cells. Perhaps such a mechanism is also operating in Schlemm's canal. The difficulty with this hypothesis is that the flow rate of aqueous humor is so low that it seems likely that shear stresses on Schlemm's canal endothelial cells would be very small.

We can investigate this further by modifying the collapsible Schlemm's canal model presented above. Since we know that canal collapse only occurs at very high IOP, let us treat the canal as rigid (non-collapsible) but provide a more realistic representation of its geometry in order to improve the estimate of the shear stress on the endothelial cells lining the canal. In particular, instead of modeling the walls of Schlemm's canal as consisting of two parallel plates, we can treat the canal as having an elliptical cross-section, with semi-minor axis a and semi-major axis b (see inset, Fig. 6.10). In this case, Equation (6.5) is replaced by [26]:

$$-\frac{\mathrm{d}p}{\mathrm{d}x} = \frac{4\mu\,Q(x)(1+\varepsilon^2)}{\pi ab^3} \tag{6.7}$$

where ε is the eccentricity of the canal cross-section, $\varepsilon = a/b$. The governing equations are now (6.4) and (6.7), which we can combine to obtain a second-order equation for $Q(x)$. To this we add the boundary conditions $Q = 0$ at $x = 0$ and $Q(x) = Q_{\text{tot}}/2N$ at $x = \pm s$, where Q_{tot} is the total aqueous outflow rate and N is the number of collector channels. The solution is

$$Q(x) = \frac{Q_{\text{tot}}}{2N}\frac{\sinh(kx)}{\sinh(ks)} \quad \text{with} \quad k^2 = \frac{4\mu\left(1+\left(a/b\right)^2\right)}{\pi ab^3 R_{\text{iw}}} \tag{6.8}$$

From knowledge of the local flow rate and a relationship between wall shear stress and pressure gradient [17] we can compute the average shear stress in the canal as a function of canal dimensions (Fig. 6.10). Allingham *et al.* [27] measured

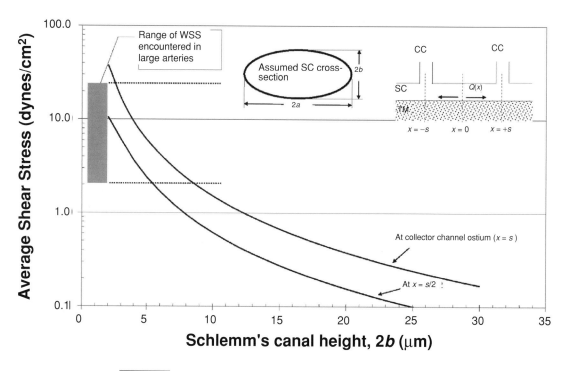

Figure 6.10

Plot of predicted shear stress acting on Schlemm's canal endothelial cells from flowing aqueous humor vs. Schlemm's canal "height", 2b (see inset). The vertical axis is the circumferentially averaged shear stress computed from the model described in the text. The insets at the top of the graph show the assumed cross-sectional shape of the canal and the terminology for computing the flow rate as a function of position in the canal, Q (x). TM, trabecular meshwork; SC, Schlemm's canal; CC, collector channel; WSS, wall shear stress. Modified from Ethier *et al.* [17], with kind permission of the authors and the Biophysical Society.

the size and cross-sectional area of Schlemm's canal, and from their data we can compute that the average inner–outer wall spacing is about 8 μm in normal eyes and 6 μm in glaucomatous eyes. The surprising conclusion is then that the shear stress is in the range 2–8 dynes/cm², not too dissimilar from that seen in large arteries. This suggests that shear stresses may have a biological effect on endothelial cells, which is supported by experimental findings that show that Schlemm's canal endothelial cells show preferential alignment [17]. It therefore seems probable that wall shear stress helps to regulate the size of Schlemm's canal, and likely has an effect on endothelial cell physiology. It is appealing to think that the cellular machinery for mechanotransduction and mechanosensing that work so well in the vascular system have simply been adapted by Schlemm's canal endothelial cells to control Schlemm's canal calibre.

What then can explain the relatively large flow resistance associated with the drainage of aqueous humor from the eye? There are two hypotheses.

1. It is known that the trabecular meshwork (or more specifically, the region adjacent to Schlemm's canal) contains high concentrations of proteoglycan-rich gels (see Section 2.4). Modeling [28] has shown that these gels could generate significant flow resistance, and recent data suggest that the turnover of these gel components is modulated by stretch-induced matrix metalloproteinase activity within the trabecular meshwork [29–31]. The hydraulic permeability of many other soft connective tissues with the body is controlled by the concentration of such proteoglycan gels [32].

2. The endothelial lining of Schlemm's canal may offer a significant barrier to flow. This cellular layer is unusual; for example, it has the highest permeability of any endothelium in the body [33], with $L_p \geqslant 4 \times 10^{-8}$ cm^2 s/g, yet it is non-fenestrated. The cells are joined by tight junctions that become more permeable as IOP increases [34] and are permeated by membrane-lined openings ("pores") that, although poorly understood, are almost certainly involved in aqueous humor transport [20].

Which of the above possibilities is correct is not yet known (see review by Ethier [20]), but the reality may lie somewhere in between. A model of the pores in the endothelial lining modulating the flow through a porous trabecular meshwork tissue [35] suggests that overall flow resistance may depend on an interaction between the endothelial pores and extracellular matrix.

6.2.3   Aqueous humor circulation in the anterior chamber

Before the aqueous humor drains out of the eye it passes through the pupil, traveling from the posterior to the anterior chamber, and then circulates in the anterior chamber. Aqueous circulation in the anterior chamber is the result of several stimuli, including blinking, accommodation (changing lens shape to alter focal length), and thermally induced natural convection. Natural convection is interesting and can have clinical implications. It occurs because the cornea is normally exposed to ambient air; consequently, the temperature at the posterior corneal surface is slightly less than body temperature. This creates a temperature gradient across the anterior chamber, so that cooler aqueous humor near the corneal surface falls and warmer aqueous humor near the iris rises. These convection patterns [36,37] tend to transport particles in vertical paths along the mid-peripheral cornea (Fig. 6.11). This phenomenon can be observed clinically in patients whose irises release abnormal amounts of pigment. Some of these pigment particles adhere to the posterior surface of the cornea, forming vertical "stripes" near the pupil known as *Krukenberg spindles*.

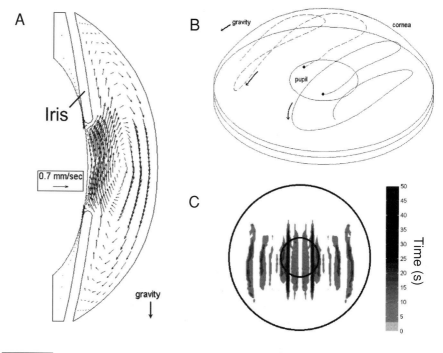

Figure 6.11

Aqueous humor circulation. (A) Computational modeling of aqueous humor circulation patterns in the anterior chamber shows fluid falling with gravity near the cornea and rising near the warmer iris. (B) This motion, together with net drainage of aqueous humor through the trabecular meshwork, leads to spiral particle paths. (C) Particles tend to travel in vertical stripes near the posterior corneal surface; here the residence times near the posterior surface of the cornea are shown. Modified from Heys and Barocas [37], with kind permission of Springer Science and Business Media.

There are several forms of glaucoma in which the elevated IOP is not a result of changes in the drainage system of the eye per se. These come under the heading of angle-closure glaucoma, a condition that occurs when the iris pivots forward and blocks access to the drainage structures in the angle of the anterior chamber. There appears to be an anatomic predisposition to this situation. The iris is extremely pliable [38], and modeling has shown interesting interactions between iris deformation and aqueous flow through the pupil and between the lens and the iris, especially when the eye is perturbed by blinking [39,40].

6.2.4 Optic nerve head biomechanics

Now that we understand IOP, and the fact that it is elevated in most forms of glaucoma, we turn our attention to the optic nerve. Recall that the retinal ganglion cell axons, responsible for carrying visual information from the retina to the brain,

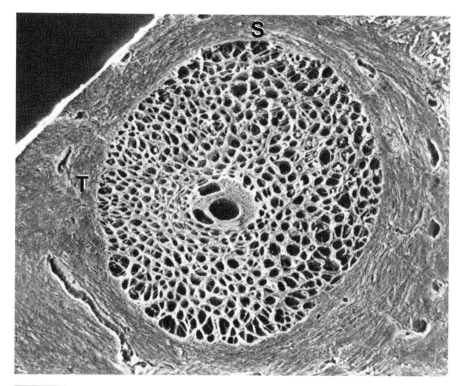

Figure 6.12

A scanning electron micrographic view of the lamina cribrosa from a normal human eye. The neural tissue has been digested away, leaving only the porous connective tissue of the laminar plates and the solid connective tissue of the sclera. After digestion, the sample was embedded and sectioned through the scleral optic nerve head. Adventitial tissue surrounding the central retinal vessels can also be seen in the center of the lamina. S, superior; T, temporal. Reprinted from Minckler [41], with permission from Elsevier.

converge from all over the retina to form the optic nerve. How does elevated IOP damage these retinal ganglion cell axons in glaucoma? We do not know the answer to this question, but there is strong evidence that a specialized tissue known as the *lamina cribrosa* plays an important role in the damage process. The lamina cribrosa is a porous connective tissue that spans the scleral canal (Fig. 6.12), mechanically supporting the retinal ganglion cells of the optic nerve as they leave the eye.

Why do we think the lamina cribrosa is important in glaucoma? To answer this question we need to know a little bit about the cellular physiology of neurons. These specialized cells can be subdivided into morphologically and functionally distinct regions, including the cell body and one or more elongated processes known as axons. The cell body contains the nucleus and is the site of protein and membrane synthesis, while the axons do not produce proteins [42]. How then can the cell supply its axons, which can be up to several meters in length, with proteins and other substances? As discussed briefly in Section 2.3, proteins and other materials are transported along axons in vesicles attached to motor proteins that "crawl" along the microtubules running within the axons. This process of axoplasmic transport is essential to maintaining the health of the axon. Early studies [43–46] demonstrated that blockage of this transport process ("axoplasmic blockade") occurs when IOP is chronically elevated, and furthermore that this blockade occurs at the level of the lamina cribrosa. Moreover, it is known that lamina cribrosa morphology is distorted in glaucoma, and that such changes can pre-date the development of vision loss [47,48]. Finally, the pattern of axon loss correlates with the density of connective tissue in the lamina cribrosa [7]. Such observations have led to much attention being focused on biomechanics of the lamina cribrosa, with the goal of understanding how elevated IOP leads to retinal ganglion cell damage.

Biomechanically, the lamina cribrosa and scleral canal are very interesting structures. The lamina cribrosa typically consists of approximately 10 cribiform plates, or lamellae, which contain collagen type IV, laminin, and elastin. Each plate is perforated by between 150 and 600 pores [49], through which the axonal bundles run. If we think of the eye as a pressurized spherical shell, then the scleral canal, which is no more than a hole in this vessel, is a site of local stress concentration. The lamina cribrosa, because it is a fairly compliant tissue spanning this canal, is expected to undergo large deformations and strains as the surrounding sclera deforms. These observations have led to the *mechanical theory of glaucomatous optic neuropathy*, which postulates that elevated mechanical stresses acting within the lamina cribrosa lead to axonal damage. Interestingly, this damage to axons may not be direct, but instead may be mediated through activation of type 1B astrocytes in the lamina cribrosa [50]. Astrocytes are a type of glial cell that function to provide support and guidance to neural cells. When glial cells become activated they proliferate, leading to a glial scar in a process known as *gliosis*. As this occurs, the activated astrocytes fail to provide trophic (i.e., nutritional and appropriate stimulatory) support to their surrounding neurons, triggering neuronal death.

There is a second theory about how retinal ganglion cells are damaged in glaucoma, called the *vasogenic theory*. It proposes that the glaucomatous insult results from insufficient vascular perfusion at the level of the lamina cribrosa, resulting in insufficient oxygen delivery (ischemic injury). Inadequate autoregulatory function in the branches of the short posterior ciliary arteries supplying the laminar region and complications in the hemodynamics of the surrounding vasculature could play a role in this process [51–54]. There is experimental evidence supporting both the mechanical and the vasogenic theories of glaucomatous damage, and it is probably the case that optic neuropathy results from a combination of both mechanisms. It is also possible that such effects could interact; for example, mechanical deformation of the lamina cribrosa could lead to ischemia via distortion of capillary beds. In any case, it should be clear from the above discussion that we need to better understand the biomechanics of optic nerve head tissues, and in particular the biomechanics of the lamina cribrosa, in order to gain insight into how glaucoma leads to visual function loss.

How can we evaluate the biomechanical environment within the lamina cribrosa? Unfortunately, the lamina cribrosa is small, relatively inaccessible, soft, and surrounded by a much stiffer tissue (sclera). This means that measurement of stress and strain in this tissue is challenging, to say the least! Most experiments have therefore just measured the deformation of the lamina in post mortem specimens, either by histological examination [55,56] or by other indirect techniques [57–59]. To determine mechanical strains, it is natural in such a situation to turn to modeling. This is not straightforward either, since the lamina is a porous, anisotropic, non-linear material with a complex geometry. However, you have to start somewhere, and it is sensible to begin by treating the lamina as a homogeneous, isotropic, linearly elastic circular plate of finite thickness subjected to a uniform pressure. Unlike the derivation presented in Section 2.8.2 for a thin elastic membrane subjected to pressure, bending stresses within the lamina cribrosa are important in supporting the pressure load and must be taken into account. Towards this end, He and Ren [60] considered the case of small deflections of a circular plate subject to tension from the surrounding sclera. Edwards and Good [61] followed a similar approach, allowing for large deflections but neglecting the tension applied by the peripheral sclera. Here we will briefly outline the approach of Edwards and Good, which is based on an energy argument arising from the classical mechanics of shells [62]. Throughout this derivation, we will exploit the fact that the thickness of the lamina cribrosa (120 to 450 μm, depending on the eye [56]) is small compared with the diameter of the lamina (typically 1.9 mm [63]).

The starting point is to write an expression for the strain energy stored in the circular disc representing the lamina cribrosa, for which there are two contributions: the strain energy of bending, V_{bending}, and the strain energy from stretching of the disc as it bows backwards, $V_{\text{stretching}}$. It can be shown [62] that for a symmetrically loaded disc of radius a, thickness h and Young's modulus E

$$V_{\text{bending}} = \pi D \int_0^a \left[\left(\frac{d^2 w}{dr^2} \right)^2 + \frac{1}{r^2} \left(\frac{dw}{dr} \right)^2 + \frac{2v}{r} \frac{dw}{dr} \frac{d^2 w}{dr^2} \right] r \, dr \qquad (6.9)$$

where $w(r)$ is the lateral deflection of the disc from the applied pressure (Fig. 6.13), v is Poisson's ratio, and D is the *flexural rigidity of the disc*, given by

$$D = \frac{Eh^2}{12(1 - v^2)}. \qquad (6.10)$$

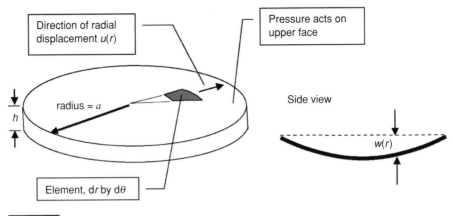

Direction of radial
displacement u(r)

Pressure acts on
upper face

radius = a

h

Side view

w(r)

Element, dr by dθ

Figure 6.13

Definition sketch for thin disc acted upon by a pressure on its upper face. The directions of the displacements u(r) and w(r) are shown. Note that the thickness of the disc, h, is exaggerated in this figure.

Equation (6.9) is not trivial to derive; in brief, it comes from considering a small element of the disc having dimensions dr by dθ, and noting that the strain energy of bending of this small element is proportional to the local curvature of the element multiplied by the internal bending moments acting on the element. For an elastic material, the internal bending moments are proportional to the local curvatures, so that the strain energy is proportional to quadratic terms involving the local curvature of the element. For a symmetrically loaded circular disc, there are two principal radii of curvature given by $(d^2w/dr^2)^{-1}$ and $[(1/r)(dw/dr)]^{-1}$, which appear in Equation (6.9).

The strain energy from stretching of this same small element is proportional to the product of the forces acting along the midplane of the disc and the strains in the disc. Again, for an elastic disc, the forces acting on the midplane are proportional to the strains, so that the quantity $V_{\text{stretching}}$ is quadratic in the strains [62]

$$V_{\text{stretching}} = \frac{\pi E h}{1 - v^2} \int_0^a \left[\varepsilon_r^3 + \varepsilon_\theta^2 + 2v\varepsilon_r\varepsilon_\theta \right] r \, dr \qquad (6.11)$$

where the two strain components ε_r and ε_θ are given by

$$\varepsilon_r = \frac{1}{2}\left(\frac{dw}{dr}\right)^2 + \frac{du}{dr} \qquad (6.12)$$

and

$$\varepsilon_\theta = \frac{u}{r} \qquad (6.13)$$

where $u(r)$ is the radial displacement of an element of the disc (Fig. 6.13).

If we knew $w(r)$ and $u(r)$ we could insert them into Equations (6.9) and (6.11) to compute the strain energy, but we do not. However, there is a creative approximate way to solve this problem, which is to make educated guesses for $w(r)$ and $u(r)$. Timoshenko [62] recommended expressions of the form:

$$w(r) = w_0 \left(1 - \frac{r^2}{a^2} \right)^2 \tag{6.14}$$

$$u(r) = r(a - r)(C_1 + C_2 r) \tag{6.15}$$

where C_1, C_2, and w_0 are constants, with w_0 representing the maximum deflection of the disc at its center. Notice that these expressions satisfy the requirements of no deflection at the edge of the disc and zero radial deflection at the center of the disc. How to determine C_1 and C_2? We can use the fact that at equilibrium the strain energy of the disc must be a minimum to write:

$$\frac{\partial}{\partial C_1}(V_{\text{bending}} + V_{\text{stretching}}) = \frac{\partial}{\partial C_1}(V_{\text{stretching}}) = 0 \tag{6.16}$$

$$\frac{\partial}{\partial C_2}(V_{\text{bending}} + V_{\text{stretching}}) = \frac{\partial}{\partial C_2}(V_{\text{stretching}}) = 0 \tag{6.17}$$

which gives expressions for C_1 and C_2 in terms of w_0 and disc properties.

The last step is to use the *principle of virtual work* to compute w_0 It is easiest to understand this principle by first considering a single particle, which we suppose is acted on by a set of forces $\mathbf{F}_1, \mathbf{F}_2, \ldots \mathbf{F}_N$. We further suppose that the particle is at equilibrium, and then assume that the particle undergoes a small movement (a *virtual displacement*), $\delta\mathbf{u}$. The virtual work done during this virtual displacement is

$$\delta\mathbf{u} \bullet \sum_{i=1}^{N} \mathbf{F}_i \tag{6.18}$$

where the virtual displacement is assumed to be small enough so that the forces are unaffected. Now it can be seen that the virtual work associated with an arbitrary displacement $\delta\mathbf{u}$ will be zero if and only if $\sum_{i=1}^{N} \mathbf{F}_i = \mathbf{0}$, which is the condition for the particle to be at equilibrium. In other words, requiring that the virtual work done on the particle be zero yields an equilibrium condition for the particle. For a collection of particles (e.g., an elastic solid) the virtual work done can be divided into that due to the external forces and that due to internal forces. The work done by internal forces is simply converted in stored elastic energy, and therefore the latter quantity is just the negative of the strain energy; consequently, in the absence of body forces, the equilibrium condition can be expressed as [64]

$$-\delta V + \int_{\Gamma} \sum_{i=1}^{N} \mathbf{F}_i \bullet \delta \mathbf{u} \, d\Gamma = 0 \qquad (6.19)$$

where the integral is taken over the bounding surface of the elastic solid, Γ, the $\mathbf{F}_i$ represent the external forces acting on the solid, and δV is the change in the strain energy owing to the virtual displacement $\delta \mathbf{u}$.

We can use this to compute the equilibrium configuration of a solid disc acted on by a uniform pressure q by noting that $\delta V = (\partial V / \partial w_0) \delta w_0$ and

$$\int_{\Gamma} \sum_{i=1}^{N} \mathbf{F}_i \bullet \delta \mathbf{u} \, d\Gamma = 2\pi \int_0^a q \, \delta w \, r \, dr = 2\pi \int_0^a q \left(1 - \frac{r^2}{a^2}\right)^2 \delta w_0 \, r \, dr$$

to write

$$\frac{\partial}{\partial w_0}(V_{\text{stretching}} + V_{\text{bending}}) = 2\pi \int_0^a q \left(1 - \frac{r^2}{a^2}\right)^2 r \, dr. \qquad (6.20)$$

The details are messy, but the net result is that an equation for w_0 as a function of the applied pressure is obtained. Edwards and Good [61] fitted the deflections of the lamina cribrosa predicted by this approach to the experimental data of Yan *et al.* [56], using the Young's modulus as a fitting parameter. Strain components throughout the lamina cribrosa can then be computed, for example by using Equations (6.12) and (6.13) for the normal strains. Edwards and Good estimated normal strains of approximately 11% and shearing strains of approximately 8% when IOP reached an elevated level of 50 mmHg [61].

The use of an analytic model for lamina cribrosa biomechanics is elegant but unfortunately requires us to make some rather restrictive assumptions, such as treating the lamina as a flat disc with no lateral tension imposed from the surrounding sclera. We can relax some of these assumptions by employing finite element modeling to study the mechanical behavior of the optic nerve head. Sigal *et al.* [65] used this approach with a variety of optic nerve head geometries (Fig. 6.14) to conclude that peak values of maximum principal strain at an IOP of 50 mmHg ranged from 5 to 8%, slightly less than but comparable to those predicted by Edwards and Good [61]. It is interesting that in in vitro experiments, strains of 5–8% induced a wide range of biological effects in neuronal cells [66–68], particularly if the strain was time varying [69]. This suggests that strains in the lamina cribrosa from elevated IOP could be damaging to the resident neural cells.

It is also noteworthy that the structure of the lamina cribrosa, where relatively stiff connective tissue elements are surrounded by fragile neural tissue, may act to effectively "amplify" strains above those computed here. For example, Bellezza

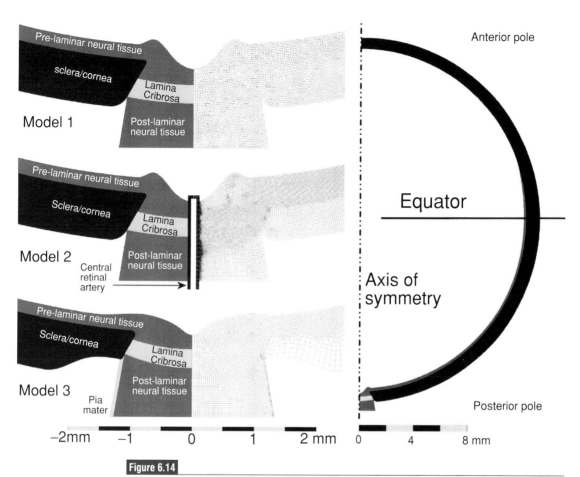

Figure 6.14

Finite element models used to study optic nerve head biomechanics in glaucoma. On the left are details of the optic nerve head region and a typical finite element mesh formed by eight-node quadrilaterals for three different optic nerve head region geometries. On the right is the full model, consisting of the eye globe with the optic nerve head tissues embedded in the sclera. From Sigal *et al.* [65]. This work is protected by copyright and it is being used with the permission of *Access Copyright*. Any alteration of its content or further copying in any form whatsoever is strictly prohibited.

et al. [70], by considering a simplified model of the lamina cribrosa consisting of a regular network of connective tissue "bridges" spanning an elliptical scleral canal, showed remarkable stress elevations in the lamina cribrosa bridges, in some cases more than 100 times the applied IOP.

6.3   Ocular blood flow

The anatomy of the circulatory system in the eye is complex to say the least. All blood is supplied to the eye globe, the lids, and the extraocular muscles by the ophthalmic artery, which is a branch of the internal carotid artery (Fig. 6.15, color plate). The eye globe itself has two main vascular systems: the *retinal* and *uveal*. The retinal circulation is a closed loop (arteries, arterioles, capillaries, venules, and veins) lying on the anterior surface of the retina (Fig. 6.16, color plate). The uveal circulation can be subdivided into two regions:

- the anterior uveal circulation supplies the iris, ciliary body (where it helps to form the aqueous humor), and the peripheral cornea
- the posterior uveal circulation feeds a specialized and quite remarkable vascular network called the *choroid*, which lies on the exterior surface of the retina (i.e., between the retina and the sclera; Fig. 6.1 color plate).

The net effect is that the retina is sandwiched between two vascular beds, receiving nourishment from both sides. Why should this complicated system be needed? It turns out that the body's scheme for detecting photons is very metabolically expensive; consequently, if there were capillary beds on only one side of the retina, the other side would be starved of oxygen. In fact, to supply the retina the choroid has the highest blood flow per perfused volume of any tissue in the body [73], with about 85% of total ocular blood flow passing through the choroid [74]. Even with this high perfusion rate, the oxygen tension in the retina falls drastically within the photoreceptor layer with distance from the choroid (Fig. 6.17).

Ensuring that enough oxygen gets to the retina is tricky: in addition to the normal difficulties of having a target tissue with time-varying oxygen needs, the entire blood supply system is in a container (the eye globe) whose internal pressure changes over time. This makes the physiology of blood flow in the eye somewhat unusual. In most of the circulatory system, flow through the capillary beds depends on vascular resistance and the pressure difference between the arterial and venous pressures, known as the *perfusion pressure*. However, in the eye, the effective perfusion pressure is the difference between the arterial pressure and IOP.

This behavior can be understood by introducing the concept of a *Starling resistor*, where fluid flows in a collapsible tube that is subject to an external pressure [76,77]. When the external pressure exceeds the pressure in the tube, the tube partially collapses until the stresses in the collapsed tube wall can support the transmural pressure difference. This collapse occurs first at the distal end of the tube, where the pressure is lowest. (Remember that the pressure decreases as we move along the tube because of viscous losses from flow in the tube.) The collapse process is very non-linear, so that a large change in vessel cross-sectional area occurs over a short distance and the pressure in the tube falls precipitously in the collapsed region. Upstream of the collapsed region, the tube pressure exceeds the external pressure, while downstream it is less than the external pressure. The collapsed region is a point of flow limitation, similar to what occurs at the throat of a nozzle in supersonic flow, or at a waterfall [77]. (When the collapsed tube is a blood vessel

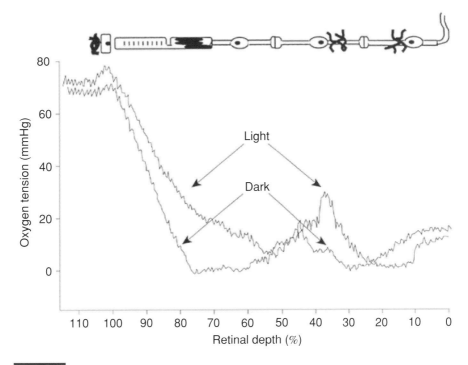

Figure 6.17

Intraretinal oxygen profiles measured across a cat retina using an oxygen electrode. The retina is shown schematically at the top, with the left edge of the figure corresponding to the outer retina and the right edge to the inner retina. The choroid lies immediately to the left of the schematic, and has an oxygen tension of approximately 70 mmHg. Notice the very steep oxygen gradient across the photoreceptor layer. Consumption of oxygen in the photoreceptors is actually higher in the dark, because of the so-called dark current, which explains the differences in the curves labeled "Light" and "Dark." Modified from Wangsa-Wirawan and Linsenmeier [75].

this is known as a "vascular waterfall.") In all of these flows, the flow is insensitive to the downstream (distal venous) pressure over a wide range of pressures; therefore, the effective driving pressure difference is the upstream pressure minus the external pressure.

In the eye, vessel collapse occurs in the veins as they pass from the pressurized interior of the eye into the sclera. Inside the ocular globe, venous pressure is greater than or equal to IOP, while in the sclera it is equal to episcleral venous pressure, typically 8–10 mmHg [21,22]. Therefore, even though the arterial pressure in the eye is somewhat lower than in the rest of the body (in the uvea, 75 mmHg systolic and 35 mmHg diastolic [78]), venous pressure in the interior of the eye is always above IOP. In fact, pressures in the capillaries of the choroid are typically 5–10 mmHg higher than IOP [79], and flow rate depends sensitively on IOP.

Since the effective perfusion pressure depends on IOP, the only way the perfusion to the retina can be kept constant as IOP changes is if the flow resistance of the retinal circulation changes. This process is known as *autoregulation*; as expected, the retinal circulation is autoregulated [74,80] so that it provides a nearly constant blood flow rate even as IOP increases up to 30 mmHg. If IOP exceeds this value, then the autoregulatory process can no longer compensate, and retinal blood supply is reduced.

It is worth remarking that collapsible tubes occur in other physiological situations. For example, collapse of the upper conducting airways as air flows through them plays a role in sleep apnea [81], snoring [82], and wheezing [83]. Veins above the heart can collapse, with the extreme case being flow in the giraffe jugular vein [84]. Blood flow through a severe stenosis, because of the Bernoulli effect, creates a very low pressure that can cause cyclic vessel collapse, which can contribute to fatigue failure of the stenotic plaque [85]. The urethra can collapse during urination, and the penetrating branches of the coronary arteries collapse during systole, when the cardiac muscle is contracting and intramuscular pressure is high.

6.4 Problems

6.1   The eye can be treated, to a first approximation, as a thin-walled elastic pressure vessel of diameter D and wall thickness t. Calculate the distensibility of the eye, $\beta = (1/V)(dV/dp)$, (compare with Equation [4.17]) as a function of D, t, and the Young's modulus of the sclera/cornea, E.

6.2   The aqueous humor circulates within the eye, flowing in at a constant rate $Q_{in} = 2$ μl/min, and draining from the eye at Q_{out}. At steady state, $Q_{out} = Q_{in}$. The eye also acts like an elastic vessel, in that its volume increases if the intraocular pressure p increases. This is expressed by

$$V = Cp \tag{6.21}$$

where V is the volume of the eye (normally about $4\,\text{cm}^3$) and C is the *compliance* of the eye (approximately 3 μl/mmHg). Finally, the outflow rate Q_{out} equals p/R, where R is the (effectively constant) resistance to outflow (about 4 mmHg min/μl).

(a)   Balancing mass, show that:

$$\frac{dp}{dt} + \frac{p - p_{ss}}{RC} = 0 \tag{6.22}$$

where p_{ss} is the steady-state pressure in the eye, equal to RQ_{in}.

(b)   If you poke your eye and increase the pressure by an amount $\delta p = 10$ mmHg, how long does it take the pressure to return to within 5% of its steady-state value?

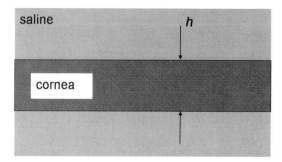

Figure 6.18

For Problem 6.3.

6.3 The cornea is a water-filled connective tissue that we will treat as being flat and of thickness h (Fig. 6.18). Because of the composition of the cornea, it traps positive ions, so that there are "excess" positive ions in the interior of the cornea compared with the surrounding fluid contacting the cornea. This is equivalent to the surface of the cornea acting like a semipermeable membrane that blocks the passage of positive ions.

   (a) When the cornea is completely dehydrated, its thickness is $h_{dry} = 220\,\mu m$, and the "excess" concentration of positive ions is 0.8 mM, compared with physiological saline. Assuming that no positive ions leave the tissue when it becomes hydrated, write an expression for the "excess" positive ion concentration as a function of corneal thickness, h.

   (b) As the cornea becomes more hydrated, it thickens and fibers in the cornea become stretched. This creates an effective positive pressure within the cornea, $p = k(h - h_o)$, with $h_o = 345\,\mu m$ and $k = 5.5$ Pa/μm. Compute the equilibrium thickness of the cornea when it is exposed to physiological saline at 37 °C and zero pressure (gauge). Note that the universal gas constant $R = 8.314$ J/(mol K). Be careful about units here: 1 mM is 10^{-3} mol/l.

6.4 In Section 6.2.1 we presented a model of Schlemm's canal as a compliant channel, as originally developed by Johnson and Kamm [23]. Show that the non-dimensional height of Schlemm's canal $\tilde{h}(x) = h(x)/h_0$, obeys the following equation:

$$\gamma^2(1 - \tilde{h}) = \tilde{h}^3 \frac{d^2\tilde{h}}{d\tilde{x}^2} + 3\tilde{h}^2 \left(\frac{d\tilde{h}}{d\tilde{x}}\right)^2 \tag{6.23}$$

where the parameter γ is given by:

$$\gamma^2 = \frac{12\mu s^2}{w h_0^3 R_{iw}} \tag{6.24}$$

and $\tilde{x} = x/s$ is the non-dimensional position in the canal. Physically, what does γ^2 represent? When γ is $\ll 1$ show that the above equation has a solution of the form $h^3(\mathrm{d}h/\mathrm{d}x) = constant$.

References

1. H. F. Krey and H. Bräuer. *Chibret Augenatlas: Eine Repetition für Ärtze mit Zeigetafeln Für Patienten* (Munich: Chibret Medical Service, 1998).
2. S. Karim, R. A. Clark, V. Poukens and J. L. Demer. Demonstration of systematic variation in human intraorbital optic nerve size by quantitative magnetic resonance imaging and histology. *Investigative Ophthalmology and Visual Science*, **45** (2004), 1047–1051.
3. J. L. Demer. The orbital pulley system: a revolution in concepts of orbital anatomy. *Annals of the New York Academy of Sciences*, **956** (2002), 17–32.
4. S. P. Bartels. Aqueous humor formation: fluid production by a sodium pump. In *The Glaucomas*, ed. R. Ritch, M. B. Shields and T. Krupin. (St. Louis, MO: Mosby, 1989), pp. 199–218.
5. E. M. Schottenstein. Intraocular pressure. In *The Glaucomas*, ed. R. Ritch, M. B. Shields and T. Krupin. (St. Louis, MO: Mosby, 1989), pp. 301–317.
6. H. A. Quigley. Number of people with glaucoma worldwide. *British Journal of Ophthalmology*, **80** (1996), 389–393.
7. H. A. Quigley. Neuronal death in glaucoma. *Progress in Retinal and Eye Research*, **18** (1999), 39–57.
8. J. Gloster. *Tonometry and Tonography* (London: J. & A. Churchill, 1966).
9. M. J. Doughty and M. L. Zaman. Human corneal thickness and its impact on intraocular pressure measures: a review and meta-analysis approach. *Survey of Ophthalmology*, **44** (2000), 367–408.
10. C. Y. Shih, J. S. Graff Zivin, S. L. Trokel and J. C. Tsai. Clinical significance of central corneal thickness in the management of glaucoma. *Archives of Ophthalmology*, **122** (2004), 1270–1275.
11. J. D. Brandt. Corneal thickness in glaucoma screening, diagnosis, and management. *Current Opinion in Ophthalmology*, **15** (2004), 85–89.
12. R. F. Brubaker. Measurement of aqueous flow by fluorophotometry. In *The Glaucomas*, ed. R. Ritch, M. B. Shields and T. Krupin. (St. Louis, MO: Mosby, 1989), pp. 337–344.

13. A. D. Beck. Review of recent publications of the Advanced Glaucoma Intervention Study. *Current Opinion in Ophthalmology*, **14** (2003), 83–85.

14. B. Becker and A. H. Neufeld. Pressure dependence of uveoscleral outflow. *Journal of Glaucoma*, **11** (2002), 545.

15. C. Linden and A. Alm. Prostaglandin analogues in the treatment of glaucoma. *Drugs and Aging*, **14** (1999), 387–398.

16. M. J. Hogan, J. A. Alvarado and J. E. Weddel. *Histology of the Human Eye* (Philadelphia, PA: W. B. Saunders, 1971).

17. C. R. Ethier, A. T. Read and D. Chan. Biomechanics of Schlemm's canal endothelial cells: influence on F-actin architecture. *Biophysical Journal*, **87** (2004), 2828–2837.

18. O. Mäepea and A. Bill. The pressures in the episcleral veins, schlemm's canal and the trabecular meshwork in monkeys: effects of changes in intraocular pressure. *Experimental Eye Research*, **49** (1989), 645–663.

19. O. Mäepea and A. Bill. Pressures in the juxtacanalicular tissue and schlemm's canal in monkeys. *Experimental Eye Research*, **54** (1992), 879–883.

20. C. R. Ethier. The inner wall of schlemm's canal. *Experimental Eye Research*, **74** (2002), 161–172.

21. R. F. Brubaker. Determination of episcleral venous pressure in the eye. A comparison of three methods. *Archives of Ophthalmology*, **77** (1967), 110–114.

22. C. D. Phelps and M. F. Armaly. Measurement of episcleral venous pressure. *American Journal of Ophthalmology*, **85** (1978), 35–42.

23. M. Johnson and R. D. Kamm. The role of schlemm's canal in aqueous outflow from the human eye. *Investigative Ophthalmology and Visual Science*, **24** (1983), 320–325.

24. F. M. White. *Viscous Fluid Flow*, 2nd edn (New York: McGraw-Hill, 1991).

25. R. F. Brubaker. The effect of intraocular pressure on conventional outflow resistance in the enucleated human eye. *Investigative Ophthalmology*, **14** (1975), 286–292.

26. R. K. Shah and A. L. London. *Laminar Flow Forced Convection in Ducts: A Source Book for Compact Heat Exchanger Analytical Data* (New York: Academic Press, 1978).

27. R. R. Allingham, A. W. de Kater and C. R. Ethier. Schlemm's canal and primary open angle glaucoma: correlation between schlemm's canal dimensions and outflow facility. *Experimental Eye Research*, **62** (1996), 101–109.

28. C. R. Ethier, R. D. Kamm, B. A. Palaszewski, M. C. Johnson and T. M. Richardson. Calculation of flow resistance in the juxtacanalicular meshwork. *Investigative Ophthalmology and Visual Science*, **27** (1986), 1741–1750.

29. J. P. Alexander, J. R. Samples and T. S. Acott. Growth factor and cytokine modulation of trabecular meshwork matrix metalloproteinase and TIMP expression. *Current Eye Research*, **17** (1998), 276–285.

30. J. M. Bradley, M. J. Kelley, X. Zhu, A. M. Anderssohn, J. P. Alexander *et al.* Effects of mechanical stretching on trabecular matrix metalloproteinases. *Investigative Ophthalmology and Visual Science*, **42** (2001), 1505–1513.

31. J. M. Bradley, J. Vranka, C. M. Colvis, D. M. Conger, J. P. Alexander *et al*. Effect of matrix metalloproteinases activity on outflow in perfused human organ culture. *Investigative Ophthalmology and Visual Science*, **39** (1998), 2649–2658.

32. J. R. Levick. Flow through interstitium and other fibrous matrices. *Quarterly Journal of Experimental Physiology*, **72** (1987), 409–437.

33. M. Johnson and K. Erickson. Mechanisms and routes of aqueous humor drainage. In *4: Principles and Practices of Ophthalmology*, ed. D. M. Albert and F. A. Jakobiec. (Philadelphia, PA: W. B. Saunders, 2000), pp. 2577–2595.

34. W. Ye, H. Gong, A. Sit, M. Johnson and T. F. Freddo. Interendothelial junctions in normal human schlemm's canal respond to changes in pressure. *Investigative Ophthalmology and Visual Science*, **38** (1997), 2460–2468.

35. M. Johnson, A. Shapiro, C. R. Ethier and R. D. Kamm. The modulation of outflow resistance by the pores of the inner wall endothelium. *Investigative Ophthalmology and Visual Science*, **33** (1992), 1670–1675.

36. C. R. Canning, M. J. Greaney, J. N. Dewynne and A. D. Fitt. Fluid flow in the anterior chamber of a human eye. *IMA Journal of Mathematics Applied in Medicine and Biology*, **19** (2002), 31–60.

37. J. J. Heys and V. H. Barocas. A Boussinesq model of natural convection in the human eye and the formation of krukenberg's spindle. *Annals of Biomedical Engineering*, **30** (2002), 392–401.

38. J. Heys and V. H. Barocas. Mechanical characterization of the bovine iris. *Journal of Biomechanics*, **32** (1999), 999–1003.

39. J. J. Heys and V. H. Barocas. Computational evaluation of the role of accommodation in pigmentary glaucoma. *Investigative Ophthalmology and Visual Science*, **43** (2002), 700–708.

40. J. J. Heys, V. H. Barocas and M. J. Taravella. Modeling passive mechanical interaction between aqueous humor and iris. *Journal of Biomechanical Engineering*, **123** (2001), 540–547.

41. D. S. Minckler. Histology of optic nerve damage in ocular hypertension and early glaucoma. *Survey of Ophthalmology*, **33**(Suppl) (1989), 401–402.

42. H. Lodish, A. Berk, S. L. Zipursky, P. Matsudaira, D. Baltimore *et al*. *Molecular Cell Biology*, 4th edn (New York: Freeman, 2000).

43. D. R. Anderson and A. Hendrickson. Effect of intraocular pressure on rapid axoplasmic transport in monkey optic nerve. *Investigative Ophthalmology*, **13** (1974), 771–783.

44. D. S. Minckler, A. H. Bunt and I. B. Klock. Radioautographic and cytochemical ultrastructural studies of axoplasmic transport in the monkey optic nerve head. *Investigative Ophthalmology*, **717** (1978), 33–50.

45. H. A. Quigley and D. R. Anderson. The dynamics and location of axonal transport blockade by acute intraocular pressure elevation in primate optic nerve. *Investigative Ophthalmology*, **15** (1976), 606–616.

46. H. A. Quigley, R. W. Flower, E. M. Addicks and D. S. McLeod. The mechanism of optic nerve damage in experimental acute intraocular pressure elevation. *Investigative Ophthalmology and Visual Science*, **19** (1980), 505–517.

47. L. A. Kerrigan-Baumrind, H. A. Quigley, M. E. Pease, D. F. Kerrigan and R. S. Mitchell. Number of ganglion cells in glaucoma eyes compared with threshold visual field tests in the same persons. *Investigative Ophthalmology and Visual Science*, **41** (2000), 741–748.

48. H. A. Quigley, E. M. Addicks, W. R. Green and A. E. Maumenee. Optic nerve damage in human glaucoma. II. The site of injury and susceptibility to damage. *Archives of Ophthalmology*, **99** (1981), 635–649.

49. H. A. Quigley and E. M. Addicks. Regional differences in the structure of the lamina cribrosa and their relation to glaucomatous optic nerve damage. *Archives of Ophthalmology*, **99** (1981), 137–143.

50. M. R. Hernandez. The optic nerve head in glaucoma: role of astrocytes in tissue remodeling. *Progress in Retinal and Eye Research*, **19** (2000), 297–321.

51. D. R. Anderson. Optic nerve blood flow. In *Optic Nerve in Glaucoma*, ed. S. M. Drance and D. R. Anderson. (New York: Kluger, 1995), pp. 311–331.

52. G. A. Cioffi. Vascular anatomy of the anterior optic nerve. In *Current Concepts on Ocular Blood Flow in Glaucoma*, ed. L. E. Pillunat, A. Harris, D. A. Anderson and E. L. Greve. (The Hague, Netherlands: Kugler, 1999), pp. 45–48.

53. S. S. Hayreh. Blood supply of the optic nerve head. A 'reality check.' In *Current Concepts on Ocular Blood Flow in Glaucoma*, ed. L. E. Pillunat, A. Harris, D. A. Anderson and E. L. Greve. (The Hague, Netherlands: Kugler, 1999), pp. 3–31.

54. E. M. van Buskirk and G. A. Cioffi. Microvasculature of the optic disc and glaucoma. In *Glaucoma: Decision Making in Therapy*, ed. M. G. Bucci. (Milan: Springer Verlag, 1996).

55. A. P. Nesterov and E. A. Egoriv. Pathological physiology of primary open angle glaucoma. In *Glaucoma*, ed. J. Cairns. (Miami, FL: Grune and Stratton, 1986), pp. 382–396.

56. D. B. Yan, A. Metheetrairut, F. M. Coloma, G. E. Trope, J. G. Heathcote *et al.* Deformation of the lamina cribrosa by elevated intraocular pressure. *British Journal of Ophthalmology*, **78** (1994), 643–648.

57. N. S. Levy and E. E. Crapps. Displacement of optic nerve head in response to short-term intraocular pressure elevation in human eyes. *Archives of Ophthalmology*, **102** (1984), 782–786.

58. N. S. Levy, E. E. Crapps and R. C. Bonney. Displacement of the optic nerve head. Response to acute intraocular pressure elevation in primate eyes. *Archives of Ophthalmology*, **99** (1981), 2166–2174.

59. D. B. Yan, J. G. Flanagan, T. Farra, G. E. Trope and C. R. Ethier. Study of regional deformation of the optic nerve head using scanning laser tomography. *Current Eye Research*, **17** (1998), 903–916.

60. D. Q. He and Z. Q. Ren. A biomathematical model for pressure-dependent lamina cribrosa behavior. *Journal of Biomechanics*, **32** (1999), 579–584.

61. M. E. Edwards and T. A. Good. Use of a mathematical model to estimate stress and strain during elevated pressure induced lamina cribrosa deformation. *Current Eye Research*, **23** (2001), 215–225.

62. S. Timoshenko and S. Woinowsky-Krieger. *Theory of Plates and Shells* (New York: McGraw-Hill, 1959).

63. J. B. Jonas, C. Y. Mardin, U. Schlotzer-Schrehardt and G. O. Naumann. Morphometry of the human lamina cribrosa surface. *Investigative Ophthalmology and Visual Science*, **32** (1991), 401–405.

64. S. Timoshenko and J. N. Goodier. *Theory of Elasticity* (New York: McGraw-Hill, 1970).

65. I. A. Sigal, J. G. Flanagan, I. Tertinegg and C. R. Ethier. Finite element modeling of optic nerve head biomechanics. *Investigative Ophthalmology and Visual Science*, **45** (2004), 4378–4387.

66. M. E. Edwards, S. S. Wang and T. A. Good. Role of viscoelastic properties of differentiated SH-SY5Y human neuroblastoma cells in cyclic shear stress injury. *Biotechnology Progress*, **17** (2001), 760–767.

67. S. S. Margulies and L. E. Thibault. A proposed tolerance criterion for diffuse axonal injury in man. *Journal of Biomechanics*, **25** (1992), 917–923.

68. B. Morrison III, H. L. Cater, C. B. Wang, F. C. Thomas, C. T. Hung *et al*. A tissue level tolerance criteria for living brain developed with an in vitro model of traumatic mechanical loading. *Stapp Car Crash Journal*, **47** (2003), 93–105.

69. D. H. Triyoso and T. A. Good. Pulsatile shear stress leads to DNA fragmentation in human SH-SY5Y neuroblastoma cell line. *Journal of Physiology*, **515**(Pt 2) (1999), 355–365.

70. A. J. Bellezza, R. T. Hart and C. F. Burgoyne. The optic nerve head as a biomechanical structure: initial finite element modeling. *Investigative Ophthalmology and Visual Science*, **41** (2000), 2991–3000.

71. H. Gray. *Anatomy of the Human Body*, 20th edn: thoroughly reviewed and re-edited by W. H. Lewis (Philadelphia, PA: Lea and Febiger, 1918).

72. H. Kolb, E. Fernandez and R. Nelson. *Webvision: The Organization of the Retina and Visual System.* Available at http://webvision.med.utah.edu/sretina.html (2005).

73. R. Collins and T. J. van der Werff. *Mathematical Models of the Dynamics of the Human Eye*, No. 34 of *Lecture Notes in Biomathematics* (Berlin: Springer Verlag, 1980).

74. A. Bill. Blood circulation and fluid dynamics in the eye. *Physiological Reviews*, **55** (1975), 383–416.

75. N. D. Wangsa-Wirawan and R. A. Linsenmeier. Retinal oxygen: fundamental and clinical aspects. *Archives of Ophthalmology*, **121** (2003), 547–557.

76. R. D. Kamm. Flow through collapsible tubes. In *Handbook of Bioengineering*, ed. R. Skalak and S. Chien. (New York: McGraw-Hill, 1987), pp. 23.1–23.19.

77. A. H. Shapiro. Steady flow in collapsible tubes. *Journal of Biomechanical Engineering*, **99** (1977), 126–147.

78. D. F. Cole. Aqueous humour formation. *Documenta Ophthalmologica*, **21** (1966), 116–238.

79. O. Mäepea. Pressures in the anterior ciliary arteries, choroidal veins and choriocapillaris. *Experimental Eye Research*, **54** (1992), 731–736.

80. J. E. Grunwald, B. L. Petrig and F. Robinson. Retinal blood flow autoregulation in response to an acute increase in blood pressure. *Investigative Ophthalmology and Visual Science*, **27** (1986), 1706–1712.

81. C. B. Stalford. Update for nurse anesthetists. The Starling resistor: a model for explaining and treating obstructive sleep apnea. *AANA Journal*, **72** (2004), 133–138.

82. L. Huang, S. J. Quinn, P. D. M. Ellis and J. E. Ffowcs Williams. Biomechanics of snoring. *Endeavour*, **19** (1995), 96–100.

83. N. Gavriely, K. B. Kelly, J. B. Grotberg and S. H. Loring. Forced expiratory wheezes are a manifestation of airway flow limitation. *Journal of Applied Physiology*, **62** (1987), 2398–2403.

84. B. S. Brook and T. J. Pedley. A model for time-dependent flow in (giraffe jugular) veins: uniform tube properties. *Journal of Biomechanics*, **35** (2002), 95–107.

85. M. Bathe and R. D. Kamm. A fluid–structure interaction finite element analysis of pulsatile blood flow through a compliant stenotic artery. *Journal of Biomechanical Engineering*, **121** (1999), 361–369.

7 The respiratory system

The function of the respiratory system is to exchange O_2 and CO_2 with the blood. To understand this system from a bioengineering viewpoint, we will first discuss the gross anatomy of the lungs and their associated structures, and then discuss the mechanics of breathing.

7.1 Gross anatomy

We divide the respiratory system into two subsystems: the *conducting airways* and the *associated structures*.

7.1.1 The conducting airways and pulmonary vasculature

The conducting airways form a fantastically complex branching tree designed to transport air efficiently into the *alveoli*, the smallest air-filled structures in the lung where blood/gas exchange takes place. Air enters through the mouth or nose then passes through (in order): the *pharynx* (the throat), the *larynx* (the voice box), and the *trachea* (the large tube passing down the neck). The trachea splits to form two *bronchi* (singular: *bronchus*), each of which feed air to one of the lungs (Fig. 7.1, color plate).

Each bronchus splits to form *bronchioles*, which, in turn, split to form smaller bronchioles, and so on (Fig. 7.2). After about 16 levels of branching, we reach the *terminal bronchioles*, which are the smallest structures that have a purely air-conducting function, that is, in which essentially no blood/gas exchange takes place (Fig. 7.3). In adult lungs, the structures distal to the terminal bronchioles consist of several generations of *respiratory bronchioles*, *alveolar ducts* and *alveolar sacs*, which collectively are known as the *acinus*; this is where the gas exchange occurs (Fig. 7.4). All of the conducting airways are lined by *pulmonary epithelial cells*, which we will see play a number of important roles in ensuring the proper function of the lungs.

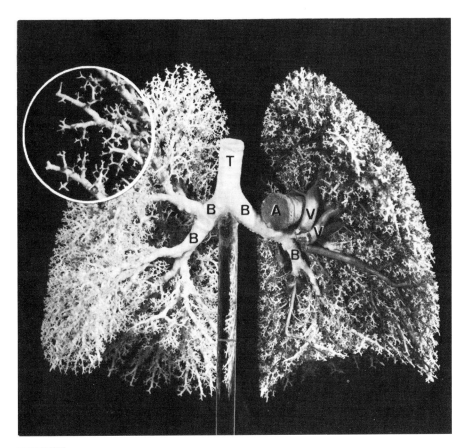

Figure 7.2

A resin cast of the airways of a pair of human lungs shows the fantastically complex geometry of the branching airways, beginning from the trachea (T) and moving down into the bronchial tree (B). Note the branching of the peripheral airways in the higher-magnification view in the inset. In the left lung, the arteries (A) and veins (V) have been filled with a slightly darker resin. Reprinted by permission of the publisher from *The Pathway for Oxygen: Structure and Function in the Mammalian Respiratory System* by E. R. Weibel, p. 273, Cambridge, MA: Harvard University Press [2]. Copyright © 1984 by the President and Fellows of Harvard College.

The alveoli (singular: *alveolus*) form after birth, by a fascinating process in which the walls of the smallest air-containing passages in the lungs become scalloped so as to enlarge their surface area. The number of alveoli gradually increases with time until about the age of six, after which the lungs are essentially fully formed (Fig. 7.5). As can be imagined, it is not easy to know exactly how many alveoli there are. Early estimates [6] suggested that there are about 300 million alveoli, with radii ranging from 75 to 300 μm, and mean radius of approximately 150 μm. More recent estimates [7] have suggested that the average number of alveoli is 480 million, with an average radius of about 100 μm. The total surface area of

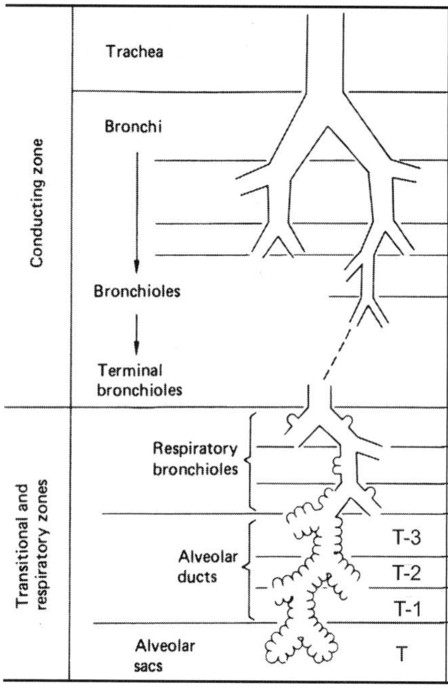

		z	Name	D (mm)	A (cm²)	Number
Conducting zone	Trachea	0	Trachea	18	2.54	1
	Bronchi	1	Mainstem bronchi	12.2	2.33	2
		2	Lobar bronchi	8.3	2.13	4
		3	Segmental bronchi	5.6	2.00	8
	Bronchioles	4	Subsegmental bronchi	4.5	2.48	16
		... 12	Bronchiole	0.95	28.8	4,096
	Terminal bronchioles	13...	Bronchiole	0.82	44.5	8,192
		16	Terminal bronchiole	0.49	225.0	65,536
Transitional and respiratory zones	Respiratory bronchioles	17	Respiratory bronchiole 1	0.40	300.0	131,072
		18	Respiratory bronchiole 2	0.38	543.0	262,144
		19	Respiratory bronchiole 3	0.36	978.0	524,288
	Alveolar ducts	T-3 20	Alveolar duct 1	0.34	1,743.0	1,048,576
		T-2 21	Alveolar duct 2	0.31	2733.0	2,097,152
		T-1 22	Alveolar duct 3	0.29	5070.0	4,194,304
	Alveolar sacs	T 23	Alveoli	0.25	7530.0	8,388,608

Figure 7.3

Schematic diagram of the organization of the airway tree, showing the different functional zones ("conducting zone," "transitional zone," and "respiratory zone") as a function of generation numbers, z. The diameter (D), cross-sectional area (A) and number of airways are shown in the colums on the right. The number of airways is actually larger than that listed here, since the tabulated values are based on a simple bifurcating model of the airways; the reality is more complex. T refers to the terminal generation, T-1 to the generation immediately preceding the terminal generation, etc. The dimensions are suitable for a lung that is inflated to three quarters of capacity. The figure at left is from Weibel [3], supplemented at right by entries from [2]. Reproduced with permission of Lippincott Williams & Wilkins.

the alveolar walls has been quoted to be as large as 140 m² [2], which is about 75 times the body's external surface area.

In addition to the conducting airways for air transport, it is necessary to have a system for delivering the blood to the alveoli so that mass transfer can take place. The lungs are supplied by the pulmonary arteries and drained by the pulmonary veins. The smallest units of the pulmonary vasculature are the pulmonary capillaries, which run as closely packed units inside the alveolar walls. The net effect is that a thin "sheet" of blood flows within much of the alveolar wall (Fig. 7.4). Interposed between the blood and the air is a thin layer of tissue, consisting of capillary endothelial cells, basement membrane, and airway epithelial cells. The total thickness of this tissue varies, with an arithmetic mean thickness of 2.22 μm and a harmonic (geometric) mean thickness of 0.62 μm [2]. The combination of an enormous alveolar surface area, a comparably large pulmonary capillary surface

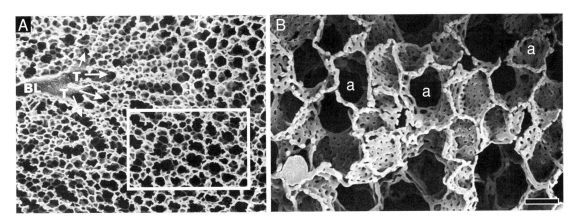

Figure 7.4

Overview of lung microanatomy. (A) Scanning electron micrograph of lung tissue, showing how a peripheral bronchiloe (BL) branches into terminal bronchioles (T), which, in turn, divide to produce respiratory bronchioles and alveolar ducts (arrows). This gazelle lung was fixed in such a way as to prevent collapse of the alveoli and terminal segments of the bronchial tree. (B) A plastic cast of the alveolar capillary network within a small group of alveoli (a), photographed after the surrounding tissue was dissolved away. The two images are not taken from the same sample or even from the same species; the white box in (A) approximately corresponds to the field of view in (B) and is simply included to give a sense of relative scale. Scale bar in panel (B) is 50 μm. (A) is reproduced from Fawcett [4] and Gehr *et al. Journal of Morphology* **168**:5-15, Copyright © 1981, and is used with permission of Wiley-Liss, Inc., a subsidiary of John Wiley & Sons, Inc. (B) is Copyright © 1990 from *Electron Microscopy of the Lung* by Schraufnagel [5] and is reproduced by permission of Routledge/Taylor & Francis Group, LLC.

area, and the very thin tissue layer between blood and air make the lung a *very* effective mass transfer device.

7.1.2 Associated structures

As well as the conducting airways, the pulmonary system includes a number of other tissues (*the associated structures*) that together are responsible for the inspiration and expiration of air. The lungs are contained within the *thoracic cage*, consisting of the *ribs, intercostal muscles*, *sternum* (breastbone), *spine, diaphragm*, and *neck and shoulder muscles*.[1] These together form a deformable "container" for the lungs. Attached to the exterior surface of the lung and to the interior surface of the thoracic cage is the *pleural membrane*. The space enclosed by this membrane is the *pleural space*, which is normally filled with the *intrapleural fluid* (Fig. 7.6). Each lung has its own pleura.

---

[1] The intercostal muscles are located between the ribs. The *diaphragm* is the large muscle at the base of the lungs, familiar to anyone who has been "winded."

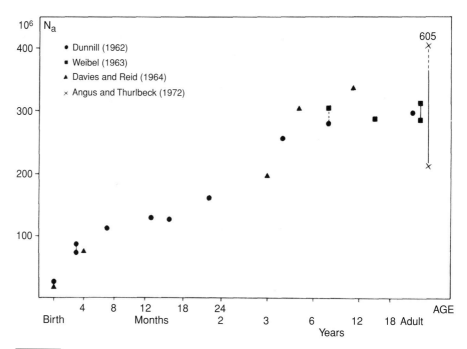

Figure 7.5

Increase in the number of alveoli with age in human lungs. There is a gradually increasing trend until about age six, after which the number of alveoli is essentially constant. The symbols show the results from different studies. The original data sources can be found in Weibel [2]. Reprinted by permission of the publisher from *The Pathway for Oxygen: Structure and Function in the Mammalian Respiratory System* by Ewald R. Weibel, p. 227, Cambridge, MA: Harvard University Press. Copyright © 1984 by the President and Fellows of Harvard College.

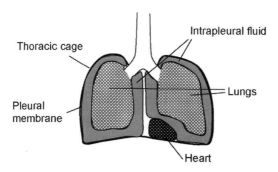

Figure 7.6

Schematic of the thoracic cage, pleural membrane, intrapleural fluid, and lungs. This is not to scale; the intrapleural fluid actually forms a very thin layer. Note that there is no communication between the right and left intrapleural fluids. After Vander *et al.* [1]. With kind permission of The McGraw-Hill Companies.

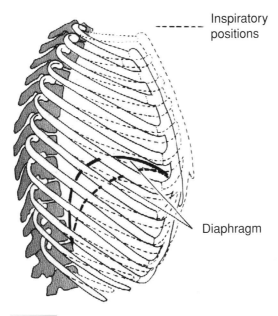

Inspiratory positions

Diaphragm

Figure 7.7

Chest wall and diaphragm motion during inspiration. The solid and dashed lines indicate expiratory and inspiratory positions, respectively. From Vander *et al.* [1]. With kind permission of The McGraw-Hill Companies.

7.2 Biomechanics of breathing

An interesting aspect of the breathing process is the fact that there are no muscles attached to the lungs to aid with inflation and deflation. Rather, the necessary forces are transmitted to the lungs through the intrapleural fluid via the mechanisms described below. This system allows a uniform inflation of the lungs while avoiding the high stresses and deformations that would result if a muscle inserted into the soft lung tissue.

Inspiration and expiration proceed by two very different mechanisms, as described below.

Inspiration. In *quiet inspiration*, the diaphragm contracts and moves downward, thereby increasing the volume of the thoracic cavity. In addition, the external intercostal muscles and the scalene muscles (at the neck) contract to move the rib cage up and out (Fig. 7.7). This also increases thoracic cavity volume, although the contribution of the intercostal and scalene muscles is minimal during quiet inspiration. Since the intrapleural fluid is essentially incompressible, an increase in thoracic cage volume must be accompanied by an increase in lung volume. The best analogy is that of a balloon inside a fluid-filled container of variable volume (e.g., a piston and cylinder). As the piston moves back and forth, the balloon's volume shrinks and grows. In *forced inspiration* the above description still applies. However, the contribution of the intercostal and scalene muscles becomes more pronounced.

Expiration. *Quiet expiration* is a passive process that relies on the natural tendency of the lung to collapse (see Section 7.3 for detailed discussion). The diaphragm, the external intercostal muscles, and the scalene muscles relax, allowing the lung and the thoracic cage to decrease in volume. *Forced expiration* is an active process. In addition to the relaxations of quiet expiration, the internal intercostal muscles contract to pivot the ribs downward and in. Also, the abdominal muscles contract, pulling down the rib cage, increasing abdominal pressure, and thereby forcing the diaphragm upwards. These mechanisms together act to reduce thoracic cavity volume, which allows a large quantity of air to be quickly expelled from the lungs.

7.3 Lung elasticity and surface tension effects

It was previously stated that the natural tendency of the lung is to collapse. Before examining why this is the case, we briefly look at what this fact implies about the intrapleural pressure.

At rest, when the subject is neither inspiring nor expiring, the pressure within the conducting airways is essentially atmospheric.[2] To counterbalance the lung's tendency to collapse, it is clear that the lung must be "blown up". In other words, the pressure in the lung must exceed the surrounding pressure, which in this case is the intrapleural pressure. Therefore, at rest, intrapleural pressure must be slightly negative. This is confirmed by measurements, with typical intrapleural pressures of -3 to -4 mmHg (gauge) at rest. During inspiration and expiration, intrapleural pressures deviate from this resting value. For example, during rapid inspiration, intrapleural pressure can fall to as low as -100 mmHg, while it can become slightly positive during forced expiration.

If the pleural cavity is opened to the atmosphere (for example in a puncture wound), then the pressure difference required to inflate the lung is lost. This causes lung collapse, or *pneumothorax*.

We now turn to the question of *why* the lung tends to collapse. There are two reasons:

- Lung tissue (or parenchyma) is naturally elastic. It contains a very high proportion of elastin (see Section 2.4). Under normal conditions, the lung is expanded

---

[2] During inspiration, alveolar pressure is slightly negative (about -1.5 mmHg) to draw air into the alveoli, and it is slightly positive during expiration.

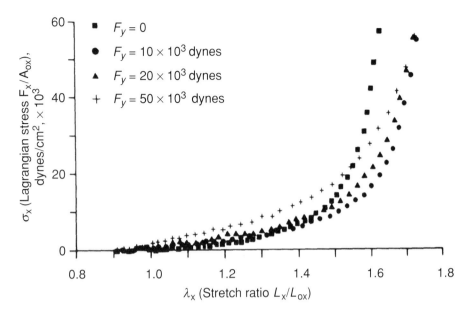

Figure 7.8

Stress–strain curve for lung parenchyma under biaxial loading conditions. For constant transverse loads (differing F_y values in dynes), the stress is plotted as a function of stretch in the x direction. Note the significant increase in tangent modulus at stretch ratios above approximately 1.5. From Vawter and Humphrey [8]. With kind permission of The McGraw-Hill Companies.

(inflated) so the lung parenchyma is under tension. The elastic restoring force from this tension acts to decrease lung volume. It is interesting to note that lung parenchyma is a strain-stiffening material (i.e., its incremental Young's modulus increases with strain, as shown in Fig. 7.8). Such behavior is typical of connective tissues and reflects the arrangement and orientation of stress-bearing fibers within the tissue (see Section 9.10.3). At low stresses, the collagen and elastin within the lung, which are the main stress-bearing materials, are randomly oriented. At high stresses they align along the direction of stretching (Fig. 7.9). Additionally, at high strain, the individual collagen and elastin fibers stiffen, because of their molecular structure, which resembles that of a helical spring (see Section 9.10.1). The net effect is that the lung becomes significantly *stiffer* as it is inflated.

• Surface tension acts to collapse the lung. To see why this is the case, consider a single alveolus filled with air. In order to maintain the viability of cells in the alveolar wall, the inner surface of the alveolus must be coated with a thin fluid layer. This creates an air/fluid interface within the alveolus, as shown in Fig. 7.10. Whenever an interface between two immiscible phases forms, energy is required to maintain that interface. The system attempts to minimize

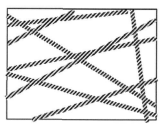

Undeformed state: low strain and low incremental Young's modulus

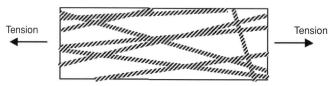

Tension

Tension

Deformed state: large strain and large incremental Young's modulus

Figure 7.9

Schematic diagram of collagen and elastin orientation in connective tissue. As fibers align in the direction of stretching, the tissue stiffens.

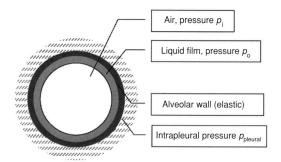

Air, pressure p_i

Liquid film, pressure p_o

Alveolar wall (elastic)

Intrapleural pressure $p_{pleural}$

Figure 7.10

A single alveolus, showing pressures in surrounding tissue, in the air, and within the liquid film lining the alveolar wall (not to scale). See text for definition of symbols.

this energy by reducing the interfacial area. This gives rise to an effective force that acts along the interface, called the *surface tension*. It is characterized by a surface tension coefficient σ, the value of which depends on the two contacting phases. In the single alveolus, a reduction in interfacial area can only be accomplished by collapsing the alveolus. To counteract this tendency, the pressure of the alveolar air (p_i) must exceed the pressure on the other side of the interface, namely the liquid film pressure p_o (see Box 7.1).

Box 7.1 Analogy of an air bubble in liquid

The situation in an alveolus is directly analogous to the case of an air bubble in an infinite volume of liquid. If the interfacial (surface) tension coefficient for the air/liquid interface is σ, then at equilibrium Laplace's law states that, for a spherical bubble of radius R, the difference between internal air pressure p_i and the pressure in the surrounding fluid, p_o, is given by

$$p_i - p_o = \frac{2\sigma}{R} \tag{7.1}$$

Using Laplace's law it is trivial to calculate the pressure difference $p_i - p_o$ in a single alveolus. Using Equation (7.1) with numerical values $R = 150\ \mu m = 1.5 \times 10^{-2}$ cm and $\sigma = 72$ dynes/cm (the latter value being appropriate for an air/water interface) yields $p_i - p_o = 9600$ dynes/cm^2, or 7.2 mmHg. At rest, when p_o is atmospheric, this result implies a liquid film pressure of -7.2 mmHg.

These data can be used to make an inference about the intrapleural pressure, $p_{pleural}$. Because of the elasticity of the lung parenchyma and alveolar wall tissue, the liquid film pressure must be *greater than* pleural pressure. Hence, we predict that $p_{pleural} < -7.2$ mmHg. However, we have stated that the measured resting value of $p_{pleural}$ is -3 to -4 mmHg, which contradicts the above calculation.

The resolution of this apparent contradiction is that we have used the wrong value of σ in Equation (7.1). The alveoli are in part lined by type II epithelial cells, which secrete a surfactant that mixes with the alveolar liquid film and greatly decreases σ for the resulting air/liquid interface. This, in turn, increases $p_{pleural}$, effectively making it easier to overcome surface tension and so to breathe. The value of σ for saline plus surfactant in contact with air has been measured (Figs. 7.11 and 7.12). The data show that the volume of σ depends on the extension of the air/liquid interface. Specifically, σ is small at low relative interfacial areas and is large at high relative interfacial areas. This means that σ is lower at the start of an inspiratory cycle than at the end. Since large values of σ imply that more negative intrapleural pressures are required for inspiration, it follows that it is easier to inspire near the beginning of an inspiratory cycle than at the end. Essentially, surface tension acts in a non-linear fashion to stiffen the lung near the end of inspiration.

A notable feature of the surface tension data is the hysteresis present: at a given relative interfacial area, the surface tension is higher when the interfacial area is increasing than when it is decreasing. This means that surface tension makes it more difficult to breathe in than to breathe out.

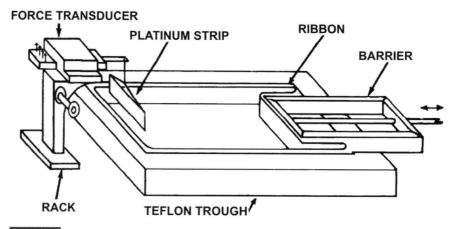

FORCE TRANSDUCER

PLATINUM STRIP

RIBBON

BARRIER

RACK

TEFLON TROUGH

Surface balance apparatus used to measure surface tension by Clements. The shallow tray is filled with saline, and the fluid to be tested is layered on top of the saline. A thin platinum strip (0.001 inch thick) is then suspended in the fluid from the arm of a very sensitive strain gauge-based force transducer. The force exerted by the fluid on the platinum strip is measured, from which the surface tension can be computed. To change the surface area of the tested fluid, a barrier can be moved along the trough. When this is done cyclically while measuring the surface tension continuously, any hysteresis effects in the surface tension versus area behavior are readily detected. Modified from Clements [9] with permission of the American Physiological Society.

The relative contributions of lung parenchymal elasticity and surface tension can be seen in Fig. 7.13, in which lungs are "inflated" either with physiological saline or air. A number of comments about these data are in order:

- When the lung is inflated with saline, there are no surface tension effects and therefore the curve labeled "2" represents the effects of parenchymal elasticity alone. The stiffening of lung parenchyma at high volumes (high strains) is evident. Also note that some hysteresis is present, as a result of viscoelastic effects in the parenchyma.
- The difference between the curves labeled "2" and "1" indicates the importance of surface tension effects. Significantly greater pressure is required to inflate the lung when surface tension effects are present. A great deal of hysteresis is evident, as expected based on the hysteresis in the value of σ shown in Fig. 7.12.

Clinically, surfactant is very important. It makes breathing much easier (less pressure difference needed to inflate the lung), so that the diaphragm and intercostal muscles can do less work with every inspiration. Premature infants are often born with reduced surfactant-producing abilities, since the lungs are one of the last organs to mature in utero (Fig. 7.14). These infants have a great deal of difficulty breathing (a condition known as *respiratory distress syndrome*) and are typically given synthetic surfactant until their type II epithelial cells mature.

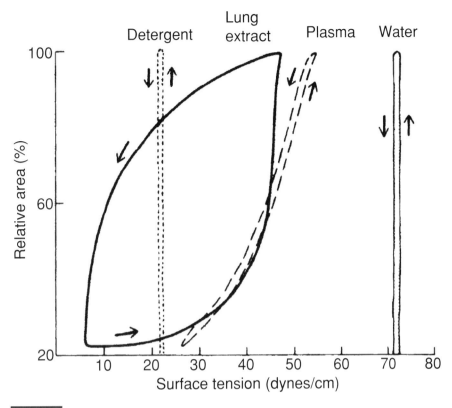

Figure 7.12

Surface tension *versus* interfacial relative area for water, plasma, detergent, and lung extract. The surface tensions of pure water and detergent are 72 and 22 dynes/cm, respectively, and show no dependence on interfacial area. Plasma and lung extract, however, have a surface tension that depends on interfacial area and also on whether the interfacial area is increasing or decreasing (arrows). This effect, leading to hysteresis in the curves, is particularly prominent for lung extract. Modified from Clements [9] with permission of the American Physiological Society.

7.4 Mass transfer

The lungs are remarkable mass transfer devices that permit blood/gas exchange. It is of interest to examine the performance of the lungs as mass transfer units, and towards this end we will consider two points of view: (i) a microscopic one, in which we focus on gas transport in a single acinus; and (ii) a macroscopic one, in which we focus on the entire lung.

When considering mass transfer at the level of the acinus, we should keep in mind that O_2 transport from air to blood consists of several steps: O_2 in the acinar air must move until it is adjacent to the thin tissue layer separating blood from air, must cross this thin tissue layer, and must then move through the blood to

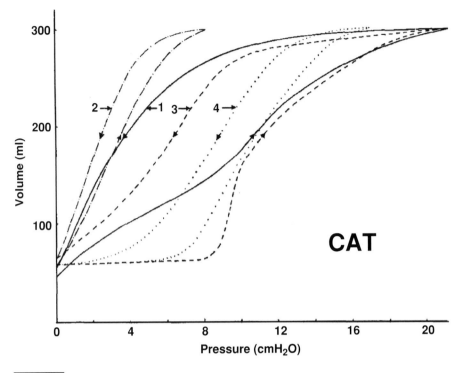

Figure 7.13

Pressure–volume relationship for inflation of an excised degassed cat lung with air or saline solution. Loop 1 is an initial inflation/deflation with air; loop 2 is an inflation/deflation with saline; loop 3 is an inflation/deflation with air after saline; and loop 4 is an air inflation/deflation after the lung was washed with detergent. Note that in all inflation–deflation cycles (but particularly when the lung is inflated with air), the deflation path differs from the inflation path. The difference between loop 1 and 2 results from the absence of surface tension effects in the saline-filled lung. After saline filling, some airways are collapsed and a significant opening pressure is required to inflate the lung, as can be seen by the initially horizontal part of the inflation paths for loops 3 and 4. Filling the lung with saline removes surfactant, as does washing with detergent, producing higher deflation pressures for loops 3 and 4. Modified from Bachofen *et al.* [10] with permission of the American Physiological Society.

bind to hemoglobin in red cells. In thinking about these processes, it is convenient to consider the "air-side" and "blood-side" mass transfer as individual processes coupled by transport across the blood–air tissue barrier. In a well-designed system, we expect that the mass transport capacity of these individual processes will be approximately matched, although we will see that it is not easy to define what "matched" means in this context.

It is common (and conceptually convenient) to think of the microscopic arrangement of alveoli and pulmonary capillaries as one in which alveoli all reside at the end of conducting airways and are more or less equivalent from a mass transport viewpoint (Fig. 7.15A, color plate). Unfortunately, this "parallel ventilation/parallel

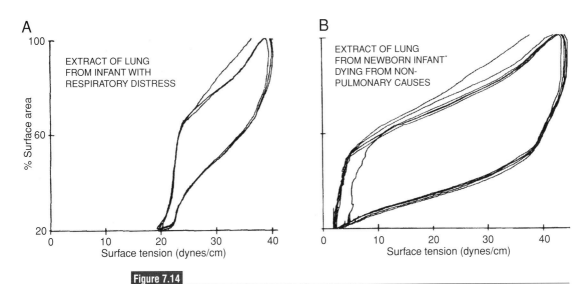

Figure 7.14

Surface tension measurements made on extracts from infant lungs using the apparatus shown in Fig. 7.11. Comparison of the measurements in lung from an infant suffering from respiratory distress syndrome (A) with those from an infant without pulmonary distress (B), and with the data shown in Fig. 7.12, shows that the minimum attainable surface tension is elevated in the lung extract from the child with respiratory distress. From Clements [9] with permission of the American Physiological Society.

perfusion" model is incorrect; in mammals, the alveoli within an acinus reside in the last six to nine generations of airways, so that oxygen is continually being extracted from acinar air as one moves deeper into the lung (Fig. 7.15B, color plate) [11]. This means that it is more appropriate to think of the acinus as a "series ventilation/parallel perfusion" system, a distinction that has important consequences for air-side mass transfer, as we will see. Let us begin our analysis of this system by considering blood-side mass transfer.

7.4.1 Blood-side acinar mass transfer

Based on the discussion above, we will treat all alveoli as being essentially equivalent from the viewpoint of blood-side mass transfer, with the exception that we will allow the driving potential (alveolar O_2 concentration) to depend on the airway generation. Some other salient features of mass transfer in a single alveolus are:

- blood flows in a sheet-like network of capillaries [12] that are in direct contact with the alveolar wall on both sides
- the thin tissue layer intervening between blood and air allows gases (e.g., O_2 and CO_2) to cross, but under normal circumstances, prevents blood from entering the alveolus; it can therefore be thought of as a semipermeable membrane.

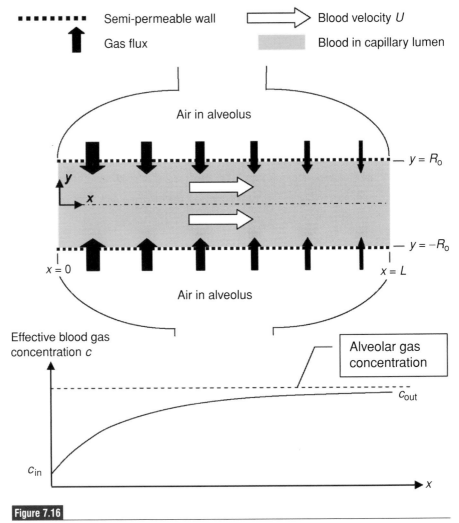

Figure 7.16

Schematic of mass transfer to an alveolar capillary (not to scale). Blood flows from left to right in the capillary, here shown as a central channel, where R_0 is the radius of a single pulmonary capillary. Alveolar air contacts the upper and lower boundaries of the channel, which are assumed to be gas permeable. The graph shows how blood gas concentration changes with axial position in the capillary, in this case approaching but not quite reaching the alveolar gas concentration.

Based on these features, we model the capillaries in contact with the alveoli as a two-dimensional channel with semipermeable walls. The height of the channel is $2R_0$, where R_0 is the radius of a single pulmonary capillary. Alveolar air is in direct contact with the upper and lower semipermeable walls, and blood flows through the channel with mean velocity U (Fig. 7.16).

We consider the transport of a gas from the alveolar air to the blood. Obviously this is relevant to the transport of O_2, but we will see shortly that the situation for

O_2 is remarkably complex, essentially because the binding of O_2 to hemoglobin depends in a non-linear manner on O_2 partial pressure and a number of other factors. Therefore, to get started, let us simply consider a generic gas whose total solubility in blood is proportional to its partial pressure. We will denote the local effective concentration of this gas in the blood by $c(x, y, t)$; it has dimensions of moles of gas per volume of blood. This gas is able to cross the semipermeable capillary walls by diffusion. Its mean concentration in the incoming blood stream is c_{in}, and in the blood leaving the capillary sheet it has mean concentration c_{out}. *We wish to solve for the blood gas concentration, $c(x, y, t)$*. A complete analysis of this problem is rather difficult, since c is in general a function of position (x and y) and time t. We will therefore simplify the problem by making several assumptions.

1. We assume that the blood is well mixed in the cross-stream (y) direction. This implies that c is not a function of y. This is actually a poor approximation, since it ignores the mass transfer boundary layer that forms along the semipermeable walls of the pulmonary capillaries. We will discuss this assumption on p. 304.
2. We assume steady-state behavior, so that c is not a function of time. This means that the pulsatile blood flow is replaced by a steady mean flow, and that the time-varying concentration of gases in the alveolar air is replaced by a steady mean concentration. This will also be discussed in more detail on p. 305.

We are left with c depending only on axial position, x. This dependence on x is central to the physics of the problem, since gas is continually leaving the alveolus, diffusing across the semipermeable membrane, and being taken up by the blood. The more rapidly the gas can diffuse across the membrane, the more quickly c will increase with x. An important parameter in this analysis is therefore expected to be the flux of gas across the walls, J, defined by

$$J = \frac{\text{moles gas diffusing across channel boundaries per unit time}}{\text{unit wall area}}. \quad (7.2)$$

The driving force for the diffusional gas flux J is the gas concentration difference between the alveolar air and the blood. Since c depends on x, this difference also depends on x, and we therefore expect J to depend on x. Consequently, we must formulate two equations to solve for the two unknowns, $c(x)$ and $J(x)$.

The first step is to perform a mass balance on the *gas* (*not* the blood) using a control volume of length δx, as shown in Fig. 7.17. Here we exploit the symmetry of the problem and consider a control volume in the upper half of the channel only. The volume flow rate of blood into this control volume (per unit depth into the page) is $Q_{blood} = U R_0$. Since there is no gas transported across the center line

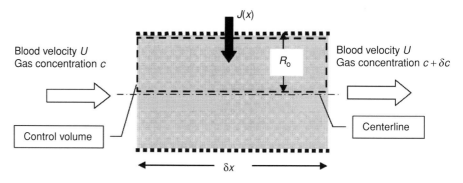

Figure 7.17

Control volume for analyzing alveolar gas transfer. Blood enters from the left at mean velocity U. Gas concentration in the incoming blood is c; in the outgoing blood it is $c + \delta c$. The control volume is δx long and one unit deep (into the page). R_0 is the radius of a single pulmonary capillary.

(symmetry!), a mass balance per unit depth into the page yields

$$c(x)Q_{\text{blood}} + J(x)\delta x = (c(x) + \delta c)\, Q_{\text{blood}}. \qquad (7.3)$$

The various terms in Equation (7.3) represent, from left to right, the mass flow rate of gas entering the left-hand edge of the control volume; the mass flow rate of gas entering the top of the control volume through the semipermeable upper channel wall; and the mass flow rate of gas leaving the right-hand edge of the control volume. Rearranging Equation (7.3) and taking the limit as $\delta x \to 0$ yields

$$J = Q_{\text{blood}} \frac{dc}{dx} = R_0 U \frac{dc}{dx}. \qquad (7.4)$$

The second step is to relate $J(x)$ to $c(x)$ by using *Fick's law*, which states that the diffusional flux, J, of a quantity is proportional to the concentration gradient of that quantity in the direction of diffusion. The proportionality coefficient is the *diffusivity*, D. In the present case, the concentration gradient is the concentration difference across the upper wall of the channel divided by the thickness of the upper wall. Taking account of the fact that the flux is in the direction from higher concentration to lower, we can therefore write the diffusional flux as

$$J(x) = D \frac{c_{\text{alv}} - c(x)}{\Delta y} \qquad (7.5)$$

where c_{alv} is the concentration of gas in the alveolar air adjacent to the upper channel wall and Δy is the thickness of the upper channel wall.

What is the meaning of the term c_{alv}? Put another way, the concentration c is defined in terms of moles of gas per volume of blood, so how can we talk about c in the alveolar air, where there is no blood? To resolve this question, we need to take a detour and discuss the concept of equilibrium between two phases. Suppose we have a gas mixture containing a gas X and that this gas mixture is allowed to come to equilibrium with a liquid in which the gas X is soluble. In the simplest case, the equilibrium concentration of gas X in solution in the liquid, c_X, will be linearly proportional to the partial pressure[3] of gas X in the mixture, p_X, i.e.,

$$c_X = \beta_X p_X \tag{7.6}$$

where the constant β_X is known as the *solubility coefficient* for gas X. For example, the solubility of O_2 in water at $37\,°C$ is approximately 1.41×10^{-9} mol $O_2/(cm^3\,mmHg)$ [2].[4] If Equation (7.6) – which is nothing other than Henry's law written in a different way – holds, then it is clear that there is a linear relationship between the concentration of gas X dissolved in a liquid and the mole fraction of gas X in the equilibrated gas phase. Hence, if all references to the gas phase are understood to be at a constant reference pressure and temperature, c_X can be thought of as a surrogate measure of the concentration of X in the gas phase. It is in this sense that we use c_{alv} in Equation (7.5). By the same logic, the reader will appreciate that we can use the partial pressure, p_X, to represent the concentration of gas X in solution in the liquid. It is common to adopt the second approach, and we will do so here. Specifically, we will now solve for the partial gas pressure $p(x)$ in the blood, with alveolar partial pressure p_{alv}, etc.

Let us now return to Equation (7.5). In general, p_{alv} is a function of position and time. However, we will assume that the alveolus is large and its contents are well mixed. Although this assumption is not particularly accurate, it permits us to approximate p_{alv} as a constant; it will be further discussed below.

The third step is to combine Equations (7.4), (7.5), and (7.6) to obtain the following first-order linear differential equation with constant coefficients for $p(x)$:

$$R_o U \frac{dp}{dx} = \frac{D}{\Delta y} [p_{alv} - p(x)]. \tag{7.7}$$

---

[3] The reader will recall that in a mixture of ideal gases, the partial pressure of gas X is equal to the mole fraction of gas X multiplied by the total pressure of the gas mixture.

[4] Sometimes solubility coefficients are given in terms of volume of gas rather than moles of gas, in which case the coefficient is known as the *Bunsen solubility coefficient*. But one must be careful when using these numbers, since the volume of gas depends on the temperature and pressure, and it should therefore be clearly stated what reference state is being used. Even worse, the solubility itself depends on temperature, so there are two temperatures involved. It is common practice in the literature to report Bunsen solubility coefficients at $37\,°C$ with the gas volume referred back to standard temperature and pressure (i.e., $0\,°C$ and 1 atm). It is also interesting that the solubility of gases in blood plasma is slightly less than in pure water. For O_2 at $37\,°C$, the plasma solubility is about 90% of that for water [13]; for CO_2 the serum:water solubility ratio is about 93% [14,15].

Since Equation (7.7) is first order, it requires one boundary condition, which is that the inlet blood gas partial pressure must equal p_{in}, i.e.,

$$p(x = 0) = p_{in}. \tag{7.8}$$

The fourth step is to solve Equation (7.7). To accomplish this it is convenient to define a dimensionless partial pressure $\hat{p}$ by

$$\hat{p} = \frac{p - p_{in}}{p_{alv} - p_{in}}. \tag{7.9}$$

Since $p(x)$ can never be greater than p_{alv}, $\hat{p}$ must always be $\leqslant 1$. Similarly, the minimum possible value of $p(x)$ is p_{in}, so that $\hat{p} \geqslant 0$. Finally, $\hat{p}$ is linearly proportional to p. Therefore, $\hat{p}$ is a convenient non-dimensional measure of concentration lying in the range $0 \leqslant \hat{p} \leqslant 1$ such that:

- $\hat{p}$ close to 1 means that blood is nearly fully loaded with gas
- $\hat{p}$ close to 0 means that blood is carrying the minimum amount of gas.

We substitute $\hat{p}$ into Equation (7.7), and after some algebraic manipulation obtain

$$\frac{d\hat{p}}{dx} + \left[\frac{D}{R_o U \Delta y} \right] (\hat{p} - 1) = 0. \tag{7.10}$$

The boundary condition is $\hat{p} = 0$ at $x = 0$. This system has solution

$$\hat{p} = 1 - e^{-x/L_{char}} \tag{7.11}$$

where L_{char} is a constant with dimensions of length equal to

$$L_{char} = \frac{U R_o \Delta y}{D}. \tag{7.12}$$

L_{char} is the e-folding length for mass transfer: that is, it is the length over which the concentration c will change by a factor of $e = 2.718 \ldots$. Equation (7.11) indicates that the blood gas partial pressure increases exponentially along the capillary. One property of the exponential function is that it goes to 0 very quickly for negative arguments, so that $\hat{p}$ approaches 1 quickly (Table 7.1). In the context of the lung this implies that the gas tension in the pulmonary capillaries rapidly approaches p_{alv}, to within 1% for $x/L_{char} > 5$. Therefore, from

Table 7.1. Normalized blood gas partial pressure, $\hat{p}$, in the alveolar model as a function of dimensionless axial position, x/L_{char}

x/L_{char}	$\hat{p}$
1	0.63
2	0.86
3	0.95
5	0.993
10	0.99996

the mass transfer viewpoint, the optimal capillary length is approximately $4L_{char}$. If the capillary is shorter, the blood cannot take up the maximum amount of gas, while a longer capillary gives no mass transfer benefit but has increased flow resistance.

In the above analysis, we have considered a gas that is being transferred from the alveolar air to the blood. Luckily, the same analysis works for the case of a gas going in the opposite direction, such as CO_2 (although see below for complexities of CO_2 transport). In that case, the definition of $\hat{p}$ is altered, but the variation of concentration with length takes the form of a decaying exponential with exactly the same characteristic length as was derived above.

It is interesting to see what optimal capillary length is predicted by this analysis. We use the following numbers, suitable for humans at rest:

- Capillary radius R_o has been morphometrically estimated to be about 4 μm $= 4 \times 10^{-4}$ cm [3,16].
- The blood speed, U, is not easy to estimate in humans. In dogs, we can estimate that the speed is approximately 0.1 cm/s [17,18], and we will assume that a similar value holds in humans. This is of the same order as blood speed in the systemic capillaries [19].
- The diffusivity, D, should include hindered diffusion in the alveolar wall, so diffusion across the alveolar wall will be slightly slower than that in water. For O_2 and CO_2, the diffusion coefficients in water at 25 °C are 2.10×10^{-5} cm^2/s and 1.92×10^{-5} cm^2/s, respectively. Weibel [2] gave a value of 1×10^{-5} cm^2/s for O_2 in connective tissue at 37 °C, and we will take this as being typical of the gas species of interest.
- The effective tissue thickness must account for local variations in thickness and the fact that gas will be transported preferentially across thinner wall regions. Weibel [2] discussed this in detail and indicated that the appropriate value is $\Delta y \approx 0.6$ μm (i.e., 6×10^{-5} cm).

These numbers give $L_{char} = 2.4 \times 10^{-4}$ cm, or 2.4 μm. The actual capillary length L is of order several hundred microns [18] to 500 μm [20], which is many times L_{char}. This analysis therefore suggests that there is significant "extra mass transfer capacity" in the pulmonary capillaries. However, we should be a little careful about this conclusion, for the following reasons:

- There is a large amount of uncertainty in some of the above parameters, arising from several sources. First, what does "capillary length" mean in a capillary network like the pulmonary capillaries, where there are extensive interconnections between capillary segments? We have taken it to be the distance from the terminal pulmonary arteriole to the pulmonary venule, but even this is sometimes not easy to define in an unambiguous manner. Second, even if we can define the relevant quantities, it is very difficult to measure them at the single capillary level, especially in humans; see [3,16,20,21] for an appreciation of some of these difficulties.
- The values taken above are averages, but there is a great deal of heterogeneity in the lung. Because blood flow and gas transport will tend to follow a path of least resistance, incorporation of such heterogeneity into models is a major challenge [22].
- We have neglected mass transfer resistances in the plasma and within red cells and have assumed that the kinetics of gas binding in the red cell are very fast. This turns out to be not quite right, as will be discussed in more detail below. The net effect of incorporating these effects would be to significantly increase L_{char}.

Putting this all together, we can say that the value for L_{char} derived above almost certainly underestimates the true value. Nonetheless, even if we were wrong by a factor of 10, our analysis still predicts that there is "extra capacity." But this is not a bad thing! Remember that the calculation above was for the resting state. During exercise, the mass transfer requirements are more severe, and it is therefore appropriate to have extra capacity for such times.[5]

Complexities associated with O_2 and CO_2 transport

A key assumption in the above analysis was that the partial pressure of the gas was directly proportional to the blood concentration of the gas. However, for the gas that we care most about – O_2 – this is quite a poor assumption. Even the situation with CO_2 is more complex than we have considered above.

---

[5] Actually, the situation during exercise is somewhat more complicated, since under-perfused pulmonary capillaries are "recruited" during exercise. Consequently, there is not a one-to-one relationship between cardiac output and U during exercise. However, the comments made above still hold in a qualitative fashion.

We will first consider the situation with O_2. Here the difficulty arises because O_2 in the blood is carried both by the aqueous phase (plasma) and by hemoglobin within the red cells. In fact, most O_2 is carried bound to hemoglobin, so Equation (7.6) must be modified to read

$$c_{O_2} = \beta_{O_2} p_{O_2} + 4c_{Hb} S(p_{O_2}) p_{O_2} \tag{7.13}$$

where the second term represents the O_2 bound to hemoglobin. The amount of O_2 carried in this manner depends on the hemoglobin concentration, c_{Hb}, and the *oxygen saturation of hemoglobin*, $S(p_{O_2})$. The factor of 4 arises because one molecule of hemoglobin can bind four molecules of oxygen. The oxygen saturation function equals the fraction of hemoglobin that is bound to oxygen, and is a pure number between 0 and 1. It can be approximated by the *Hill equation*

$$S(p_{O_2}) = \frac{\left(p_{O_2}/p_{50}\right)^n}{1 + \left(p_{O_2}/p_{50}\right)^n} \tag{7.14}$$

where empirically $n = 2.7$ and p_{50}, the partial pressure at which hemoglobin is 50% saturated by O_2, is approximately 27.2 mmHg. Equation (7.14) shows that O_2 blood concentration depends on O_2 partial pressure in a highly non-linear manner (Fig. 7.18), so it is no longer straightforward to relate gas-phase partial pressures to blood-phase gas concentrations. Additionally, O_2 saturation depends on blood pH and CO_2 concentration (the *Bohr effect* [2]), both of which vary with position in the pulmonary capillary. The net effect is that O_2 binds hemoglobin much more avidly near the entrance to the pulmonary capillaries than would be expected by linear theory, which is helpful in that it facilitates rapid uptake of O_2 by blood in the lungs.

Because of these complications, there is a non-exponential relation between p and x for O_2 (Fig. 7.19A). However, the *scaling* in the above analysis (i.e., the dependence of L_{char} as given by Equation [7.12]) is still approximately valid. For a more complete discussion of O_2 mass transfer, the reader is referred to the books by Weibel [2] and Cooney [15].

The situation for CO_2 transport is slightly simpler. CO_2 is about 24 times more soluble in the aqueous phase than is O_2, and it is carried in the blood in three main forms: as a dissolved gas in the plasma (7% of the total), bound to hemoglobin (30%), and in the form of bicarbonate ions in the red cells (63%) [15]. The net effect is that there is less non-linearity in the relationship between CO_2 partial pressure and CO_2 concentration in blood than there is for O_2, and hence the partial pressure distribution more closely approximates an exponential for CO_2 (Fig. 7.19B).

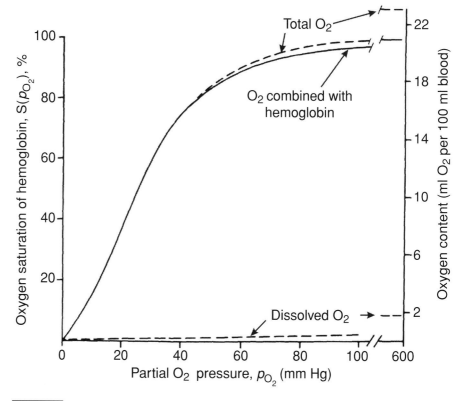

Figure 7.18

Oxygen-binding characteristics of blood. The lower dashed line represents the O_2-carrying capacity of plasma (right scale); it can be seen that at a typical alveolar O_2 partial pressure of 100 mmHg, plasma can only carry about 0.24 ml of O_2 at standard temperature and pressure (STP) per 100 ml plasma. The S-shaped solid line shows the carrying capacity of hemoglobin (right scale) and the O_2 saturation function (left scale). The upper dashed line shows the total O_2-carrying capacity of blood (right scale), which is about 20.8 ml O_2 at STP per 100 ml blood. It can be appreciated how critical the O_2-carrying function of hemoglobin is, since without it the blood simply could not carry enough O_2 to peripheral tissues without having an enormous cardiac output. From Weibel [2], based on West, *Respiratory Physiology. The Essentials*, Williams & Wilkins, 1974, with permission.

Assumptions of the blood-side model, and more sophisticated models

At this point we should revisit our assumptions. There were two main ones.

The first assumption was that the alveolar wall was the major mass transfer barrier, so we could neglect mass transfer effects in the capillary, thereby neglecting the dependence of c on y. There are three such capillary-side mass transfer effects for O_2: transfer through the plasma to the red cell surface, transport within the red cell, and the kinetics of O_2 binding to hemoglobin. The latter two are often grouped together in the literature.

Classical analysis would suggest that we should consider the mass transfer boundary layer in the blood to estimate the resistance for O_2 transport through

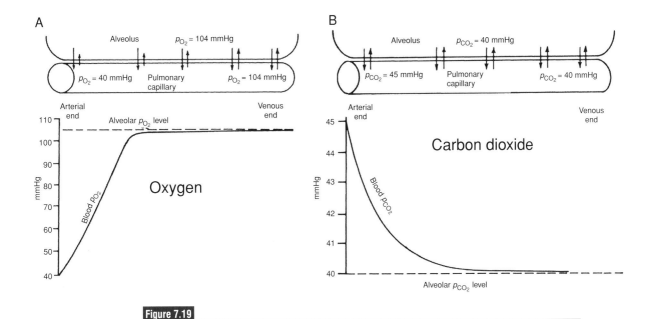

Figure 7.19

Distributions in a pulmonary capillary of O_2 (A) and CO_2 (B). Note that the distribution of O_2 is not an exponential, for reasons described in the text. For CO_2, there is an exponential decay of partial pressure (p_{CO_2}) with axial distance down the capillary. In this plot, the gas concentration is expressed in terms of the gas partial pressure for component i, p_i, given by $p_i = $ mole fraction i × total pressure. This can also be written as $p_i = p_{tot} c_i / (\Sigma_j c_j)$, where p_{tot} is the total pressure, c_i is the molar concentration of gas species i, and Σ_j refers to a sum over all species present. From Guyton [23], based on calculations reported in Milhorn and Pulley [24].

the plasma. However, this is not really appropriate, since red cells are passing through the capillary in the near vicinity of the wall, hence disturbing this mass transfer boundary layer. It is more appropriate to think of a simple diffusional process across the plasma layer between the capillary wall and the red cell. One can then think of the plasma layer as being an extension of the capillary wall, so the effective thickness Δy becomes the sum of these two layers. Weibel *et al.* [20] estimated the thickness of this composite layer to be approximately 1.1 μm, which approximately doubles the value of L_{char} obtained from the above analysis.

More serious is the issue of O_2 transport and binding within the red cell. A detailed analysis of this effect is beyond the scope of this book, but Weibel has considered these factors in a model of pulmonary capillary mass transfer that implicitly (but not explicitly) includes the variation of O_2 tension along the length of the capillary [2]. His numbers, while only approximate, suggest that mass transfer effects in the red cell can actually dominate the overall mass transfer process, and the effective value of L_{char} would be three to five times larger than our computed value if these effects were taken into account.

The second assumption was that the system could be described as steady, so that neither c nor c_{alv} depended on time. The magnitude of the error introduced by this assumption depends on the time scale for mass transport across the alveolar wall compared with the time scale of changes in alveolar O_2 and CO_2 concentration. If the mass transport is fast, then the process is quasi-steady, and the above model will be approximately correct if all concentrations, fluxes, etc. are taken as time-averaged quantities. The time scale for transport in our model is the time required to diffuse across the alveolar wall. The physics of the diffusion process show that this is $\Delta y^2 / 2D$, which is about 10^{-4} s. This is very fast compared with the time scale of changes in alveolar O_2 and CO_2 concentration, which is the breathing period (about 5 s). Therefore, this assumption is acceptable. Furthermore, at least for the case of O_2, the variation in alveolar O_2 concentration is rather modest, ranging from about 100 to 105 mmHg over the breathing cycle [2]. Considering that the O_2 tension in the blood entering the capillaries is only approximately 40 mmHg, it is acceptable to replace the alveolar O_2 partial pressure by its mean value.

We can see that there is room to develop more sophisticated models than the one described above. For example, it turns out that the hematocrit in the pulmonary capillaries is important; since most O_2 is taken up into red cells, and red cells do not fill the entire capillary, only part of the capillary wall is effectively "available" for O_2 transport. This can be accounted for by computing an effective capillary wall area weighted by the presence of red cells [20] or by more complex models where individual red cells are considered [22,25,26].

In closing, we would like to mention another way to think about mass transfer at the capillary level. If we consider the entire lung, we expect that the transport rate of O_2 (or any other gas) from air to the blood, Q_{O_2}, can be written as

$$Q_{O_2} = D_{LO_2}(pA_{O_2} - \bar{p}C_{O_2}) \tag{7.15}$$

where pA_{O_2} is the O_2 partial pressure in the alveolar air, $\bar{p}C_{O_2}$ is a suitable mean O_2 partial pressure in the pulmonary capillaries, and D_{LO_2} is known as the *pulmonary diffusing capacity*. In a 70 kg healthy adult human, experimental measurements show that D_{LO_2} is about 30 ml O_2/(min mmHg) at rest and 100 ml O_2/(min mmHg) during heavy exercise [2]. Now, in principle, it should be possible to relate the pulmonary diffusing capacity to the characteristics of the capillaries in the lung. The calculation proceeds in a manner similar to that of a single capillary, except that it is now the entire capillary surface area in the lung that is relevant. The rate of uptake of O_2 by hemoglobin – which we did not account for in our analysis – can be estimated from measurements on whole blood. The computed value of D_{LO_2} obtained in this manner is close to, but somewhat larger than, the measured value. For full details, the reader is referred to the wonderful book by Weibel [2], and its update [20].

7.4.2 Air-side acinar mass transfer

Now let us shift our attention to mass transfer on the air side. Thanks to the large diffusivity of O_2 in air (approximately $0.2\,cm^2/s$ [27]), diffusion on the length scale of a single alveolus is relatively rapid, with a diffusion time of only about $0.01\,s$ across a $300\,\mu m$ alveolus. This would suggest that the air side should contribute little mass transfer resistance to the overall air/blood mass transfer process. However, the situation is a little more complex than this. We must recall that air transport in the pulmonary airways occurs by a combination of convection and diffusion: transport in the larger airways is convection dominated, while diffusion becomes increasingly important as we progress to the smaller airways. It turns out that diffusion takes over from convection somewhere around airway generation 17 in humans [11]. That means that O_2 has to diffuse all the way from generation 17 to the terminal alveoli. We will refer to this as a "gas exchange unit;" it corresponds to one eighth of an acinus. With reference to Fig. 7.15B (color plate), we can see that the relevant length scale for diffusive O_2 transfer is therefore not an alveolar diameter but rather the size of the gas exchange unit.

To understand air-side mass transfer better, let us consider the fate of an O_2 molecule in an acinus. It is transported by convection to the proximal end of the gas exchange unit (generation 17) and then diffuses deeper into the airways. Along the way, it has some probability of coming close to the alveolar walls and entering the blood. If the rate at which O_2 is transported along the acinus is very slow compared with the uptake rate into the blood, it can be seen that the O_2 concentration will significantly decrease as one moves distally towards the terminal alveoli. This phenomenon is known as *screening* [11] and is undesirable, since it means that distal alveoli "see" reduced O_2 concentration and therefore are not working optimally. To understand this phenomenon better, we can define a *screening length* Λ, which can be interpreted as the characteristic size of a gas exchange unit at which screening begins to occur. The question then becomes what the ratio of Λ to the actual size of the gas exchange units is.

Based on the above discussion, we expect that the screening length Λ will depend on the diffusivity of O_2 in air, D_{O_2}, and on the ease with which O_2 can cross the acinar wall to enter the blood [11]. The latter quantity can be characterized by the permeability of the acinar wall to O_2, W_{O_2}, defined via

$$N_{O_2} = W_{O_2}\, A\, \Delta c_{O_2} \qquad (7.16)$$

where N_{O_2} is the rate at which O_2 is transported across the wall of the gas exchange unit (moles O_2 per unit time), A is the surface area of the wall of the gas exchange unit, and Δc_{O_2} is the O_2 concentration difference between alveolar air and blood.

N_{O_2} has dimensions of length/time, and hence dimensional analysis shows that $\Lambda W_{O_2}/D_{O_2}$ is a Π-group. Just as in Sections 4.3.4 and 5.1.2, we can then argue that this single Π-group should be constant, which implies that the screening length can be estimated by

$$\Lambda \sim \frac{D_{O_2}}{W_{O_2}} \tag{7.17}$$

We can estimate W_{O_2} from the properties of the acinar wall (thickness, diffusivity of O_2 in connective tissue, solubility of O_2 in water) just as in Section 7.4.1. Sapoval *et al.* [11] reported relevant parameters for several species and derived a value of Λ of approximately 28 cm in humans.

We need to compare this value of Λ with the characteristic size of the gas exchange unit to determine whether or not screening occurs. The relevant characteristic size turns out not to be the flow pathway length in the gas exchange unit; instead, we have to take account of the convoluted surface of the acinar wall, and hence the relevant length scale is [11]

$$L \sim \frac{A}{D} \tag{7.18}$$

where D is the diameter of a circle that encloses a single gas exchange unit. In humans, L is approximately 30 cm, which almost exactly equals Λ. This close agreement between Λ and L suggests that the geometry of the acinus is such that screening is minimal, and hence that air-side O_2 transfer is optimized. A more detailed calculation [11] showed that some screening occurs, but its effects are not severe.

This is a very satisfying result, but what happens during exercise, when O_2 needs become greater and more O_2 crosses the acinar wall per unit area? The above analysis suggests that screening *would* occur in such a situation, which would be very undesirable. However, we must recall that ventilation rate increases under such conditions, with a corresponding increase in air velocities. The net effect is that air is convected further into the acinus, so that the point at which diffusion takes over is pushed deeper into the lung (Fig. 7.20). This means that the effective length scale of the gas exchange unit L decreases to match the decreased screening length Λ. The interested reader is referred to the papers of Sapoval and coworkers [11,28,29] for more details.

7.4.3 Whole lung mass transfer

We now shift attention from the alveolar level to consider gas transfer for the entire lung. To do so, we need first to understand a few basic facts about gas volumes transferred in and out of the lungs during normal breathing.

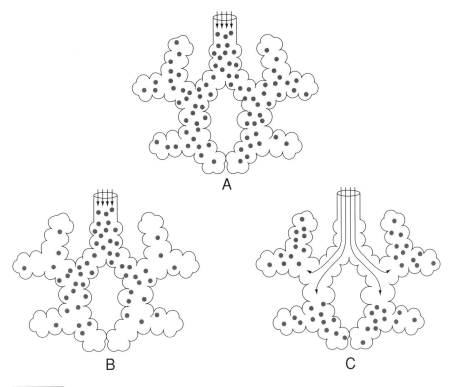

Figure 7.20
Schematic description of different ways in which the acinar gas exchange unit might function. The arrows symbolize convection; the end of the arrows is the transition point after which diffusional transport takes over. The dot represents O_2 molecules. (A) The diffusivity of O_2 is very large and essentially no screening occurs; consequently O_2 concentrations in the terminal alveoli are close to those in the large airways. (B) Here screening is occurring; significant amounts of O_2 are removed from air before it reaches the terminal alveoli. (C) In exercise, increased flow rates shift the convective/diffusive transition point to more distal airways and screening is minimized. From Sapoval *et al.* [11]. Copyright 2002 National Academy of Sciences, U.S.A.

Total lung volume is approximately 6 liters. However, only a small fraction (500 ml at rest) of this total volume is exchanged with each breath. The volume exchanged per breath is the *tidal volume*, and it is exchanged approximately 12 times per minute. In addition to the tidal volume, other relevant terms are listed below (see also Fig. 7.21).

Residual volume. The volume of air remaining in the lungs after maximal expiration, normally approximately 1200 ml.

Expiratory reserve volume. The volume of air that can be exhaled after normal exhalation of one tidal volume, normally approximately 1200 ml.

Inspiratory reserve volume. The volume of air that can be inhaled after normal inhalation of one tidal volume, normally approximately 3100 ml.

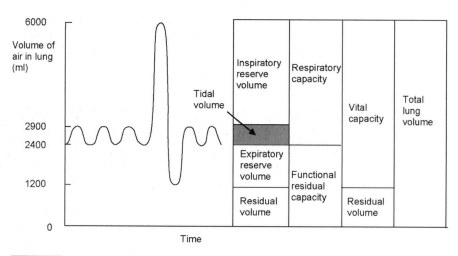

Lung volumes and capacities. The left part of the figure shows an idealized spirometer tracing for three normal breathing cycles, a maximal inspiration and expiration, and two more normal breathing cycles. When the subject inspires, the spirometer trace moves up; with expiration it moves down. See text for definition of terms on the right side of the figure. After Vander *et al.* [1]. Reproduced with kind permission of The McGraw-Hill Companies.

Vital capacity. The sum of tidal volume plus inspiratory reserve volume plus expiratory reserve volume. It is the lung's working maximum volume; note that it is less than the total lung volume.

With this background, we now consider overall gas transfer in a normal lung at rest. We expect the exhaled air to contain more CO_2 and less O_2 than the inspired air and wish to quantify these concentration differences. This requires that we know the body's rates of CO_2 production and O_2 consumption at rest, which are related to overall metabolic rate:

- at rest, the body's O_2 consumption rate is 250 ml/min at standard temperature and pressure (STP, or 1 atmosphere and 273 K)
- at rest the CO_2 production rate is 200 ml/min at STP.

It is conventional to refer all volumes to body temperature and pressure (BTP), which is taken as 1 atmosphere and 310 K. At BTP, the O_2 consumption and CO_2 production rates are 284 and 227 ml/min, respectively.

For purposes of this calculation, we will assume that the ambient air has the composition shown in Table 7.2, corresponding to a dry day. In all calculations below we will assume that air is a mixture of perfect gases. For such a mixture, the molar fraction of component i, n_i, is related to the partial pressure (p_i) and partial

Table 7.2. *Air composition in inspired and expired tidal volumes. All volumes in the table are referenced to BTP. The assumed molar fractions are in column two, corresponding to dry ambient air (trace gases are not shown). The corresponding partial pressures and partial volumes (in a 500 ml inspired tidal volume) are shown in columns three and four. The partial volumes in the expired tidal volume are shown in column 5, calculated as described in the text.*

Gas	Molar fraction (%)	Partial pressure (mmHg)	Partial volume in one:	
			Inspired tidal volume (ml)	Expired tidal volume (ml)
N_2	78.62	597	393.1	393.1
O_2	20.84	159	104.2	80.5
CO_2	0.04	0.3	0.2	19.1
H_2O	0.50	3.7	2.5	32.5
Total	100.00	760	500	525.2

volume (V_i) of component i by

$$n_i = \frac{p_i}{p_{tot}} = \frac{V_i}{V_{tot}} \tag{7.19}$$

where p_{tot} and V_{tot} are the total pressure and total volume (respectively) of the gas mixture. Note that the total pressure and total volume obey

$$p_{tot} = \sum_j p_j \tag{7.20}$$

$$V_{tot} = \sum_j V_j \tag{7.21}$$

where the sum is over all gases in the mixture.

Given the ambient air composition in column two of Table 7.2, it is straightforward to use Equation (7.19) to compute the partial pressures and partial volumes in a 500 ml inspired tidal volume (columns three and four). The partial volumes in column five are calculated as follows:

- Nitrogen is not metabolized, and so its partial volume does not change.
- Oxygen is consumed at the rate of (284 ml/min)/(12 breaths/min), or 23.7 ml/ breath. Therefore, the expired air contains $104.2 - 23.7 = 80.5$ ml O_2.
- Carbon dioxide is produced at the rate of (227 ml/min)/(12 breaths/min), or 18.9 ml/breath. Therefore, the expired air contains $0.2 + 18.9 = 19.1$ ml CO_2.

Table 7.3. Comparison of calculated and measured tidal volume compositions. Calculated compositions are from column five of Table 7.2, measured values are from Cooney [15]. With kind permission of Taylor & Francis.

Gas	Composition (%)	
	Calculated	Measured
N_2	74.8	74.5
O_2	15.4	15.7
CO_2	3.6	3.6
H_2O	6.2	6.2

- The water mass balance is slightly more complex. The essential fact is that the expired air is, to a very good approximation, fully humidified. Therefore, the partial pressure of water vapour in expired air equals the saturation pressure of water at 310 K, which is 47 mmHg. Since partial pressures are directly proportional to partial volumes, we can then write

$$\frac{\text{volume water}}{\text{volume dry gases}} = \frac{\text{partial pressure water}}{\text{partial pressure dry gases}} = \frac{47 \text{ mmHg}}{(760 - 47) \text{ mmHg}}. \quad (7.22)$$

In Equation (7.22), we use the fact that the total pressure in the expired tidal volume is 760 mmHg, consistent with the fact that all values in Table 7.2 are referenced to BTP. Adding the volumes in column five of N_2, O_2, and CO_2 gives a dry gas volume of 492.7 ml. Hence, Equation (7.22) can be used to compute

$$\text{volume water} = \frac{47}{760 - 47} \, 492.7 \text{ ml} = 32.5 \text{ ml}, \quad (7.23)$$

which is entered in Table 7.2 in column five. The total expired volume (at BTP) is then calculated to be 525.2 ml.

The accuracy of this calculation can be checked by comparing the calculated expired air composition with measured values. Table 7.3 shows that the agreement is very good.

It is interesting to note that Table 7.2 indicates that the expired volume exceeds the total inspired volume.[6] At first sight this seems paradoxical, but it can be explained by the fact that mass is being added to the expired air by the lungs.

---

[6] This volume "discrepancy" is not as large as might be expected from Table 7.2. Recall that all volumes in Table 7.2 are at BTP. However, the actual pressure during inspiration is slightly less than atmospheric; consequently, the actual inspired tidal volume within the lungs is slightly greater than 500 ml. Similarly, during expiration the alveolar pressure is slightly greater than 1 atm so that the actual expired tidal volume within the lungs is slightly less than 525.2 ml.

(The reader will recall that volume fluxes do not have to balance, but mass fluxes do.) Inspection of Table 7.2 shows that a great deal of the volume "discrepancy" between inhaled and exhaled air results from water addition by the lungs, and in fact the lungs are a major site of water loss. For the ambient air conditions of Table 7.2, 30.0 ml (at BTP) of water are lost per breath. Assuming that the water vapor is a perfect gas, this corresponds to

$$
n = \frac{pV}{RT}
$$
$$
= \frac{1 \text{ atm} \times 30.0 \text{ ml}}{82.05 \times 10^{-3} \text{ liter atm/(mole K)} \times 310 \text{ K}} \frac{10^{-3} \text{ liter}}{\text{ml}}
$$
$$
= 1.18 \times 10^{-3} \text{ mol } H_2O \tag{7.24}
$$

This corresponds to a loss of 0.0212 g water per breath. Over the course of one day, breathing at an assumed rate of 12 breaths/min, this produces a net daily loss of 366 g water, which is approximately 15% of the body's total daily water loss from all sources.

7.5 Particle transport in the lung

The air that enters the conducting airways during breathing can carry particles with it. These particles can be useful, for example when aerosol droplets containing inflammation-reducing steroids are inhaled by a child to treat his or her asthma. Unfortunately, they can also be harmful, such as occurs when asbestos fibers (or other toxic materials) are inhaled by workers. A critical question in relation to particulate inhalation is where the particles are deposited within the airways. For example, several systems for the delivery of inhaled insulin to diabetics are now being tested [30]; in this application, it is important to get as many of the drug-containing particles as possible down into the small airways so as to exploit the large surface area available for insulin transfer into the blood.

A number of authors have considered how particles are transported within the airways; Grotberg has provided a recent review [31]. The most critical issue is the ability of the particle to follow the curves and bends of the airways as it is carried into the lung by the air. We can get some insight into this by considering a small spherical particle of diameter d_p, moving at velocity $\mathbf{u_p}$ in a flow with local velocity $\mathbf{U}$. We will consider the case where the Reynolds number based on particle diameter is $\ll 1$, so that the fluid drag force on the particle is given by Stokes' law [32]

$$
\text{drag} = 3\pi\mu d_p(\mathbf{U} - \mathbf{u_p}) \tag{7.25}
$$

where μ is fluid viscosity. In its simplest form, the equation of motion for the particle is therefore [33]

$$m\frac{d\mathbf{u_p}}{dt} = 3\pi\mu d_p(\mathbf{U} - \mathbf{u_p}) \qquad (7.26)$$

where m is the mass of the particle. The reader can verify that Equation (7.26) can be cast into the dimensionless form

$$St\frac{d\mathbf{u_p^*}}{dt^*} = (\mathbf{U^*} - \mathbf{u_p^*}) \qquad (7.27)$$

where velocities with asterisks have been made dimensionless with respect to a reference fluid velocity U; t^* has been made dimensionless with respect to a reference time scale for the flow, τ; and the *Stokes number* (St) is given by

$$St = \frac{m}{3\pi\mu d_p\tau} = \frac{\rho_p d_p^2}{18\mu\tau} \qquad (7.28)$$

where ρ_p is the density of the particle. Equation (7.27) shows that we will obtain two very different types of limiting behavior depending on the magnitude of the Stokes number. If $St \gg 1$, it can be seen that the particle's velocity will not change with time; consequently, the particle will travel in a straight line and will eventually run into an airway wall. If, however, $St \ll 1$, the particle velocity $\mathbf{u_p}$ will always be very close to the local fluid velocity $\mathbf{U}$; in this case, the particle will be able to "follow" the flow very well and will tend not to run into walls. The Stokes number, therefore, tells us something about the ability of a particle to follow fluid path lines within the flow.

From the definition of the Stokes number (Equation [7.28]), we can see that a large Stokes number will result if the particle is dense or, more significantly, if it is large. The prediction is therefore that smaller particles should make it further into the lungs than larger ones, and that that penetration efficiency should be a non-linear function of particle diameter. In fact, for larger particles, as many as 45% of all particles are deposited in the passages of the nose and mouth, depending on the inhalation conditions [31]. This is usually beneficial, since we prefer to stop particulate contaminants from entering the lower airways, but it is undesirable when we are trying to deliver a drug or other therapeutic substance into the lung. The strategy in such cases is to try to make the particles as small as possible. For particles that are 2–10 μm in diameter, which typically have Stokes numbers of the order of 0.005 to 0.5, the particles are able to enter the main bronchi, but even then significant numbers are trapped after several generations of airway splitting, with preferential deposition at the airway bifurcations along the carinal ridges [31].

Despite the filtering function provided by the mouth and nose, the large volumes of air that move in and out of the lungs mean that significant numbers of airborne

Figure 7.22

Micrographs showing the epithelial lining of a small bronchus, consisting of three cell types: ciliated (C), goblet (G), and basal (B). The transmission electron micrograph at left shows macrophages (M) lying on the ciliated cells, as well as underlying fibroblasts (F) and connective tissue fibers (cf). The scanning electron micrograph at right shows the extensive cilia (C) and a goblet cell secreting a mucous droplet. Scale markers in both panels: 5 μm. Reprinted by permission of the publisher from *The Pathway for Oxygen: Structure and Function in the Mammalian Respiratory System* by E. R. Weibel, p. 237, Cambridge, MA: Harvard University Press, Copyright © 1984 by the President and Fellows of Harvard College.

particles enter the airways and adhere to the large area presented by the airway and acinar walls. Even for "inert" particles, continued normal lung function requires a mechanism to deal with these materials. What does the body do with these particles? It has several coping strategies.

1. The conducting airway walls are coated with a thin layer of mucus, which causes particles to stick to the walls where they impact. Mucus is constantly being synthesized by *goblet cells* lining the airways and is then moved upwards

towards the throat by the action of *cilia*, small hair-like projections of the epithelial cells lining the airway walls (Fig. 7.22). These cilia beat back and forth at a regular frequency to create a "ciliary escalator" that delivers mucus (with attached and embedded particles) to the pharynx, where it is usually swallowed with little attention. Only when excessive amounts of mucus are present in the airways, such as during an upper respiratory tract infection or during respiration in a dusty environment, do we notice the mucus and have to assist the cilia by coughing.

2. Some particles (the smallest ones) pass into the alveoli, where they adhere to the alveolar walls. Cilia are not present at this level in the lung, so the particles must be dealt with locally. These particles are internalized by macrophages lying on the pulmonary epithelial surface in a process known as *phagocytosis*. Depending on the composition of the particle, it is then broken down to a greater or lesser extent in the macrophage by lytic enzymes, and the breakdown products are stored in the cell in membrane-delimited spaces called *residual bodies*.

7.6 Problems

7.1 Consider a small spherical bubble of radius R.

(a) Show that the energy required to expand this sphere by a *small* amount ΔR is $2\sigma \Delta V / R$. Here ΔV is the increase in volume and σ is the interfacial tension.

(b) Estimate the time-averaged power required to overcome alveolar surface tension during normal breathing. Take $R = 150$ μm, $\sigma = 25$ dynes/cm, and breathing rate $= 12$ breaths/min.

(c) Repeat this calculation for the cat, where $R = 50$ μm and the tidal volume is 20 ml. Compare this calculated value with a *rough* estimate of the power obtained from Fig. 7.23. (Take beginning of normal inspiration to occur at 100 ml.) Is surface tension or lung tissue elasticity the dominant restoring force in the cat lung?

7.2 This question is concerned with the energy required to inflate the lung. Specifically, we wish to know what fraction of the total inflation energy is used to overcome alveolar surface tension forces, and what fraction is used to overcome parenchymal elasticity.

(a) How much energy is required to inflate a spherical bubble from radius R_1 to radius R_2? The surface tension coefficient is constant and equal to σ. Hint: think in terms of a pressure–volume relationship.

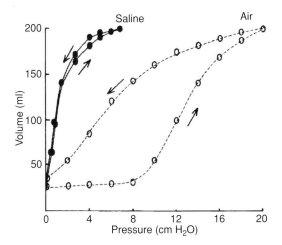

Figure 7.23

For Problem 7.1. A pressure–volume curve for a cat's lungs as they were inflated and then deflated, first with air and then with saline. Arrows indicate direction of inflation/deflation. Modified from Radford [34] with permission of the American Physiological Society.

(b)   Considering surface tension effects only, how much energy is required to inflate all 300 million alveoli in the lungs? Assume that the total alveolar volume before inspiration is 2.5 liters, that the tidal volume is 500 ml, that all alveoli are identically sized spheres, and that the effective surface tension coefficient during inspiration is constant and equal to 35 dynes/cm.

(c)   Idealized data from the air inflation of a pair of lungs are plotted in Fig. 7.24. Based on your calculations from parts (a) and (b), plot (to scale) on the same graph the pressure–volume curve expected from surface tension effects only. What fraction of the total energy required to inflate the lungs is from surface tension effects?

7.3   A balloon is surrounded by a tank of liquid at negative pressure and is connected to the atmosphere by a tube of length L and cross-sectional area A (Fig. 7.25). The pressure inside the balloon p oscillates above and below atmospheric pressure causing small changes in the balloon volume V. The elasticity of the balloon is characterized by its compliance C, defined by $\Delta p = \Delta V / C$

(a)   Derive a second-order differential equation for $\Delta V(t)$, assuming that (i) the pressure differential along the tube accelerates the air in the tube and is not used to overcome entrance, exit, or tube losses; and (ii) the air density ρ is constant. From the equation, show that the natural frequency of the system is $\sqrt{A/\rho L C}$.

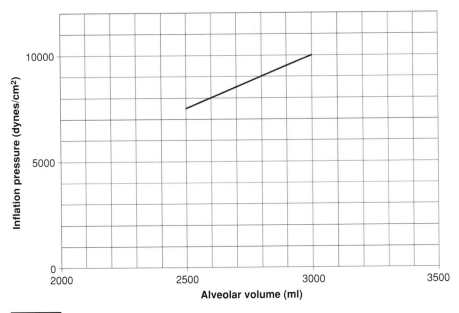

Figure 7.24

For Problem 7.2. Note that inflation pressure equals alveolar pressure minus intrapleural pressure.

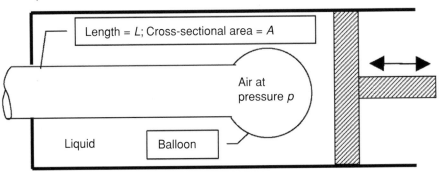

Figure 7.25

For Problem 7.3.

(b)   For a 70 kg man, A/L is approximately 0.001 m. The equivalent value
      for a 12 kg dog would be approximately $[12/70]^{1/3}$ of the value, or
      5.6×10^{-4} m. The compliance of dog lungs is approximately
      0.029 l/cmH$_2$O. Estimate the natural frequency of a dog's breathing
      using the formula developed in (a). Measurements indicate that dogs
      with a body mass of 12 kg pant at about 5.3 Hz. Comment briefly on any
      differences between your answer and the measured frequency.

Lung inflation curves

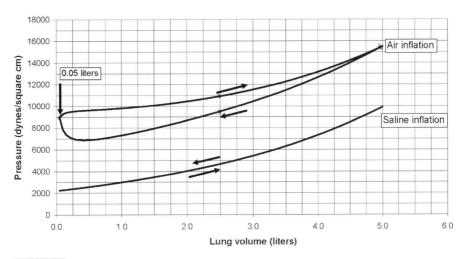

Figure 7.26

For Problem 7.4.

7.4   A lung is inflated with water and then with air. The pressure–volume curves for these two inflation procedures are shown in Fig. 7.26, with the right-pointing arrow representing inflation and the left-pointing arrow representing deflation. Assume that the lung has 150×10^6 identical alveoli, and that alveoli make up a constant 85% fraction of total lung volume. Based on these curves, graph the relationship between surface tension coefficient (in dynes/cm) and alveolar radius (in microns). Your graph should be *quantitatively* correct. This is best accomplished by choosing some key points from Fig. 7.26, transforming them to suitable values on your graph, and then interpolating by sketching.

7.5   Figure 7.12 shows that a solution containing lung extract exhibits hysteresis in its surface tension versus area relationship. In other words, the surface tension is higher during inflation of the lung than during deflation.

(a)   By recalling that mechanical work can be expressed as $\int p\,dV$, show that the work required to inflate all the alveoli in the lung against the effects of surface tension can be written as

$$Work = \int \sigma \; dA \qquad (7.29)$$

where σ is the surface tension coefficient, A is the aggregate surface area of all alveoli in the lung, and the integral is carried out from minimum surface area (start of inspiration) to maximum surface area (end of

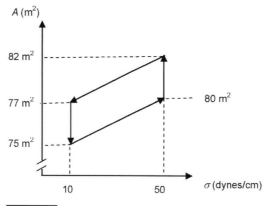

For Problem 7.5.

inspiration). To show this result, you may assume that the pressure outside the alveoli is constant and equal to 0 (gauge). You will find it convenient to recall that the volume and surface area of a sphere are $V = 4\pi R^3/3$ and $S = 4\pi R^2$. This implies that $dV = 4\pi R^2 dR$ and $dS = 8\pi R\, dR$.

(b) Assume that the surface tension versus area curve for the entire lung over one breathing cycle can be approximated by the shape in Fig. 7.27. Using this information, determine how much energy is dissipated in surface tension hysteresis effects during one breathing cycle.

7.6 Consider a fluid layer A surrounded on both sides by a fluid, B. The fluid layer can be thought of as a membrane. A species S is diffusing across this membrane, and has concentrations c_1^B and c_2^B (in the fluid B) on the two sides of the membrane, as shown in Fig. 7.28. It often happens that the solubility of material S inside the membrane is different than its solubility in bulk solution B. We therefore define a partition coefficient k as

$$k = \frac{\text{concentration of S in material A (at equilibrium)}}{\text{concentration of S in material B}}. \qquad (7.30)$$

Hence $k < 1$ means that S is less soluble in the membrane, and $k > 1$ means S is more soluble. Write down an expression for the flux across the membrane in terms of c_1^B, c_2^B, D_A (diffusion coefficient of S in A) and Δy_A. Sketch the concentration profile. What is the effective diffusion coefficient value with partitioning, D_{eff}?

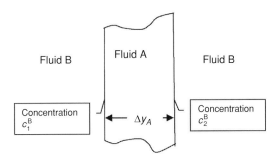

Figure 7.28

For Problem 7.6.

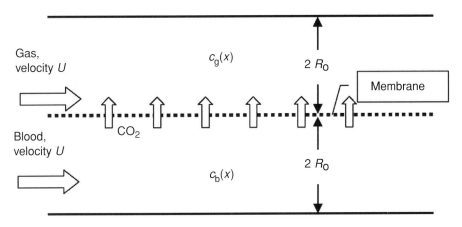

Figure 7.29

For Problem 7.7.

7.7 A membrane oxygenator is designed to supply O_2 and to remove CO_2 from blood during surgery. Blood and a gas mixture flow from left to right in two channels separated by a semipermeable membrane (Fig. 7.29). Blood gives up CO_2 to the gas stream continuously along the channel, so that both the blood and gas CO_2 concentrations are functions of position, x. Assume that the channel width and average velocities are the same for both blood and gas streams.

(a) If $c_g(x)$ is the CO_2 concentration in gas stream; $c_b(x)$ is the CO_2 concentration in blood; U is the mean velocity of both blood and gas streams; D_{eff} is the effective diffusion constant of CO_2 in the membrane and Δy is the membrane thickness; show that the CO_2 concentrations satisfy:

$$\frac{dc_b}{dx} = -\frac{c_b - c_g}{L_{char}} \quad (7.31)$$

where

$$c_b + c_g = constant. \tag{7.32}$$

Make and state appropriate assumptions.

(b) If the CO_2 concentration in the gas is zero at $x = 0$ and is c_b^0 in the blood at $x = 0$, at what x/L_{char} is c_b reduced to 60% of c_b^0?

7.8 In Section 7.4.1, we modeled gas transfer between blood in a pulmonary capillary and an alveolus, taking account of the fact that the concentration of gas in the blood varied with position. However, we did not account for the time-dependent nature of the mass transfer process, which is the subject of this question. Suppose that the CO_2 concentration *everywhere* in a capillary can be expressed as some spatially averaged value c_{cap}. Because blood from the right heart is continually flowing into the capillary, c_{cap} can be approximated as being constant in time. Assume that at time zero, fresh air (effective CO_2 concentration of 0) enters the alveolus, that at every instant the air in the alveolus is well mixed and does not communicate with air in the terminal bronchioles, and that all air in the alveolus is expelled and replaced by fresh air with every breath (breathing period $= T$).

(a) *Qualitatively* sketch a labeled graph of alveolar CO_2 concentration vs. time for the above assumptions.

(b) If the total surface area for blood/gas exchange associated with one alveolus is A, and the volume of the alveolus is V, derive an expression for how the CO_2 concentration in the alveolus changes with time. State assumptions. You will want to start with a mass balance in the alveolus.

7.9 A membrane oxygenator is being designed as part of a heart–lung bypass machine. It must be able to transfer 200 ml/min of O_2 into blood flowing at 5 l/min. Assume the blood enters the oxygenator with an effective O_2 concentration of 0.1 ml O_2/ml blood.

(a) With what O_2 concentration should the blood leave the oxygenator? You can solve this question easily by thinking about an overall mass balance.

(b) One design is to make the oxygenator as a "stack" containing many "units", as shown Fig. 7.30. Each unit consists of a channel filled with flowing blood, an O_2-filled channel, and flat membranes separating the channels. The membranes are $10 \, cm \times 10 \, cm$ by $5 \, \mu m$ thick, and the height of each blood-containing channel is $1 \, cm$. The O_2-containing channels are filled with 100% O_2, which is equivalent to a blood concentration of 0.204 ml O_2/ml blood. How many membrane units are needed to supply the required oxygen? The value for D_{eff} of O_2 in the membranes is measured as $10^{-6} \, cm^2/s$.

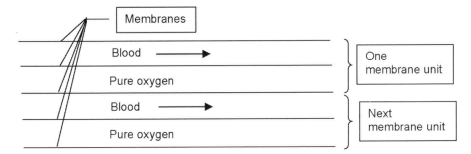

Figure 7.30

For Problem 7.9.

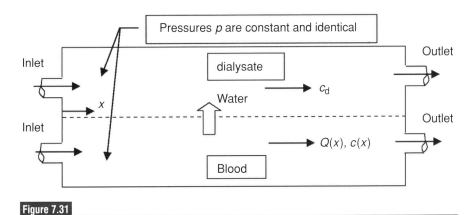

Figure 7.31

For Problem 7.10.

7.10 A dialyzer is shown in Fig. 7.31. The dialysate solution has glucose added to it so that it is hypertonic (has an osmotic pressure greater than that of blood). This causes water to be drawn into the dialysate, even though the pressures in the blood and dialysate are identical. We wish to determine how much water will be removed from the blood plasma in this unit. For purposes of this question, we will denote the volume flow rate of blood per unit depth into the page by $Q(x)$, the molar concentration of osmotically active components in the blood by $c(x)$, and the molar concentration of osmotically active substances in the dialysate by c_d. We will assume that *only* water can cross the membrane. We will also assume that blood and dialysate obey van't Hoff's law for osmotic pressure.

(a) Briefly state in words why c and Q change with axial location, x.

(b) Show that the blood flow rate per unit depth into the page satisfies

$$\frac{\mathrm{d}Q}{\mathrm{d}x} = L_p RT [c(x) - c_d] \tag{7.33}$$

where L_p is the membrane permeability (volume flow of water across the membrane per unit area per unit osmotic pressure difference).

(c) Show that the product of $c(x)$ and $Q(x)$ is constant. Hint: consider the mass flow rate of osmotically active components in the blood per unit depth into the page.

(d) If the flow rate of dialysate is large, the concentration c_d can be treated as a constant. Under this approximation, show that $Q(x)$ satisfies

$$\ln\left[\frac{Q_r - Q_0}{Q_r - Q(x)}\right] + \frac{Q_0 - Q(x)}{Q_r} = \frac{L_p\, RT\, c_0\, Q_0}{Q_r^2}x \qquad (7.34)$$

where Q_0 and c_0 are the values of $Q(x)$ and $c(x)$ at the inlet, and Q_r is a constant defined as $Q_0 c_0/c_d$. Hint:

$$\int \frac{Q}{Q_r - Q}dQ = -Q_r \ln\left[\frac{Q_r - Q}{Q_r}\right] - Q + \ constant. \qquad (7.35)$$

(e) If the total blood flow rate entering the dialyzer is 250 ml/min, the dialyzer is 10 cm deep into the page, and the maximum allowable water loss from the blood is 10 ml/min, what length L should the dialyzer be? The dialysate osmotic concentration, c_d, is 0.32 mol/l, the inlet blood osmotic concentration is 0.285 mol/l, the membrane permeability is 1×10^{-8} cm/(s Pa), and the working temperature is 310 K. The universal gas constant R is 8.314 J/(mol K).

7.11 Consider the transport of oxygen in *stagnant* blood. More specifically, suppose that a large container filled with deoxygenated whole blood is suddenly placed in an atmosphere of pure oxygen at concentration c_o, where c_o is constant (Fig. 7.32). Because of the very non-linear nature with which hemoglobin binds O_2, we will approximate this situation as one in which a "front" forms and propagates: above this interface the hemoglobin is oxygenated and there is free O_2 in solution in the plasma, while below this interface the free O_2 concentration is zero and the hemoglobin is deoxygenated. The front location, z_f, advances with time.

(a) Use Fick's law for diffusion of the free O_2 to argue that the concentration profile above the front is linear, and that the flux of O_2, J, is:

$$J = \frac{D}{z_f}c_o \qquad (7.36)$$

where D is the diffusivity of oxygen in whole blood. For purposes of

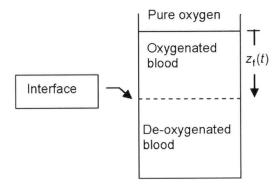

For Problem 7.11.

this portion of the question, you may assume that the front is slowly moving compared with the time scale for diffusion.

(b) Assuming that 4 moles of O_2 bind rapidly and irreversibly to 1 mole of hemoglobin at the front, and that the concentration of deoxygenated hemoglobin below the front is c_{Hb}, show that the front position satisfies:

$$z_f = \sqrt{\frac{2Dc_0t}{4c_{Hb} + c_0/2}} \qquad (7.37)$$

Hint: perform an unsteady mass balance on the O_2 in the container. Do not forget to take into account O_2 that is bound irreversibly to hemoglobin.

7.12 Consider mass transfer from pulmonary capillaries to a single alveolus. Suppose that, in addition to the normal 0.6 µm thick tissue layer between the blood and the air, scar tissue has formed that is 1 µm thick. The effective diffusivity of CO_2 in this scar tissue is 0.7×10^{-6} cm^2/s. When CO_2 has to cross both the "normal" tissue and the scar tissue, the mass transfer efficiency of the alveolus is reduced. Considering the entire alveolus, compute the net percentage reduction in blood-to-air CO_2 transfer due to the scar tissue for a 50 µm long capillary. *Note*: you do not need to re-derive equations; you should be able to make some simple modifications to equations in the text to get what you need. Remember that when two mass transfer barriers are in series, their mass transfer resistances add. You may use parameter values (except for capillary length) given in Section 7.4.1.

7.13 This question is concerned with mass transfer in the whole lung.

(a) If you switch from breathing air to breathing xenon, what is the *minimum* number of breaths after which the concentration of xenon in your lungs is 99%, by volume? Assume that no xenon diffuses out of the lungs, that perfect mixing occurs in the lungs, and that the xenon environment is so large that the exhaled air does not change the xenon concentration of 100%.

(b) If you switch back to an air environment and breathe normally, after how many breaths is the xenon concentration in your lungs reduced to 0.1%, by volume?

7.14 You decide to cure your hiccups by breathing into a paper bag. Assume that the bag is initially filled with 500 ml air (referenced to BTP) having the ambient air composition given in Table 7.2. Assume a breathing rate of 12 breaths/min, a CO_2 production rate of 235 ml/min at BTP, and an O_2 consumption rate of 284 ml/min at BTP. You can also assume that you begin by inhaling all the air in the bag (i.e., tidal volume is 500 ml for the first breath), and that on each subsequent breath you increase your tidal volume so as to completely empty the bag on inhalation. Compute the CO_2 concentration (as a percentage) in the bag after exhalation on the 10th breath. Although not particularly realistic, you can assume that the normal amount of CO_2 is transferred from the lungs to the expelled air with every breath, even though the CO_2 concentration in the bag is continually increasing.

7.15 You are standing on the top of Mount Everest (elevation 29 028 ft; 8708 m) where atmospheric pressure is 235 mmHg and the ambient temperature is 0 °C. The air composition is 78.6% N_2, 20.8% O_2, 0.04% CO_2, and 0.50% H_2O. Treat the air as an ideal gas.

(a) If you remove your oxygen set, how many times per minute must you breathe so as to satisfy your O_2 requirements of 284 ml/min at BTP? (Recall that BTP is 1 atmosphere and 37 °C). Assume that you take in tidal volumes of 1000 ml (at ambient conditions) and that you can transfer only 30% of the O_2 into your blood.

(b) Assuming that you could breathe that fast, what would the composition of the expired air be? Your CO_2 production rate is 227 ml/min at BTP, and expired air is fully humidified. (Partial pressure of water vapour at 37 °C and 100% relative humidity is 47 mmHg.)

7.16 On a particular day, the air has the composition shown in Table 7.4. You may assume a 530 ml tidal volume (at BTP), a breathing rate of 12 breaths/min, an O_2 consumption rate of 295 ml/min at BTP, and a CO_2 production rate of 235 ml/min at BTP.

Table 7.4. For Problem 7.16

Gas	Partial pressure (mmHg)
N_2	594
O_2	156
CO_2	0.3
H_2O	9.7
Total	760

Table 7.5. For Problem 7.17

Gas	Molar fraction in ambient air (%)
N_2	75.85
O_2	20.11
CO_2	0.04
H_2O	4.00
Total	100.00

The 530 ml tidal volume can be broken down into two parts: 150 ml of dead space air, and 380 ml of alveolar air. The dead space refers to the portion of the conducting airways where no blood/gas exchange takes place, such as the mouth, trachea, etc. Assume that no CO_2 or O_2 exchange takes place in the dead space, but that the expired dead space air is at 100% relative humidity (at BTP). What is the composition of the expired alveolar air, and of the expired dead space air, expressed as percentages by volume of N_2, CO_2, O_2, and H_2O?

7.17 During an experiment, a subject breathes in normally from the atmosphere but breathes out into a special device that collects the water in each exhaled breath. At the end of the experiment, the collected mass of water is 1.299 g. During the experiment, the ambient air had the composition shown in Table 7.5. How many breaths did the subject take during the experiment? Make and state necessary assumptions. You do not need to re-derive any formulae appearing in the text.

7.18 While a subject was exercising vigorously the measurements in Table 7.6 were made. All measurements are referenced to *BTP*. Note that tidal volume was 1000 ml (not 500 ml) and the breathing rate was 25 breaths/min. What were the subject's O_2 consumption rate and CO_2 production rate (ml gas at *STP* per minute)?

Table 7.6. For Problem 7.18

Gas	Molar fraction (%)	Partial pressure (mmHg)	Partial volume in one:	
			Inspired tidal volume (ml)	Expired tidal volume (ml)
N_2	78.62	597	786.2	786.2
O_2	20.84	159	208.4	158.4
CO_2	0.04	0.3	0.4	37.3
H_2O	0.50	3.7	5.0	64.7
Total	100.00	760	1000	1046.6

Table 7.7. For Problem 7.19

Gas	Molar fraction (%)
N_2	78.62
O_2	20.84
CO_2	0.04
H_2O	0.50
Total	100.00

7.19 An adult mouse has a respiratory rate of $\sim$163 breaths/min and a tidal volume of $\sim$0.15 ml at BTP (37 °C) [36]. The average O_2 consumption for one strain of male adult mice is 45.5 ml O_2/hour at STP [25]. For the inspired air composition shown in Table 7.7, compute the expired air composition. You may assume that the ratio of CO_2 production/O_2 consumption is the same for humans and mice.

7.20 In respiratory analysis, it is simplest to assume that both lungs are identical for purposes of gas exchange. This effectively implies that the air in the two lungs can be pooled together for purposes of quantitative analysis. However, in some cases, the volume of air entering each lung and the blood perfusion to each lung is different. This means that the amount of CO_2 added to each lung (per breath) will be different. For purposes of this question, we will assume that the number of moles of CO_2 added to a lung is proportional to the blood perfusion to that lung.

(a) Suppose that for a given subject the tidal volume and CO_2 concentration in expired air are measured to be 540 ml and 3.5%, respectively, and the dead space volume is known to be 160 ml. (See Problem 7.16 for definition of dead space volume.) If it is known that the blood perfusion to the right lung is twice that to the left lung, while the tidal volume entering the right lung is 1.6 times that of the left, compute the alveolar CO_2 concentrations in each lung.

(b) What criterion ensures that the alveolar CO_2 concentration is equal for both lungs?

References

1. A. J. Vander, J. H. Sherman and D. S. Luciano. *Human Physiology: The Mechanisms of Body Function*, 4th edn (New York: McGraw-Hill, 1985).

2. E. R. Weibel. *The Pathway for Oxygen: Structure and Function in the Mammalian Respiratory System* (Cambridge, MA: Harvard University Press, 1984).

3. E. R. Weibel. *Morphometry of the Human Lung* (New York: Academic Press, 1963).

4. D. W. Fawcett. *Bloom and Fawcett: A Textbook of Histology* (Philadephia, PA: W.B. Saunders, 1986).

5. D. E. Schraufnagel (ed.) *Electron Microscopy of the Lung* (New York: Marcel Dekker, 1990), p. 298.

6. E. R. Weibel and D. M. Gomez. Architecture of the human lung. Use of quantitative methods establishes fundamental relations between size and number of lung structures. *Science*, **137** (1962), 577–585.

7. M. Ochs, J. R. Nyengaard, A. Jung, L. Knudsen, M. Voigt *et al.* The number of alveoli in the human lung. *American Journal of Respiratory and Critical Care Medicine*, **169** (2004), 120–124.

8. D. L. Vawter and J. D. Humphrey. Elasticity of the lung. In *Handbook of Bioengineering*, ed. R. Skalak and S. Chien. (New York: McGraw-Hill, 1987), pp. 24.1–24.20.

9. J. A. Clements. Surface phenomena in relation to pulmonary function. *Physiologist*, **5** (1962), 11–28.

10. H. Bachofen, J. Hildebrandt and M. Bachofen. Pressure–volume curves of air- and liquid-filled excised lungs: surface tension in situ. *Journal of Applied Physiology*, **29** (1970), 422–431.

11. B. Sapoval, M. Filoche and E. R. Weibel. Smaller is better – but not too small: a physical scale for the design of the mammalian pulmonary acinus. *Proceedings of the National Academy of Sciences USA*, **99** (2002), 10411–10416.

12. S. S. Sobin, Y. C. Fung, H. M. Tremer and T. H. Rosenquist. Elasticity of the pulmonary alveolar microvascular sheet in the cat. *Circulation Research*, **30** (1972), 440–450.

13. C. Christoforides, L. H. Laasberg and J. Hedley-Whyte. Effect of temperature on solubility of O_2 in human plasma. *Journal of Applied Physiology*, **26** (1969), 56–60.

14. W. H. Austin, E. Lacombe, P. W. Rand and M. Chatterjee. Solubility of carbon dioxide in serum from 15 to 38 °C. *Journal of Applied Physiology*, **18** (1963), 301–304.

15. D. O. Cooney. *Biomedical Engineering Principles* (New York: Marcel Dekker, 1976).

16. B. M. Wiebe and H. Laursen. Human lung volume, alveolar surface area, and capillary length. *Microscopy Research and Technique*, **32** (1995), 255–262.

17. R. L. Capen, L. P. Latham and W. W. Wagner, Jr. Comparison of direct and indirect measurements of pulmonary capillary transit times. *Journal of Applied Physiology*, **62** (1987), 1150–1154.

18. N. C. Staub and E. L. Schultz. Pulmonary capillary length in dogs, cat and rabbit. *Respiration Physiology*, **5** (1968), 371–378.

19. C. G. Caro, T. J. Pedley, R. C. Schroter and W. A. Seed. *The Mechanics of the Circulation* (Oxford: Oxford University Press, 1978).

20. E. R. Weibel, W. J. Federspiel, F. Fryder-Doffey, C. C. Hsia, M. Konig *et al.* Morphometric model for pulmonary diffusing capacity. I. Membrane diffusing capacity. *Respiration Physiology*, **93** (1993), 125–149.

21. W. Huang, R. T. Yen, M. McLaurine and G. Bledsoe. Morphometry of the human pulmonary vasculature. *Journal of Applied Physiology*, **81** (1996), 2123–2133.

22. A. S. Popel. A finite-element model of oxygen diffusion in the pulmonary capillaries. *Journal of Applied Physiology*, **82** (1997), 1717–1718.

23. A. C. Guyton. *Textbook of Medical Physiology*, 4th edn (Philadelphia, PA: W. B. Saunders, 1971).

24. H. T. Milhorn, Jr. and P. E. Pulley, Jr. A theoretical study of pulmonary capillary gas exchange and venous admixture. *Biophysical Journal*, **8** (1968), 337–357.

25. A. O. Frank, C. J. Chuong and R. L. Johnson. A finite-element model of oxygen diffusion in the pulmonary capillaries. *Journal of Applied Physiology*, **82** (1997), 2036–2044.

26. A. A. Merrikh and J. L. Lage. Effect of blood flow on gas transport in a pulmonary capillary. *Journal of Biomechanical Engineering*, **127** (2005), 432–439.

27. E. L. Cussler. *Diffusion: Mass Transfer in Fluid Systems*, 2nd edn (New York: Cambridge University Press, 1997).

28. M. Felici, M. Filoche and B. Sapoval. Diffusional screening in the human pulmonary acinus. *Journal of Applied Physiology*, **94** (2003), 2010–2016.

29. E. R. Weibel, B. Sapoval and M. Filoche. Design of peripheral airways for efficient gas exchange. *Respiratory Physiology and Neurobiology*, **148** (2005), 3–21.

30. D. R. Owens, B. Zinman and G. Bolli. Alternative routes of insulin delivery. *Diabetic Medicine*, **20** (2003), 886–898.

31. J. B. Grotberg. Respiratory fluid mechanics and transport processes. *Annual Review of Biomedical Engineering*, **3** (2001), 421–457.

32. F. M. White. *Viscous Fluid Flow*, 2nd edn (New York: McGraw-Hill, 1991).

33. L. A. Spielman. Particle capture from low-speed laminar flows. *Annual Review of Fluid Mechanics*, **9** (1977), 297–319.

34. E. P. Radford, Jr. Recent studies of mechanical properties of mammalian lungs. In *Tissue Elasticity*, ed. J. W. Remington. (Washington, DC: American Physiological Society, 1957), pp. 177–190.

35. E. L. Green (ed.) for the Jackson Laboratory. *Biology of the Laboratory Mouse*, 2nd edn (New York: Dover, 1966).

36. M. R. Dohm, J. P. Hayes and T. Garland, Jr. The quantitative genetics of maximal and basal rates of oxygen consumption in mice. *Genetics*, **159** (2001), 267–277.

8 Muscles and movement

Together, muscles and bones confer structure and the capacity for motion to the body. Any analysis of locomotion (Ch. 10) must therefore have as a basis an understanding of muscle and bone mechanics.

There are three types of muscle, each having particular characteristics.

1. **Skeletal muscle.** Skeletal muscles comprise 40 to 45% of total body weight. They are usually attached to bones via tendons, are responsible for locomotion and body motion, and are usually under voluntary control.
2. **Smooth muscle.** Smooth muscle is typically found surrounding the lumen of "tubes" within the body, such as blood vessels, the urinary tract, and the gastrointestinal tract. Smooth muscle is responsible for controlling the caliber (size) of the lumen and also for generating peristaltic waves (e.g., in the gastrointestinal tract). Control of smooth muscle is largely involuntary.
3. **Cardiac muscle.** Cardiac muscle makes up the major bulk of the heart mass and is sufficiently unique to be considered a different muscle type. It is under involuntary control.

In this chapter we will concentrate on skeletal muscle only.

There are three types of skeletal muscle, classified according to how the muscles produce ATP, which is used during the contraction process. The two basic ATP-production strategies are:

1. **Aerobic.** In aerobic respiration, ATP is produced by the breakdown of precursors in the presence of O_2. This is a high efficiency pathway but cannot proceed if O_2 is not present.
2. **Anaerobic.** In anaerobic respiration (also known as glycolysis), ATP is produced without O_2 present. This pathway is less efficient than aerobic respiration and produces the undesirable by-product *lactic acid*. Accumulation of lactic acid in muscle tissue produces the characteristic ache that follows too strenuous a workout.

Available now. LINEAR MOTOR. Rugged and dependable: design optimized by worldwide field testing over an extended period. All models offer the economy of "fuel-cell" type energy conversion and will run on a wide range of commonly available fuels. Low stand-by power, but can be switched within milliseconds to as much as 1 kW mech/kg (peak, dry). Modular construction, and wide range of available subunits, permit tailor-made solutions to otherwise intractable mechanical problems.

Choice of two control systems:

1. Externally triggered mode. Versatile, general-purpose units. Digitally controlled by picojoule pulses. Despite low input energy level, very high signal-to-noise ratio. Energy amplification 10^6 approx. Mechanical characteristics: (1 cm modules) max. speed: optional between 0.1 and 100 mm/sec. Stress generated: 2 to 5×10^5 N/m^2.
2. Autonomous mode with integral oscillators. Especially suitable for pumping applications. Modules available with frequency and mechanical impedance appropriate for:
   • Solids and slurries (0.01–1.0 Hz).
   • Liquids (0.5–5 Hz): lifetime $2–6 \times 10^9$ operations (typical) $3–6 \times 10^9$ (maximum) – independent of frequency.
   • Gases (50–1,000 Hz).

Many optional extras e.g. built-in servo (length and velocity) where fine control is required. Direct piping of oxygen. Thermal generation. Etc.

Good to eat.

Figure 8.1

Announcement of a lecture on muscle, presented by Professor D. R. Wilkie to the Institution of Electrical Engineers in London. Modified from Lehninger [1]. Reproduced with kind permission of W. H. Freeman.

The three skeletal muscle types are

1. **Slow oxidative.**   Also known as slow twitch muscle, these cells produce ATP by aerobic respiration. These muscle fibers are used for endurance activities.
2. **Fast oxidative.**   These are also known as intermediate twitch fibers. In these cells, ATP is normally produced aerobically, but production can be fairly easily switched over to glycolysis. These muscle fibers are used for activities of moderate duration and moderate strength.
3. **Glycolytic.**   Also known as fast twitch fibers, these cells produce ATP anaerobically. They are used for high-intensity, high-strength activities.

In any given muscle, there will be a mixture of the above three types. Depending on the application, different muscle fibers can be recruited within the muscle.

8.1 Skeletal muscle morphology and physiology

Skeletal muscle is a remarkably efficient and adaptable tissue (see Fig. 8.1). Skeletal muscles are typically relatively long and thin, often described as *spindle shaped*. In

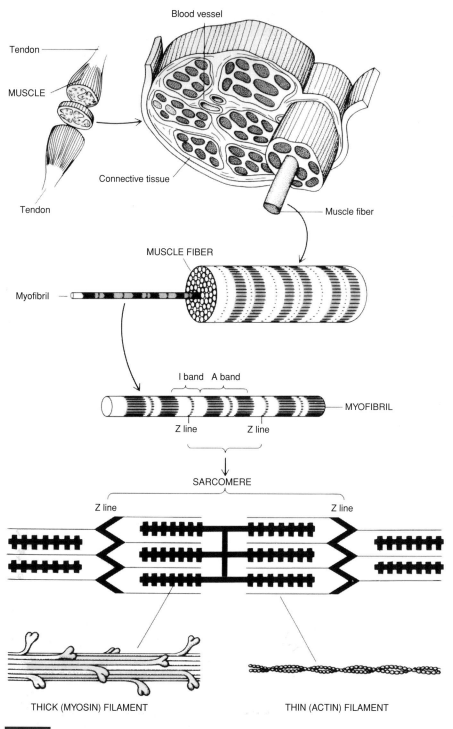

Figure 8.2

Overview of skeletal muscle structure, showing progressively finer levels of organization. From Vander *et al.* [2]. Reproduced with kind permission of The McGraw-Hill Companies.

cross-section, skeletal muscle is seen to consist of blood vessels, connective tissues, and muscle cells, also known as *muscle fibers* (Fig. 8.2). Individual muscle fibers are long rod-shaped cells, 10 to 100 μm in diameter, and up to 30 cm in length! Typically, they are multinucleated and contain many mitochondria.[1] However, the most prominent constituent of the muscle fiber is the *myofibril*. Myofibrils are long rod-shaped elements that are responsible for causing muscle cell contraction. Morphologically, myofibrils have a regular banded structure, the repeat unit of which is called a *sarcomere* (Fig. 8.3). This periodicity is a consequence of the internal structure of the myofibril, which is composed of a definite arrangement of even smaller filaments:

- *thick filaments*, or *myosin*, which have a diameter of 12 to 18 nm
- *thin filaments*, or *actin*, which have a diameter of 5 to 8 nm (see Section 2.3.1).

The myosin and actin are arranged as shown in Fig. 8.2. The essential features of this arrangement are that the myosin and actin filaments interpenetrate, and that the myosin filaments have cross-bridges which are able to attach temporarily to the actin filaments (Fig. 8.4).

Knowledge of sarcomere morphology is necessary to understand the *sliding filament model*, which describes the details of how sarcomeres effect muscle contraction (Fig. 8.5, color plate). It is known that in the presence of ATP, myosin cross-bridges are able to bind to complementary sites on the actin filaments, thereby forming the so-called *actin–myosin complex*. This binding utilizes energy, which is provided by cleavage of the high-energy bond in ATP to give ADP plus PO_3^{2-}. Loss of the phosphate from the complex causes the myosin heads to "tip," thereby "dragging" the actin filament along the myosin filament. Once the myosin head has reached its maximally "tipped" position, it is able to bind another ATP molecule, which causes the myosin head to detach from the actin and return to its base configuration.

So long as ATP is available, the net effect is similar to that of a ratchet, in which myosin heads attach, tip to drag actin, detach, tip again, etc. When repeated many times, and by myosin heads throughout the muscle, the net effect is to shorten each sarcomere and thereby cause the muscle to contract. This is a mechanism for the direct conversion of chemical energy (stored in ATP) into motion.

As for all systems in the body, muscle contraction is controlled, in this case by *troponin* and *tropomyosin*. These are two proteins that normally act together to cover the cross-bridge binding sites on actin, thereby preventing the formation of the actin–myosin complex and myofibril contraction. However, when calcium

---

[1] The reader will recall from Section 2.2 that mitochondria are the organelles responsible for energy production in cells. Since muscle motion requires substantial amounts of ATP, it is logical that muscle fibers should be heavily invested with mitochondria.

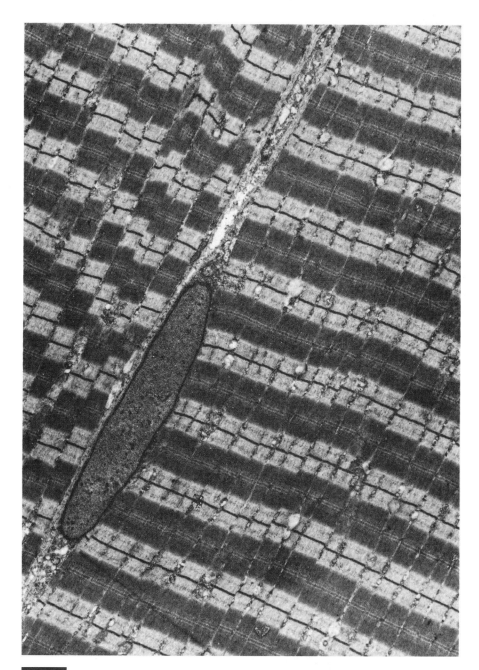

Figure 8.3

Transmission electron micrograph of skeletal muscle. The sample has been treated to make the sarcomeres more evident, allowing us to visualize their regular repeat structure. Note the fairly uniform diameter of the myofibrils and the nucleus located at the edge of the myofibril bundles. Corresponding bands of adjacent myofibrils are usually in register across the muscle fiber, as shown in the right portion of the image. They are out of alignment in the upper left region; this is probably a specimen preparation artefact. From Fawcett [3] with permission of Hodder Education.

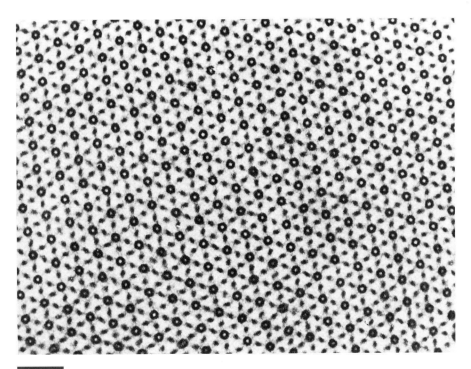

Figure 8.4

Electron micrograph of insect flight muscle in cross-section, showing actin filaments (small dashes) interpenetrating into myosin filaments (circles). Notice the beautifully regular filament architecture. From Fawcett [3] and Prof. Hans Ris, with kind permission.

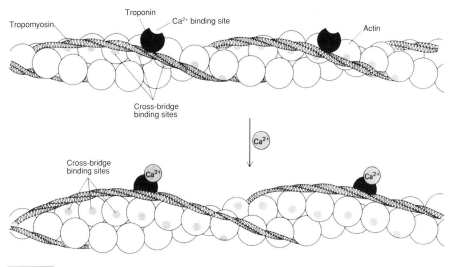

Figure 8.6

Role of troponin and tropomyosin in exposing actin-binding sites, and thereby controlling myofibril contraction. Binding of Ca^{2+} to troponin moves the tropomyosin, exposing binding sites, and allowing myosin cross-bridges to bind to actin. From Vander *et al.* [2]. Reproduced with kind permission of The McGraw-Hill Companies.

ions (Ca^{2+}) bind to troponin, tropomyosin undergoes a conformational change that exposes the binding sites on actin (Fig. 8.6). The presence of free Ca^{2+} in the cell (at concentrations >1 μmol/l) will induce myofibril contraction,[2] and Ca^{2+} availability is therefore able to provide a control mechanism for muscular contraction. The above discussion naturally prompts the question of how Ca^{2+} levels within the myofibrils are controlled. To answer this question, it is necessary to describe briefly the process of *neurotransmission* and the anatomy and physiology of the *motor end plate*. We will start with neurotransmission.

Nerve cells must communicate with muscle cells and with other nerve cells. This is accomplished through biochemical means, via the release of compounds (*neurotransmitters*) from the signaling nerve cell to the recipient muscle or nerve cell. The main site of such communication is a specialized junction known as a *synapse*. Typically, the process begins with the arrival of an *action potential* – an electrochemical signal that propagates as a traveling wave along the axons of neurons – at a synapse.[3] This causes neurotransmitters to be released into the space between the signaling and recipient cells (the *synaptic cleft*). The neurotransmitters diffuse across the synaptic cleft, are recognized by the recipient cell (in our case, muscle), and activate processes in the recipient cell. A single nerve fiber can synapse with as few as several muscle cells, or as many as several thousand muscle cells; muscles that require fine motor control have fewer muscle fibers per nerve. A synapse between a nerve and a muscle cell is also known as a motor end plate, and the collection of all muscle cells controlled by a single nerve fiber is known as a *motor unit* (Fig. 8.7). Nerves that control muscles are known as *motor neurons*.

The arrival of an action potential at the motor end plate, and subsequent neurotransmitter release and uptake by the muscle cells, causes Ca^{2+} to be released from storage vesicles into the cytoplasm of the muscle cell. The release and spread of Ca^{2+} is facilitated by the *sarcoplasmic reticulum*, which is essentially a connected network of channels through which Ca^{2+} can be distributed within the cell (Fig. 8.8). There is a delay of approximately 3–10 ms between arrival of the action potential and the resulting release of Ca^{2+} in the muscle cell. This is known as the *latency period* between stimulation and contraction. Because of its central role in signaling within the cell, cytoplasmic Ca^{2+} concentration is tightly controlled by the cell. Once Ca^{2+} is released within the cell to cause muscular contraction, the cell immediately begins to re-sequester the ions in storage vesicles. As Ca^{2+} is removed from the cytoplasm by this process, the contraction stops

---

[2] The presence of ATP is also required; normally, however, there is sufficient ATP present that this is not a limiting factor.
[3] See Appendix for further discussion of how action potentials are generated and propagated.

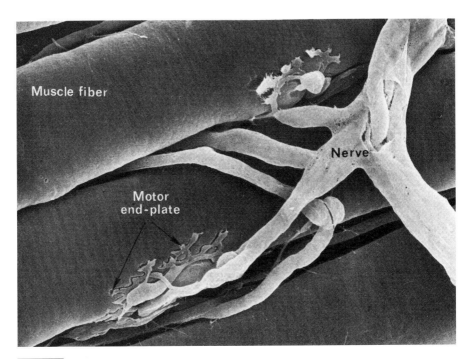

Scanning electron micrograph of a motor nerve and two end-plates on adjacent muscle fibers. From Desaki and Uehara [5]. With kind permission of Springer Science and Business Media.

and the muscle returns to its original length. The net effect is that a single action potential produces a transient contraction of the motor unit, known as a *twitch* (Fig. 8.9).

8.1.1   Isotonic versus isometric contraction

Suppose that we isolate and excise a single muscle and wish to characterize its behavior. There are two basic types of test that can be performed on it (Fig. 8.10):

- **isotonic:** in an isotonic test, a known (constant) tension is applied to the muscle and its length is recorded
- **isometric:** in an isometric test the muscle is held at a constant length and the force generated by the muscle during contraction is recorded.

Consider the response of a muscle in an isometric test. In response to a single action potential, the muscle will develop tension as soon as the latent period has passed.

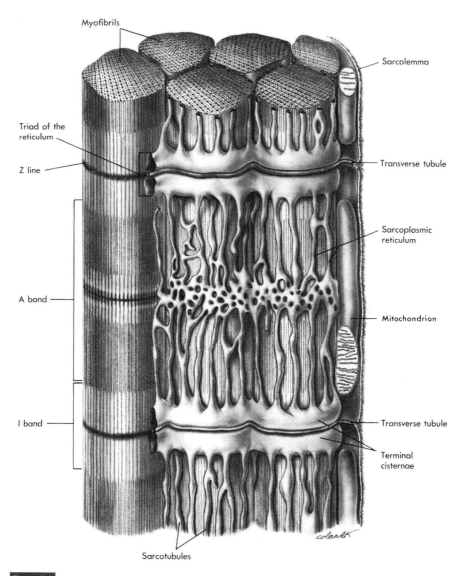

Myofibrils

Sarcolemma

Triad of the reticulum

Z line

Transverse tubule

Sarcoplasmic reticulum

A band

Mitochondrion

I band

Transverse tubule

Terminal cisternae

Sarcotubules

Figure 8.8

Schematic diagram of sarcoplasmic reticulum in skeletal muscle. The network of longitudinal sarcotubules, transverse tubules, terminal cisternae, and sarcoplasmic reticulum permits rapid and efficient distribution of Ca^{2+} within the muscle. This schematic is specific for amphibian skeletal muscle; mammalian skeletal muscle has two transverse tubules per sarcomere, located at the A–I junctions. From Fawcett [3], based on *The Journal of Cell Biology*, 1965, 25, pp. 209–231. Reproduced by copyright permission of the Rockefeller University Press.

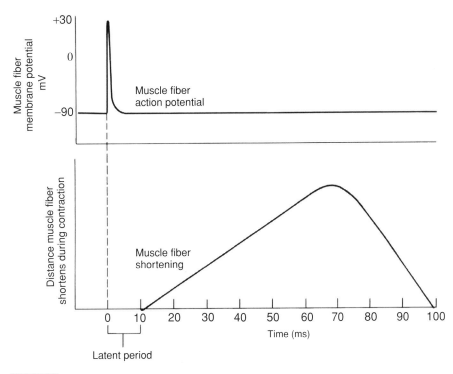

Figure 8.9

Twitch of a muscle fiber in response to a single action potential. Note the latent period between action potential arrival and the development of mechanical contraction. From Vander *et al.* [2]. Reproduced with kind permission of The McGraw-Hill Companies.

Intuitively, we expect that the developed tension will reach some maximum value then decline to the resting state (upper panel of Fig. 8.11).[4] The duration of the twitch is approximately 100 ms.

The situation is somewhat different in the isotonic test. There can be no muscle motion until sufficient force is developed to overcome the applied tension. Since the developed tension does not reach its maximal value instantaneously, this implies that the larger the applied tension, the longer a delay there will be until the muscle begins to shorten. Thus, in the isotonic case, the latency period increases with load (applied tension), as shown in the lower panel of Fig. 8.12. By the same reasoning, the period during which the muscle can shorten during an isotonic experiment

---

[4] The reader may wonder how it is possible for the muscle to "contract" even though it is being held at constant length. It must be remembered that muscle consists of muscle cells and connective tissue. In response to stimulus, the muscle cells shorten. Since the muscle length is constant, the connective tissue must extend. This is discussed in more detail in Section 8.2.

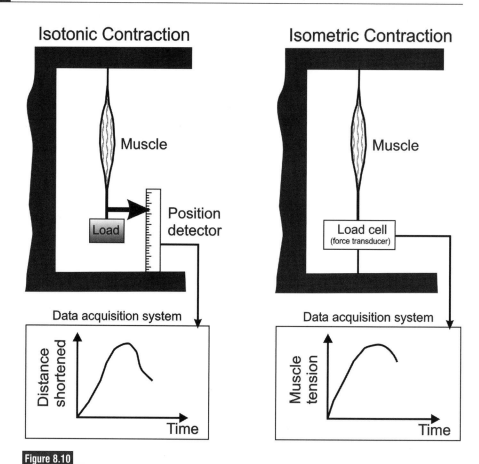

Schematic description of experimental set up used to measure isotonic and isometric muscle contractions.

is shorter when the applied tension is large. Therefore, the distance shortened decreases with increasing applied load in an isotonic test (Fig. 8.12).

To this point, we have only discussed muscle response to single action potentials. However, in practice a second action potential will usually arrive before the first twitch has finished. In this case, the two twitches sum together to produce a greater contraction than could be achieved from a single twitch, a condition known as *unfused tetanus* (Fig. 8.13). If a rapid train of action potentials is applied to the motor unit, the twitches sum together to yield an approximately constant contraction in a state known as *fused tetanus*. Increasing the stimulation frequency beyond this point does not lead to an increase in developed tension in the muscle. The tension developed in a muscle during fused tetanus is known as *tetanic tension*.

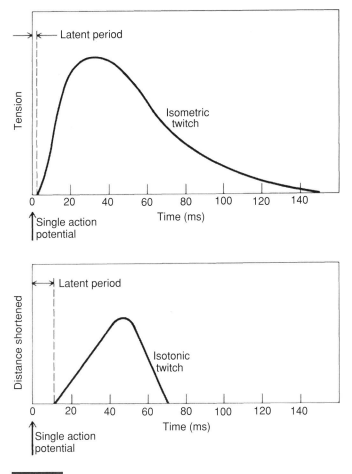

Latent period in isometric and isotonic twitches. The mechanical response after a single action potential for a skeletal muscle fiber is shown. From Vander *et al.* [2]. Reproduced with kind permission of The McGraw-Hill Companies.

8.2 Muscle constitutive modeling

The tension which a muscle fiber can generate depends on a number of factors.

Length. There is found to be an optimal length for muscle fibers; that is, in an isometric experiment, maximum tetanic tension is generated at some optimal length l_0. When the muscle is stretched to beyond this length, the actin and myosin filaments overlap less, with the result that less tetanic tension can be developed at this length. Conversely, when the muscle is shortened to less than its optimal

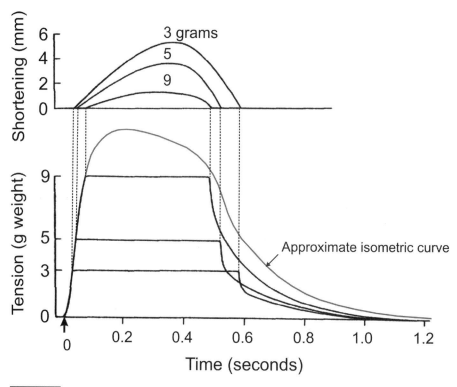

Figure 8.12

Tracings of distance shortened (top graph) and developed tension (bottom graph) versus time in frog sartorius muscle for isotonic twitches against loads of 3, 5 and 9 g. The muscle was stimulated by a single action potential at time zero (arrow). Notice how the developed tension increases while the muscle length stays unchanged until the tension reaches the applied load, after which point the muscle begins to shorten. Also shown in the bottom panel is a curve for an isometric contraction (light line). Isotonic twitch curves are replotted from Jewell and Wilkie [6] with permission of Blackwell Publishing Ltd; the isometric twitch curve is extrapolated from data in the same article but for a different muscle and should therefore be considered to be only approximate.

length, the actin and myosin filaments interfere with one another, also reducing the tetanic tension. This produces a load–length relationship, which is graphed in Fig. 8.14. In the body, muscles always work in a length range where developed tension is at least half of maximum.

Load.   Suppose that a muscle is subjected to high-frequency stimulation for an extended period, so that it goes into tetanus. If we hold the muscle isometrically and then suddenly release it, the velocity of contraction V can be measured for different loads placed on the muscle, P. For some large load, the muscle will be unable to move the load, and the contraction velocity will be zero. On the other extreme, for small loads, it will be able to contract rapidly. This leads to the load–velocity relationship shown in Fig. 8.15. A. V. Hill [9] showed that this relationship could

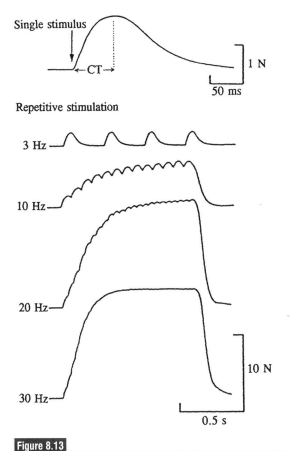

Single stimulus

CT

1 N

50 ms

Repetitive stimulation

3 Hz

10 Hz

20 Hz

10 N

30 Hz

0.5 s

Figure 8.13

Tension developed by a human extensor hallucis brevis muscle versus time for a range of stimulus frequencies. The top trace shows a twitch in response to a single stimulus under isometric conditions. The twitch contraction time (CT) is the interval between the beginning of tension development and the peak tension. The lower four traces show the tension developed under progressively higher stimulation frequencies (3 Hz to 30 Hz), progressing from unfused to fused tetanus. Note the different scales for the upper trace and the lower four traces. From McComas [7], with kind permission of the author.

be described by the equation

$$(a + P)(b + V) = b(P_0 + a) \tag{8.1}$$

where a, b_0 and P_0 are constants, with P_0 equal to the applied load when V is zero.

Time.   We have previously noted that the tension developed during an isometric twitch is time dependent. This results from several factors. First, the development of tension by the myofibrils in response to an action potential is time dependent. Second, a whole muscle consists of more than just myofibrils, including significant amounts of connective tissue, blood vessels, etc. Connective tissue, in particular, has a significant amount of elasticity and also acts to damp motion;

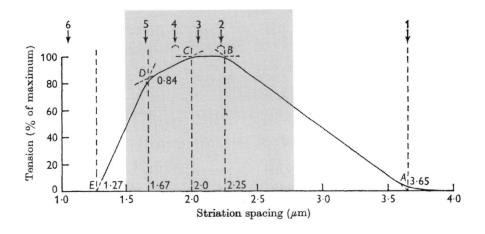

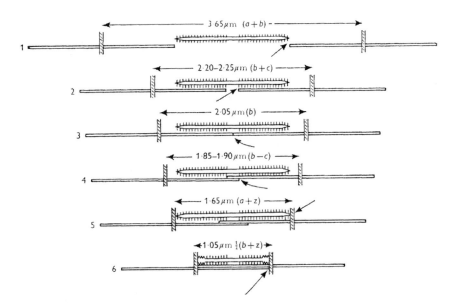

Figure 8.14

Load–length relationship for skeletal muscle. The vertical axis in the top panel is isometric tetanic tension in an isolated frog skeletal muscle, expressed as a percentage of the maximum attainable tension; the horizontal axis is striation spacing (in micrometers), a measure of sarcomere length. Maximum attainable tension occurs at an optimal sarcomere length of 2.13 μm. The shaded region represents the range of lengths (approximately 70% to 130% of the optimal sarcomere length) that occur normally in the body. The diagrams of sarcomere geometry for different sarcomere lengths at the bottom correspond to labels 1–6 in the graph. The lengths shown on the lower panel are measured from one Z-line to the next Z-line (see Fig. 8.2). Modified from Gordon *et al.* [8] with permission of Blackwell Publishing Ltd.

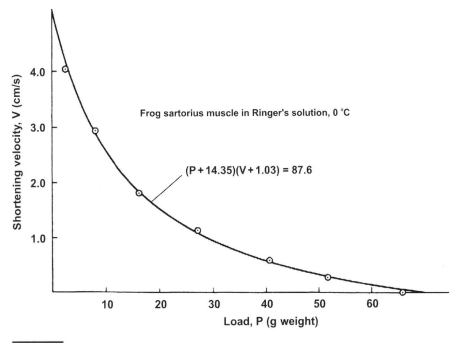

Frog sartorius muscle in Ringer's solution, 0 °C

$$(P + 14.35)(V + 1.03) = 87.6$$

Figure 8.15

Load–velocity relationship for isotonic tetanized muscle under different loadings. Data are obtained from frog skeletal muscle (sartorius muscle), and the solid line is a plot of Equation (8.1) with load P expressed in g weight and V expressed in cm/s. Modified from Hill [9], with permission of the Royal Society; this paper is recommended to the reader as a wonderful example of experimental ingenuity and problem solving.

consequently, it modifies the force output of the myofibrils (Fig. 8.16). This can be made more specific by considering a whole muscle to consist of three mechanical components:

- **a contractile, or active, element**, consisting of the actin–myosin complex, which produces a force rapidly after stimulation
- **a spring, or elastic component**, representing the elasticity of the connective tissue within the muscle
- **a dashpot**, representing the effects of frictional dissipation within the muscle; frictional dissipation occurs as different portions of the muscle slide relative to one another and as fluid within the muscle is sheared.

The reader will recognize the similarities between this lumped parameter approach and the lumped parameter modeling of cell biomechanics (Section 2.6.1), and we will use this model again when we come to connective tissue biomechanics (Section 9.10.5). Here the unique feature is the active element.

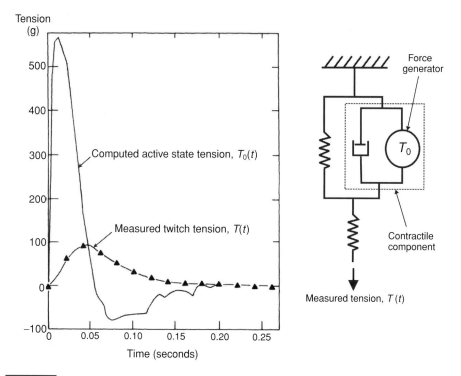

Figure 8.16

The graph shows the measured time course of muscle tension, $T(t)$, during an isometric twitch of a frog gastocnemius muscle at 24 °C (triangles). The curve labeled "computed active state tension" is an estimate of the tension generated by the active component (the myofibrils) of the muscle, computed using the model in the right panel. This model incorporates a damping element (dashpot) and two elastic elements, the properties of which were estimated from independent experiments. Notice the very significant difference between the time course and magnitude of the tension generated by the force generator and the tension delivered to the surroundings at the ends of the muscle. Adapted from McMahon, T. A.; *Muscles, Reflexes, and Locomotion* [10], based on data from [11]. © 1984 by T. A. McMahon. Reprinted by permission of Princeton University Press.

To examine the time course of muscle shortening further, consider an idealized model of muscle consisting of the three components identified above, as shown schematically in Fig. 8.17. The myofibrils (actin–myosin complex, or active element) are idealized as a force generator that produces a constant tension T_0 for a finite duration C upon stimulation (Fig. 8.17).[5] The spring is assumed to be linearly elastic, with spring constant k_0. The dashpot is also assumed to be linear, so that the force produced by the dashpot is given by

$$F_{dash} = \eta_0 \frac{dx_1}{dt} \qquad (8.2)$$

---

[5] Examination of Fig. 8.16 shows that this is not strictly the case, but this assumption at least maintains the essential feature that the active state tension is transient in response to a single stimulus.

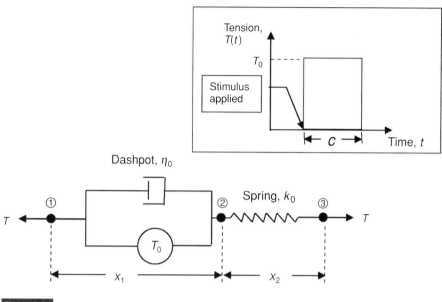

Figure 8.17

Three element model for muscle. x_1 is the length of the dashpot portion of the muscle (assumed equal to the length of the force generator) and x_2 is the length of the elastic portion of the muscle. See the text for the description of the characteristics of the dashpot and spring. Inset shows the force generated by the active element as a function of time following a single stimulation.

where η_0 is the dashpot damping constant and x_1 is defined in Fig. 8.17. Note that the force exerted by the dashpot on the surroundings is such to oppose motion.

As was the case for the lumped parameter models representing cell biomechanics (Section 2.6.1), the arrangements of components in this three element model is somewhat arbitrary; for example, there is no reason *a priori* why the damping element should be in parallel with the active element, or why the elastic element should be in series. In fact, alternative models have been used [10]. However, this model is as valid as any other three element arrangement and yields results that are qualitatively correct.

The tension T transmitted by the muscle to its surroundings will depend on T_0, C, K_0, and η_0. We assume that the initial tension on the muscle is zero, and that the muscle has negligible mass. Using these assumptions, we can ask what $T(t)$ is for the above arrangement during an isometric contraction. We answer this question by first noting that if the contraction is isometric, we have $x_1 + x_2 = constant$. Differentiation of this expression yields

$$\frac{dx_1}{dt} + \frac{dx_2}{dt} = 0. \tag{8.3}$$

Splitting the model at node ②, and drawing a free body diagram for the spring immediately shows that the internal tension T_{int} between the spring and the remainder of the model satisfies:

$$T - T_{int} = m_{spring}\, a_{spring}. \tag{8.4}$$

However, since we have assumed that the mass of the muscle is small, the right-hand side of Equation (8.4) can be neglected and the internal tension must equal $T(t)$. The constitutive equation for the spring can therefore be written as

$$T = k_0\big(x_2 - x_2^0\big) \tag{8.5}$$

where x_2^0 is the relaxed length of the spring.

A similar argument applied to forces acting on the dashpot/active element combination shows that the sum of forces produced by the dashpot and the active element must be equal to $T(t)$. Assuming that the active element begins to contract at time $t = 0$, and taking proper account of the orientation of forces, this implies that

$$T = T_0 + \eta_0 \frac{dx_1}{dt} \quad \text{for } 0 \leqslant t \leqslant C. \tag{8.6}$$

$$T = \eta_0 \frac{dx_1}{dt} \quad \text{for } t > C. \tag{8.7}$$

Consider first the period $0 \leqslant t \leqslant C$: combining Equations (8.3), (8.5) and (8.6) yields

$$\frac{T - T_0}{\eta_0} + \frac{1}{k_0}\frac{dT}{dt} = 0. \tag{8.8}$$

This is a first-order, linear, ordinary differential equation for $T(t)$. Its solution, taking account of the assumption that the initial tension in the muscle was zero, is

$$T(t) = T_0\left[1 - e^{-k_0 t/\eta_0}\right] \quad \text{for } 0 \leqslant t \leqslant C. \tag{8.9}$$

This shows that the force increases exponentially towards the maximal value of T_0. A similar derivation for the period $t > C$ gives a solution

$$T(t) = T_0\left[1 - e^{-k_0 C/\eta_0}\right] e^{-k_0(t-C)/\eta_0} \quad \text{for } t > C. \tag{8.10}$$

This produces a force versus time curve as shown in Fig. 8.18. It can be seen that the output of the active element is modified extensively before being transmitted to the surroundings.

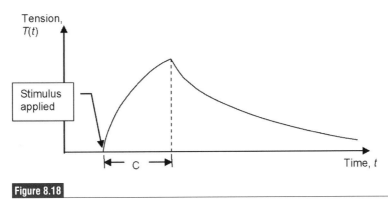

Figure 8.18

Force output predicted from the three element model for muscle, as given by Equations (8.9) and (8.10).

8.3 Whole muscle mechanics

We now turn to the study of forces produced by whole muscles in practical situations within the body. Any study of whole muscle mechanics must examine two main factors: first, the arrangement of muscle fibers within a particular muscle, and second, the arrangement of that muscle with respect to the bones it attaches to. We will consider the first issue in this section; interactions between muscles and bones will be considered in Section 8.4.

8.3.1 Parallel versus pinnate muscle types

In many muscles, the muscle fibers are simply aligned with the axis of the muscle. In such cases, the force generated by the muscle during contraction, F, is the force per unit area generated by muscle fibers, f, multiplied by the cross-sectional area of the muscle, A. With reference to Fig. 8.19:

$$F_{\text{parallel}} = fA = 2awf. \tag{8.11}$$

There are a number of alternative muscle fiber arrangements. One of these is known as a *pinnate* (or *pennate*) *arrangement*, in which the muscle fibers are oriented obliquely with respect to the axis of contraction of the muscle. For example, in Fig. 8.20, a *bipinnate* muscle is schematically shown, consisting of two pinnate groups. An example of a bipinnate muscle that many readers will be familiar with is in the pincer of a crab or lobster, where the interface between the two muscle halves is a hard connective tissue. Humans also have pinnate muscles; one example

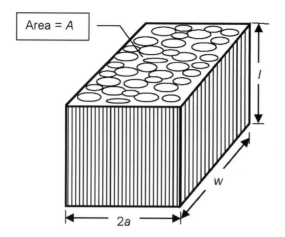

Figure 8.19

Portion of a parallel fiber muscle. The muscle fibers are aligned vertically, and the muscle cross-sectional area A is $2aw$. The length of the muscle fibers is l.

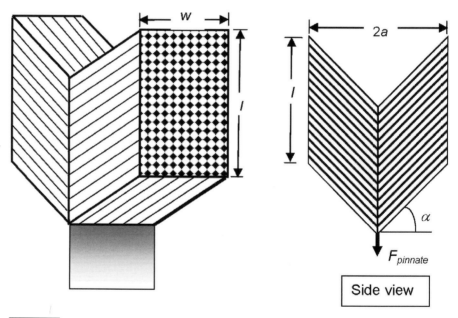

Figure 8.20

Schematic drawing of a portion of a bipinnate muscle. This muscle has the same volume as that shown in Fig. 8.19, but the fibers are inclined at an angle $90° - \alpha$ with respect to the direction in which tension is delivered.

is the *rectus femoris*, the major muscle on the front of the femur. As discussed by Fung [12], it is straightforward to show that the force generated by a bipinnate muscle, F_{pinnate}, is given by

$$F_{\text{pinnate}} = 2wfl\cos\alpha\sin\alpha. \tag{8.12}$$

It is of interest to compare the force generated by the bipinnate arrangement with that generated by the parallel arrangement. To make the comparison meaningful, we must ensure that the bipinnate and parallel muscles have the same total volume, which inspection of Figs. 8.19 and Fig. 8.20 will confirm. From Equations (8.11) and (8.12)

$$\frac{F_{\text{pinnate}}}{F_{\text{parallel}}} = \frac{l\cos\alpha\sin\alpha}{a}. \tag{8.13}$$

The term $\cos\alpha\sin\alpha$ is always < 1, but the ratio l/a can be very large. If this is the case, then Equation (8.13) shows that the force generated by the bipinnate arrangement can be larger than that generated by the parallel arrangement, for a given muscle mass. Further, the bipinnate muscle does not expand laterally as it contracts, while a parallel muscle arrangement does. Consequently, there are advantages to a pinnate arrangement. However, there is also a disadvantage: for a given shortening velocity of the muscle fibers, the bipinnate muscle will shorten more slowly than the parallel arrangement. These points are discussed in greater detail by McMahon ([10], pp. 54 ff.).

8.4 Muscle/bone interactions

Skeletal muscle is always attached to bone, and forces developed during muscular contraction are usually transmitted to the "outside world" via the bone. As an example, consider running: muscles attached to bones within the leg contract in such a way that the feet exert a force on the ground which propels the runner forward. The specific geometry of the muscle/bone arrangement clearly influences the forces transmitted to the surroundings, the speed of motion of the bones that the muscles attach to, and the stresses within the bones involved. We examine these issues below.

8.4.1 Foreleg motion in two species

We begin with an example in which foreleg motion in two different animals is compared. Specifically, we examine rotation of the *humerus*, the upper bone in the foreleg, about the shoulder joint in the armadillo and the horse. Humeral rotation is primarily accomplished by the *teres major muscle*, which inserts into the humerus

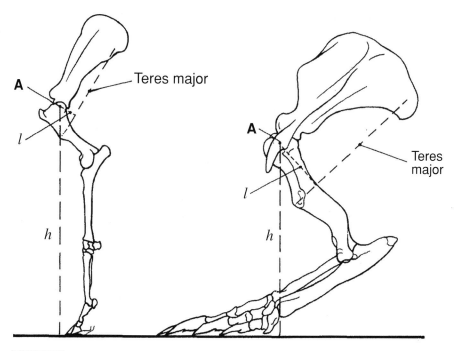

Teres major

A

Teres major

Figure 8.21

Anatomy of the bones in the foreleg from the horse (left) and the armadillo (right). The forelegs have been scaled so that they appear the same size. The point labeled "A" is the center of rotation of the humerus. The distance l is measured perpendicular to the line of action of the teres major muscle. From Young [13]. Reproduced with kind permission of Oxford University Press and Prof. J. Maynard Smith.

and the *scapula* (shoulder blade), as shown in Fig. 8.21. We wish to determine the mechanical advantage of the teres major muscle in these two species: that is, the ratio of forces transmitted to the foot for a given amount of force in the muscle. This will depend on the distance between the foot and the center of rotation, h, and the normal distance between the center of rotation and the line of action of the teres major, l. We assume that the knee joint is locked, and call:

F_m = force developed in the teres major muscle
F_g = force exerted on the ground by the foot.

If we further assume that the inertia of the foreleg can be neglected, then the sum of moments about the center of rotation A must be zero. In this case, it directly follows that the force transmitted to the ground by the foot, F_g, is given by

$$F_g = \frac{l}{h} F_m. \tag{8.14}$$

Equation (8.14) shows that the mechanical advantage of the teres major muscle is given by the ratio l/h. The larger this ratio, the greater the force delivered to the foot per unit force generated in the teres major muscle. However, the larger this ratio, the slower the foot will move for a given velocity of contraction of the teres major. For animals that are specialized for digging and burrowing, such as the armadillo, large forces must be delivered to the foot. It is expected that such animals will have a relatively large l/h. By comparison, animals that are specialized for running, such as the horse, require a foot that moves quickly. This is accomplished by a small l/h ratio. Measurements show that l/h is approximately 1/4 for an armadillo and 1/13 for a horse, in agreement with the above discussion.

The armadillo is further specialized to be able to deliver large forces to the foot. Examination of Fig. 8.21 shows that the armadillo has a proportionately much larger *olecranon process*, the bony protrusion behind the "elbow" joint. This is the point of insertion of the *triceps muscle*, which is responsible for straightening the front knee joint. Thus, a large olecranon process provides a large lever arm, which, in turn, leads to relatively larger forces acting to straighten the front leg.

A corollary of the above discussion is that animals like the armadillo accelerate to top running speed faster than an animal like a horse. This seems counter-intuitive, until we recognize that the top speed for the armadillo is also much less than that for the horse, so that the absolute acceleration of the horse is greater.

8.4.2 Flexion of the elbow

We now examine flexion (bending) of the elbow in humans, with particular emphasis on the stress this places on the bones in the forearm. As shown in Fig. 8.22, there is one bone in the upper arm (*the humerus*) and two bones in the forearm (*the radius* and *the ulna*). There are numerous bones in the wrist and hand; however, for the purposes of this analysis we will treat the wrist and hand as a rigid assembly and primarily concentrate on stresses in the radius and ulna.

We wish to estimate the stresses in the forearm caused by holding a weight W in place against the force of gravity. We will assume that the forearm is in supination,[6] as shown in Fig. 8.22. Analysis of this problem first requires that the geometry of the relevant musculoskeletal elements be quantified. Hsu *et al.* [14] defined the length of the radius bone as the distance from the proximal articulating surface to the distal "tip" of the styloid process (see Fig. 8.22), and measured a mean radius bone length of 25.8 cm in eight human arms. Using a similar approach, Murray *et al.* [15] report a mean length of 23.9 cm for 10 human arms. We will take a length of 24 cm as being typical for the radius bone. We will further assume that the weight acts through a line 6 cm distal to the tip of the styloid process, corresponding approximately to the middle of the palm. Finally, we will use the

---

[6] When the forearm is in *supination*, it is rotated so that the palm is facing upwards. The opposite configuration is *pronation*, with the palm facing down.

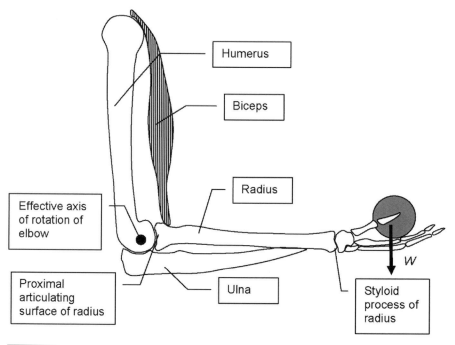

Humerus

Biceps

Radius

Effective axis
of rotation of
elbow

Proximal
articulating
surface of radius

Ulna

W

Styloid
process of
radius

Figure 8.22

Bones of the upper and lower arm, as seen from the lateral side. In the analysis described in the text, a weight *W* is
assumed to be supported against the force of gravity as shown. The distal portion of the ulna is hidden behind the radius.

anthropometric data of Murray *et al.* [15] to take the mean length of the humerus as
32 cm.

The effective axis of rotation of the elbow is not fixed but varies with the degree
of elbow flexion [16]. However, for purposes of this example, we will assume a
fixed axis of rotation that is 1 cm to the left of the proximal articulating surface of
the radius, and 1 cm superior to the distal end of the humerus (Fig. 8.22).

Turning now to the muscles responsible for elbow flexion, we will consider four
major muscles (Fig. 8.23), as listed below.

1. **The biceps.** This muscle arises from the scapula at two points of origin;
   muscle from these two locations fuses to give the biceps. Distally, the biceps
   insert primarily into the radius, with a small attachment to the ulna.
2. **The brachialis.** This muscle arises from the anterior surface of the humerus,
   approximately halfway to the elbow. Distally, it attaches to the ulna.
3. **The brachioradialis.** This muscle arises on the lower lateral portion of the
   humerus and inserts into the radius near the wrist. The muscle itself is fairly
   short, but the tendon attaching to the radius is quite long.

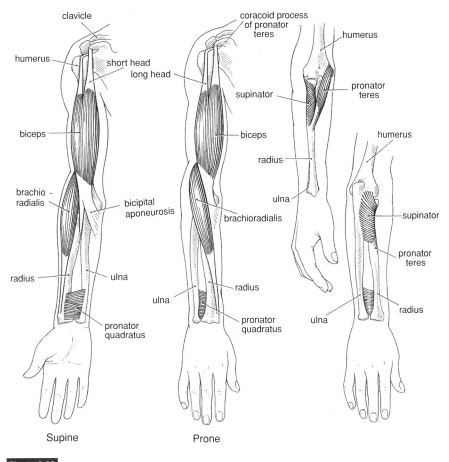

clavicle

humerus

short head
long head

biceps

brachio-
radialis

bicipital
aponeurosis

radius

ulna

pronator
quadratus

coracoid process
of pronator
teres

humerus

supinator

biceps

radius

ulna

brachioradialis

ulna

radius

pronator
quadratus

humerus

pronator
teres

radius

ulna

humerus

supinator

pronator
teres

radius

ulna

Supine   Prone

Figure 8.23

Selected muscles responsible for motion of the forearm in humans. Not all muscles responsible for elbow flexion are shown in this image, which also includes muscles responsible for forearm rotation (pronation/supination). From Young [13]. Reproduced with kind permission of Oxford University Press.

4. **The extensor carpi radialis longus (ECRL).** This muscle arises from the lateral aspect of the humerus very near the elbow and inserts on the second metacarpal (wrist) bone. Since the hand articulates with the radius, this can be mechanically approximated as an insertion into the radius.

Relevant anatomic characteristics of these muscles are shown in Table 8.1. The muscle origin and insertion locations are obviously approximate, since the attachment of the muscles is spread over a finite area of bone; to determine a single value for the origin or insertion locations we have used the centroid of the

Table 8.1. Origin and insertion points and other characteristics of the four major muscles participating in elbow flexion. L_H is the distance from the effective center of rotation for the elbow to the muscle origin location on the humerus; L_F is the corresponding measurement for the insertion location on the forearm. PCSA is the physiologic cross-sectional area of the muscle (see text). $\theta = \tan^{-1}(L_H/L_F)$ is the angle that the muscle makes with respect to the horizontal when the forearm is in the position shown in Fig. 8.22. "Inserts into" refers to which bone the muscle inserts into in the forearm.

Muscle	L_H (cm)	L_F (cm)	PCSA (cm^2)	θ($^\circ$)	Inserts into
Biceps	31[a]	8	12.3	76	Radius
Brachialis	10	5	13.0	63	Ulna
Brachioradialis	8	24	2.9	18	Radius
Extensor carpi radialis longus	3	25[b]	3.6	7	Radius

[a] The biceps does not originate from the humerus. Its effective point of origin is taken at the top of the humerus.
[b] The ECRL inserts into the wrist; its effective insertion point is taken as the end of the radius.

origin or insertion region as estimated from anatomic drawings. The physiological cross-sectional area (PCSA), defined as the muscle volume divided by the muscle fiber length measured normal to the fiber axis, is particularly important for us. Because it takes into account muscle fiber orientation, it is a measure of the number of sarcomeres in parallel in the muscle. This area is therefore considered to be an indirect measure of the muscle's force-generating capacity [17,18]. Murray *et al.* [18] reported PCSA values for elbow flexors measured on fixed cadaver limbs, but if their numbers are used in the calculation below a rather low value for the supported weight is obtained, possibly because tissue fixation causes shrinkage and therefore underestimates cross-sectional area. If the PCSA values of Murray *et al.* are multiplied by a factor of 2.4 so as to match the PCSA values reported by Klein *et al.* [19] obtained from magnetic resonance imaging of the biceps, brachialis, and brachioradialis of young males, then the values are consistent with other measures; for example, the resulting biceps' PCSA computed in this manner is 12.3 cm^2, which is comparable to the maximum anatomic cross-sectional area of 16.8 cm^2 measured in 585 subjects by Hubal *et al.* [20].

Neglecting the mass of the forearm and the moment due to the elbow joint, we can draw a free body diagram for the forearm (Fig. 8.24). Denoting the force generated per unit PCSA of muscle by f, and assuming that this quantity is the same for all muscles (more on this later), we can sum moments about the effective

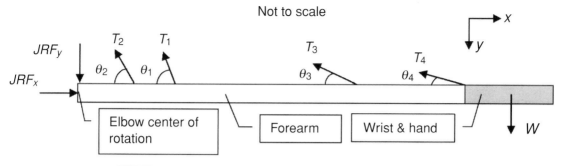

Figure 8.24

Free body diagram of lower arm, showing forces resulting from muscle insertions (T_i, i = 1 . . . 4), the two components of the joint reaction forces at the elbow (JRF_x and JRF_y), and the supported weight (W). The indices 1–4 refer to the biceps, brachialis, brachioradialis, and extensor carpi radialis longus muscles, respectively.

center of rotation of the elbow to obtain

$$f \sum_{i=1}^{4} L_{F_i} [PCSA]_i \sin \theta_i = W L_{tot} \tag{8.15}$$

where L_{tot} is the distance from the elbow to the line of action of the weight W, and the subscript i = 1 . . . 4 refers to each of the four muscles in Table 8.1. Here we assume that the joint reaction force (JRF) acts through the center of rotation for the elbow.

Equation (8.15) can be used to solve for the weight W if f and L_{tot} are specified. Measurements indicate that f can be as high as 28 N/cm^2 (40 lbf/in^2) for short durations.[7] However, sustained voluntary force development gives values of f slightly lower than this. For example, Kent-Braun and Ng [21] measured values of f between 17.1 and 20.7 N/cm^2 for the ankle dorsiflexor muscle in human subjects of different age. McMahon noted that f is approximately 20 N/cm^2 (29 lbf/in^2) for skeletal muscle in a wide range of species, independent of body size [10]. We will take f as 20 N/cm^2. Based on the geometric assumptions that we have made, L_{tot} is 31 cm, and the supported weight W is computed as $W = 120.1$ N, equivalent to a mass of 12.2 kg or a net moment of 37.2 N m about the elbow's center of rotation. Recall that this is the weight supported by a single arm and that it neglects the weight of the forearm itself.

There have been a number of assumptions made in arriving at this result. For example, the idea that all muscles exert the same force is an oversimplification; research has shown that the force exerted by muscles cooperating in a single task differs from muscle to muscle, and it has been argued that the force distribution is such to minimize the effort [22]. Moreover, the force that can be generated by a

---

[7] Other interesting statistics for muscle: the power density of muscle when doing medium-term repetitive tasks is approximately 40 W/kg muscle. The corresponding efficiency of energy conversion from foodstuffs is approximately 20%.

given muscle depends on the length of the muscle (Fig. 8.14), which varies with elbow flexion angle. This means that the force generated by the muscle is not just proportional to the PCSA but also depends in a complex way on the degree of elbow flexion. Murray *et al.* [18] have considered these effects and have presented a more complete analysis for elbow flexion, although for a slightly different forearm orientation than we have used here. The interested reader is referred to Murray [18] for more details; here we will stick with our simple assumption of a uniform value for f for all muscles. Finally, our treatment of the elbow as a simple hinge with reaction forces acting through the center of rotation is somewhat unrealistic, but a more complete analysis is beyond the scope of this book.

Given all of the above assumptions, how does our computed moment of 37.2 N m compare with experimental data? Wang and Buchanan [23] measured the moment generated during elbow flexion with the forearm in the neutral position (halfway between pronation and supination, as if the subject were holding the handle of a full coffee mug), obtaining values between 60 and 90 N m, which is about double the predictions of our calculations. Klein *et al.* [19] measured the force exerted at the wrist during elbow flexion with the forearm in supination and $100°$ of flexion, obtaining a mean value of 331 N in 20 young male subjects. This corresponds to a moment of roughly 80 N m, again about double what our computation predicts. This discrepancy may be because we have neglected several smaller muscles that help to stabilize and flex the elbow joint, because of small postural differences between our calculation and the experimental conditions, or because of the assumptions listed in the previous paragraph (including the value for f). However, the value that we have computed is not far different from experimental data and agrees with everyday experience. (Most people could support a 12 kg mass in their hand for a while.) We see that even something as biomechanically "simple" as computing the weight that can be supported in a person's hand is not so straightforward!

Returning to our calculation, we can now ask what joint reaction force this weight W will generate at the elbow. Referring to Fig. 8.24, we can sum forces in the vertical direction to obtain the vertical joint reaction force at the elbow, JRF_y as:

$$\mathrm{JRF}_y = \sum_{i=1}^{4} f \, [\mathrm{PCSA}]_i \sin \theta_i - W = 377.6 \text{ N}, \qquad (8.16)$$

which is approximately three times greater than the supported weight. A similar force balance in the x direction gives $\mathrm{JRF}_x = 304.3$ N. Both of these values are significantly larger than the supported weight, and this is typical of most joints in the body where the muscles are working at a "mechanical disadvantage," i.e., through a short lever arm. This means that the muscles must generate significant forces to support loads, which invariably translates into large joint reaction forces.

What magnitude of bending stress does this force distribution generate in the bones of the forearm? This turns out to be a very difficult question to answer; here we present a simplified way to tackle this problem. The first step is to compute the bending moment distribution in the forearm. The reader will recall that if one makes an imaginary cut through a loaded beam so as to separate the beam into two parts, then consideration of equilibrium for each part shows that there must be internal forces and moments acting on this cutting face (Fig. 8.25; upper panel). The internal moment is known as the *bending moment*, while the internal forces consist of a normal force and (in three dimensions) two shearing forces. The bending moment and internal forces change with position and can be computed by a static force and moment balance on either of the two parts of the beam for different imaginary cutting plane positions. In the case of the forearm, which we have approximated as being loaded by point forces, the bending moment distribution consists of linear segments (why?), and is as shown in Fig. 8.25. Inspection of this diagram shows that the maximum bending moment occurs 8 cm from the elbow and has magnitude 2323 N cm. Note that this result does not depend on assumptions about the architecture or shape of the bones in the forearm; rather, it follows directly from a force and moment balance.

Now we come to the tricky part: if there were only one bone in the forearm, a reasonable way to proceed would be to assume that this fictitious bone could be approximated as an Euler–Bernoulli beam.[8] In this case, the axial stress due to the bending moment, $M(x)$, and compressive internal force, $F_x(x)$, is given by

$$\sigma(x, y) = \frac{M(x)y}{I_z} + \frac{F_x(x)}{A} \tag{8.17}$$

where y is the distance from the neutral axis of the bone, A is the cross-sectional area of the bone, and I_z is the areal moment of inertia for the bone cross-section about the z-axis,

$$I_z = \int_A y^2 \, dA \tag{8.18}$$

where the integral in Equation (8.18) is carried out over the beam cross-section.

---

[8] There are six key assumptions underlying Euler–Bernoulli beam theory: (1) the beam is long and slender; (2) the beam cross-section is constant along its axis; (3) the beam is loaded in its plane of symmetry, i.e., there are no torsional loads on the beam; (4) deformations remain small; (5) the material is isotropic and (6) plane sections of the beam remain plane. Problematic in this context are assumptions (2), (3), (5), and, as we will see, (6). The reader is cautioned that the following material is therefore only approximate.

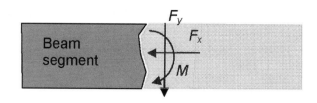

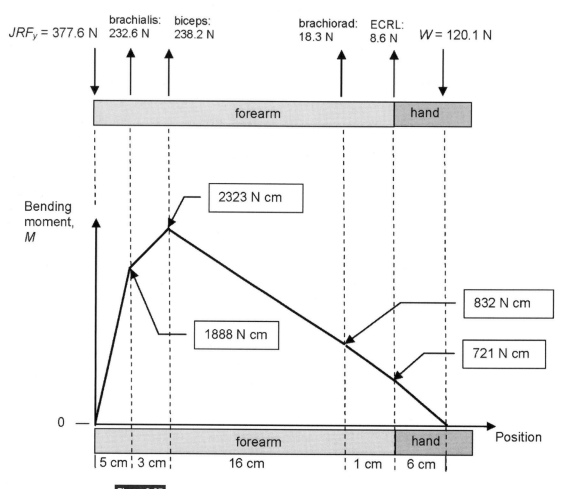

Figure 8.25

Bending moment distribution in the forearm. The upper portion of the figure shows a generic beam cut at an arbitrary location, with the internal bending moment and force components acting on the dark-shaded portion of the beam superimposed. The middle panel shows the vertical components of forces acting on the forearm, arising from muscles, joint reaction forces at the elbow, and the supported weight. The bottom panel shows the bending moment distribution in the forearm, computed as described in the text. Distances are not to scale. ECRL, extensor carpi radialis longus; brachiorad, brachioradialis.

It is instructive to use Equation (8.17) to consider what would occur if the radius bone were the only bone in the forearm and therefore carried the forearm's entire load. To work with Equation (8.17) we need geometric information about the radius bone, which is complicated by the fact that this bone has a non-circular cross-section that changes with axial position. To keep the problem tractable we will simply consider one axial location and treat the bone as an equivalent hollow beam of circular cross-section with inner and outer diameters, D_i and D_o respectively. Burr and Martin [24] used a combination of microscopy and calliper measurements to assess polar moment of inertia, J, and "diameter" in 86 human radius bones at a cross-section located 37.5% of the length of the radius from the distal end. We can use their average value for the polar moment of inertia in males (0.35 cm^4) to compute an effective outer diameter for the radius bone, as follows. The *medullary canal* (the hollow central core of the bone) has a diameter about half that of the outer bone diameter, so by using the formula $J = \pi(D_o^4 - D_i^4)/32$ with the ratio of inner to outer diameters $D_i/D_o = 1/2$ we obtain $D_o = 1.40 \text{ cm}$. This value is surprisingly close to the average outer "diameter" of about 1.4 cm (for males) measured by callipers in the anterior–posterior direction. We therefore approximate the equivalent outer diameter as 1.4 cm and the inner diameter as 0.7 cm. Using these values we can compute $I_z = \pi(D_o^4 - D_i^4)/64 = 0.177 \text{ cm}^4$ and $A = \pi(D_o^2 - D_i^2)/4 = 1.15 \text{ cm}^2$.

Returning to Equation (8.17), the ratio of the maximum value of the first term to the value of the second term is $8M(x)D_o/[(D_o^2 + D_i^2)F_x(x)]$, where we have used the fact that that the maximum value of y for a beam of circular cross-section is $D_o/2$. Since we will ultimately be concerned with extreme values of the terms in Equation (8.17), we will use this formula for the ratio and plug in the maximum value of $M(x)$ as 2323 N cm (Fig. 8.25) and the maximum value of $F_x(x)$ as 304 N. Using $D_o = 1.4$ cm, we then compute that the ratio of the first to the second term on the right hand side of Equation (8.17) can be as large as 35, which suggests that the maximum stress induced by bending is much larger than that induced by axial force, at least at the location where $M(x)$ is maximal. Therefore, it seems reasonable to ignore stresses in the bone owing to axial forces, and we will henceforth consider the role of bending only.

Of course, there is not just one bone in the forearm, and now we must consider the role of both the radius and the ulna. Equation (8.17) is based on the assumptions of Euler–Bernoulli beam theory, which fail for the forearm since it has two bones in it. (This is an example of a *truss beam*, for which it is no longer guaranteed that plane sections remain plane.) What we can say is that the radius and the ulna will each carry a part of the total bending moment. If we denote by subscripts r and u the portions carried by the radius and ulna, we can state

$$M(x) = M_u(x) + M_r(x) \tag{8.19}$$

Based on our arguments above, it seems reasonable to neglect the contribution of F_x to stress in the bone, so that we can write:

$$\sigma_r(x, y) = \frac{M_r(x)y_r}{I_{z,r}} \tag{8.20}$$

and

$$\sigma_u(x, y) = \frac{M_u(x)y_u}{I_{z,u}} = \frac{(M(x) - M_r(x))y_u}{I_{z,u}} \tag{8.21}$$

where y_r and y_u in Equations (8.20) and (8.21) are measured from the neutral axis of the respective bone.

As formulated, we cannot solve this problem since it is under-determined. However, as we will see in Section 9.5, bone remodels in response to stress, and it therefore seems plausible that the radius and ulna might be sized so that the maximum stresses predicted by Equations (8.20) and (8.21) would be approximately equal. Making this assumption, we can write:

$$\max\left\{\frac{M_r(x)y_r}{I_{z,r}}\right\} = \max\left\{\frac{(M(x) - M_r(x))y_u}{I_{z,u}}\right\} \tag{8.22}$$

Using the fact that the maximum value of y occurs at $D_o/2$, Equation (8.22) can be rewritten as

$$\max\{M_r(x)\} = \frac{\max\{M(x)\}}{1 + \left(D_{o,r}I_{z,u}/D_{o,u}I_{z,r}\right)} \tag{8.23}$$

At the location where the bending moment is maximum, the ulna has approximately the same diameter as the radius. Assuming that the respective medullary canal diameters are also the same for the two bones at this location, the area moments of inertia for the two bones are the same. We can therefore set the term in parentheses in the denominator of Equation (8.23) to 1, which implies that the bending moment in the radius is about half of that in the forearm. Using Equation (8.20) we can then write:

$$\max\{\sigma_r(x, y)\} = \frac{16 D_{o,r}}{\pi\left(D_{o,r}^4 - D_{i,r}^4\right)} \max\{M(x)\} \tag{8.24}$$

Substituting numerical values gives a maximum σ_r of 4599 N/cm^2 ($\approx$ 46 MPa). As we will see in Section 9.3.1 (Table 9.3), the yield stress for the type of bone found in the radius (cortical bone) is approximately 130 MPa in tension and 200 MPa in compression. Therefore, the computed maximum bending stress is about 23–35% of the yield stress for bone. The Young's modulus for longitudinal loading is approximately 17 GPa, so this loading corresponds to a peak surface strain of approximately 2700 microstrain. In Section 9.5, we will see that peak surface strains are typically in the range 2000–3000 microstrain [25], which agrees quite well with the strain levels we predict with this analysis.

Obviously the analysis that we have presented above has made many simplifications. A complete treatment would consider how forces are transmitted between the radius and the ulna, which, in turn, depends on the distribution of bone-on-bone contact, muscles, and connective tissue between these bones. The radius and ulna are in direct contact over a small region at their distal ends, and over a region of 1–2 inches near their proximal ends. There is also a connective tissue that lies between these bones called the *interosseous membrane*. Studies have shown that when the forearm is loaded along its long axis there is significant force transmission through the interosseous membrane [26–28], but it is unclear how much force transmission there is when the arm is supporting a weight as shown in Fig. 8.22. It is clear that force transmission at the elbow and in the forearm is complex (see for example Werner and An [29]); perhaps some brave biomechanics student reading this book will undertake a more complete analysis as part of a senior year thesis project?

8.4.3 Biomechanics of the knee

A third example that illustrates the importance of muscle–bone interactions in force transmission is the knee. The biomechanical loads placed on the knee joint are severe. This is reflected in the high incidence of knee injuries among athletic injuries: "Studies indicate that the knee is the most commonly injured area of the body and possibly accounts for 45% of all sports injuries" [30]. There are three bones involved in the knee joint proper:

- the femur, the large bone in the upper leg
- the tibia, the larger of the two bones in the lower leg
- the patella, or kneecap.

The strength and stability of the joint depends to a very large degree on the tendons and ligaments of the knee, in contradistinction to a joint such as the hip, which depends more on a close fit between the bones.

The femur has two "bulbs," or *condyles*, at its distal end, known as the lateral and medial femoral condyles. These rest in slightly concave portions on the upper surface of the tibia, known as the *tibial condyles*. The main structures ensuring that the femur and tibia remain in contact are the *cruciate ligaments*, so named because they appear to cross over one another. The anterior cruciate emanates from the anterior midplane of the tibia and attaches to the posterior aspect of the lateral femoral condyle. The posterior cruciate ligament emanates from the posterior tibial

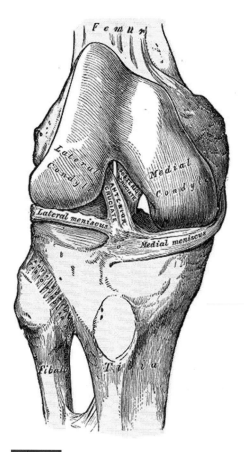

Figure 8.26

Some of the bones, ligaments, tendons, and cartilage in the human knee. This is a right knee seen from the anterior side. Note the anterior and posterior cruciate ligaments between the femur and the tibial plateau. Reprinted with permission from *Gray's Anatomy* [31].

midplane and attaches to the anterior aspect of the femoral condyle (Fig. 8.26). The arrangement of the cruciate ligaments is such that the weight is carried by progressively superior aspects of the femoral condyle during knee flexion (Fig. 8.27). There are also numerous muscles and ligaments outside the capsule of the knee that act to confer stability on the knee.

The superior tibia and the inferior femur have a cartilaginous covering, and a layer of cartilage known as the *meniscus* interposes between the femur and the tibia. The entire knee is enclosed in a fluid-filled capsule. This results in a joint in which relative motion is supported on a thin fluid film underlain by cartilage. This arrangement has very low wear and friction, and allows the healthy knee to be flexed many millions of times over the course of a lifetime.

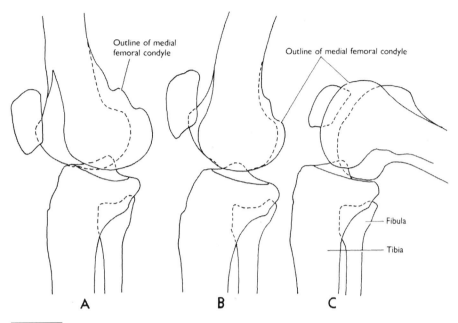

Outline of medial femoral condyle

Outline of medial femoral condyle

Fibula

Tibia

A   B   C

Figure 8.27

Bones of the right knee joint in humans, showing motion of bones during three phases of knee flexion. The tracings are taken from radiographs with the foot firmly fixed on the ground, so that there is no movement of the tibia and fibula and the full effects of rotation are visible in the femur. The tracings are viewed from the medial side, with the parts hidden by the more medial structures being shown by broken lines. (A) Position of full extension with the femur fully rotated medially and the knee joint locked. (B) Slight flexion. Note that there has been considerable lateral rotation of the femur, so that the outlines of the condyles are nearly superimposed. This movement occurs at the very outset of flexion. (C) Considerable flexion of the knee with further lateral rotation of the femur on the tibia. From Romanes [32]. Reproduced with kind permission of Oxford University Press.

A second important structural system acts to extend the knee joint and comprises the patella, *patellar ligament* (sometimes also called the *patellar tendon*), and *quadriceps*. The quadriceps are the large set of four muscles running along the anterior surface of the thigh. They extend tendons that anastomose (join) to form the quadriceps tendon, which, in turn, inserts into the superior aspect of the patella. Emanating from the inferior patella is the patellar ligament, which attaches to a small tuberosity on the front of the tibia (Fig. 8.28). The patella itself slides in a groove on the anterior femoral surface known as the *femoral intercondyler groove*. This groove, as well as the posterior surface of the patella, are covered with cartilage to allow low-friction sliding motion.

On the posterior surface of the knee there are a number of muscles that act to flex the knee joint. One consequence of this arrangement of flexing and extending systems is that a large compressive force can be generated between the patella and the femur. This is particularly true during maneuvers such as standing from a

Table 8.2. Measured femoro-patellar contact loads during squatting, for physiological joint angles. Values in the last column are femoro-patellar contact forces, and should be multiplied by three to get in vivo loads. Values are mean for n = 12 knees (±95% confidence intervals). Reprinted from Huberti and Hayes [35]. With kind permission of Elsevier.

Knee flexion angle (°)	Contact area (cm²)	Contact area as percentage of total articular area (%)	Average contact pressure (MPa)	Resultant contact force (N)
20	2.6 ± 0.4	20.5	2.0 ± 0.4	497 ± 90
30	3.1 ± 0.3	24.9	2.4 ± 0.6	573 ± 125
60	3.9 ± 0.5	30.4	4.1 ± 1.4	1411 ± 331
90	4.1 ± 1.1	32.2	4.4 ± 1.0	1555 ± 419

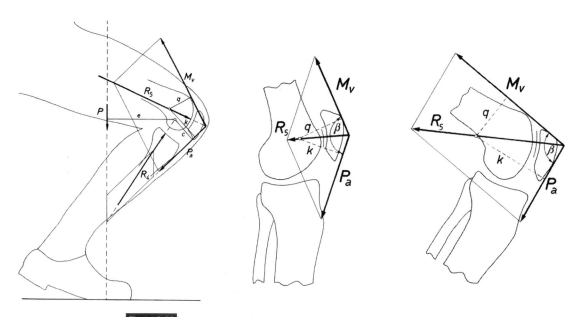

Figure 8.28

Diagram of forces acting in and on the knee during flexion. The left drawing shows an overview, while the middle and right drawings show close-up views for two different extents of knee flexion. As the lever arms q and k and the angle β change, the femoro-patellar contact force can vary substantially. In particular, when β is small (for example, during squatting), this force (R_5) can be very large (see Table 8.2). P is the portion of body weight supported by the leg; P_a is the force in the patellar tendon; M_v is the force exerted on the patella by the quadriceps; R_4 is the femoro-tibial contact force; R_5 is the compressive force between the femur and the patella; e is the lever arm of force P; c is the lever arm with which force P_a acts on the femoro-tibial joint; k is the lever arm with which force P_a acts on the patello-femoral joint; q is the lever arm with which force M_v acts on the patello-femoral joint. Modified from Maquet [33]. With kind permission of Springer Science and Business Media.

Table 8.3. Calculated femoro-patellar contact loads during walking, based on the model shown in Fig. 8.28. Values in the last column are femoro-patellar contact forces, and should be multiplied by three to get in vivo loads. Values are mean for $n = 12$ knees ($\pm 95\%$ confidence intervals). See Fig. 8.28 for explanation of symbols. From Maquet [33]. With kind permission of Springer Science and Business Media.

Phases	Angle $\beta_{2,4}$ of Fischer	Force R_5 (kg)	Patello-femoral contact area (cm^2)	Average compressive articular stresses (kg/cm^2)
12	23° 33′	219.0	5.56	39.37
13	28° 35′	126.8	6.07	20.92
14	24° 45′	126.6	5.90	21.13
15	18° 31′	33.5	4.88	6.86

squat, when the knee is flexed (Fig. 8.28). For example, Huberti and Hayes [34] estimated an in vivo contact force of approximately 1500 N at a flexion angle of 20°, increasing to 4600 N at a flexion angle of 90° (Table 8.2). Even during normal walking the forces can be significant: Maquet [33] estimated femoro-patellar contact forces of between 300 and 2000 N during normal walking (Table 8.3). This translates into compressive stresses of the order of 2–4 MPa on the femoro-patellar contact surface. It is clear that the cartilage on the posterior patella and anterior femur is subject to a great deal of wear and tear.

8.5 Problems

8.1 In your own words, describe the main events occurring between the arrival of an action potential at a motor neuron end plate and contraction of the corresponding muscle. Use 250 words or less.

8.2 Shown in Fig. 8.29 is a cross-sectional view through muscle, showing actin and myosin filaments. Knowing that muscle can generate a maximum force of 20 N/cm^2, determine the maximum force exerted by each myosin filament. Make and state appropriate assumptions.

8.3 A highly idealized version of part of the tension–length relationship for cardiac muscle is graphed in the left portion of Fig. 8.30. This relationship effectively determines the pumping behavior of the left ventricle, as follows. Increased blood volume within the left ventricle causes stretching of the ventricular wall muscle fibers, which, in turn, causes the contraction of the ventricle to be more forceful. In this way, the left ventricular blood ejection

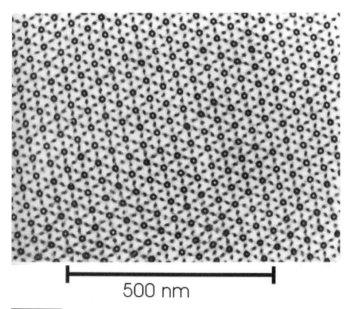

Figure 8.29

For Problem 8.2. Modified from Fawcett [3] and Prof. Hans Ris, with kind permission.

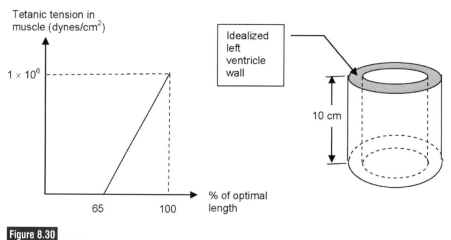

Figure 8.30

For Problem 8.3.

pressure, and thus also the ejected blood volume, increase in response to increased presystolic ventricular volume.

Using this information, plot (to scale) the left ventricular blood ejection pressure as a function of presystolic ventricular volume. (The ejection pressure is the maximum pressure achieved during the isovolumetric phase of contraction.) For purposes of this question, you may assume that the left

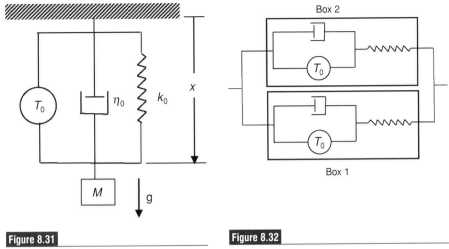

Figure 8.31

For Problem 8.5.

Figure 8.32

For Problem 8.6.

ventricle is a thin-walled right circular cylinder with constant wall thickness of 0.7 cm (see right portion of Fig. 8.30). Valves are located in the top of the cylinder, and the bottom and top of the cylinder are passive (i.e., do not participate in active contraction). The internal diameter of the ventricle when maximum muscle tension occurs is 6 cm. You may assume that tetanic tension is developed during the isovolumetric phase.

8.4   A certain muscle is known to behave according to the three element model presented in Section 8.2, with an effective dashpot damping coefficient of $\eta_0 = 2.5\,\text{N s/m}$. When stimulated with a single twitch in an isometric experiment, it produces 80% maximal tension after 40 ms. While keeping the same muscle length, the muscle is then put in series with a spring having k_0 of 200 N/m. What tension is measured in a new isometric experiment 20 ms after a twitch?

8.5   A muscle is supported from a fixed point and has a mass M attached to it (Fig. 8.31). Assume that the muscle can be modeled using a three element model, *noting that the arrangement of elements is different than used in Section 8.2.* Call the muscle length x, and denote the value of x before the muscle begins to contract by x_0. At time $t = 0$, the active component of the muscle begins to contract and produces a constant tension T_0 for duration C. This causes the mass to rise, i.e., causes x to decrease with time.
(a)   Treating the muscle as massless, show that $x(t)$ is given by

$$x - x_0 = -\frac{T_0}{k_0}\left[1 + \frac{r_2\,e^{r_1 t} - r_1\,e^{r_2 t}}{r_1 - r_2}\right] \qquad (8.25)$$

where

$$r_1 = -\frac{\eta_0}{2M}\left[1 + \sqrt{1 - 4k_0 M/\eta_0^2}\right]$$

$$r_2 = -\frac{\eta_0}{2M}\left[1 - \sqrt{1 - 4k_0 M/\eta_0^2}\right].$$

(b) If $T_0 = 15\,$N, $k_0 = 500\,$N/m, $M = 1\,$kg, $\eta_0 = 100\,$N s/m, and $C = 0.1\,$s, calculate how far the mass M will have risen at the end of the contraction (i.e., at $t = C$).

8.6 In Section 8.2 the response of the spring–dashpot–active element model of muscle to a single stimulus was examined. Here we will look at overall muscle response to multiple stimuli. Consider the model for muscle shown in Fig. 8.32, where each of boxes 1 and 2 contains a spring–dashpot–active element system identical to that discussed in the text.

The muscle is held isometrically. At $t = 0$, a stimulus arrives and the active element in box 1 contracts, while the components in box 2 are unaffected. At $t = C/2$ a second stimulus arrives that causes the active element in box 2 to contract without affecting the components of box 1. What is the total developed tension within the muscle at time $t = C$? Use the values $\eta_0 = 0.06\,$dyne s/cm, $k_0 = 0.3\,$dyne/cm, $C = 0.4\,$s, and $T_0 = 4\,$dyne.

8.7 In the text a three element model for muscle was presented, consisting of a linear spring (constant k_0), a dashpot (constant η_0), and a contractile element that generates a tension T_0 for a period C after being stimulated (Fig. 8.17). Predict the muscle length L for all $t > 0$ in response to a single stimulus at time $t = 0$ during an *isotonic* experiment with imposed tension $T_1 < T_0$. The muscle length before stimulus is L_0 and upon relaxation the muscle cannot stretch past L_0. Sketch your predictions on a graph of L versus t.

8.8 Prove that the force generated (per unit depth), F, by a pinnate muscle arrangement is

$$F = 2fl \sin \alpha \cos \alpha \tag{8.26}$$

where f is the force generated per unit cross-sectional muscle area, l is the pinnate segment length, and α is the muscle fiber angle.

8.9 What total axial force can be generated by a pinnate muscle having the form and dimensions shown in Fig. 8.33? The force, f, developed by muscle per unit cross-sectional area may be taken as $20\,$N/cm^2.

8.10 Let us revisit the biomechanical differences between the horse and armadillo forelegs, as discussed in Section 8.4.1. In addition to the interspecies difference in the l/h ratio, the horse is larger and more massive than the

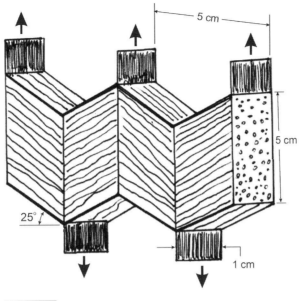

For Problem 8.9.

armadillo. The purpose of this question is to take these factors into account when computing the acceleration of these two animals as they start running from rest. Suppose that a length L characterizes the size of each animal. This implies that:

- each animal's body mass is proportional to L^3, i.e., body mass $= k_1 L^3$
- each animal's teres major muscle has a cross-sectional area proportional to L^2, i.e., muscle cross-sectional area $= k_2 L^2$
- each animal's teres major muscle has a contraction velocity proportional to L, i.e., contraction velocity $= k_3 L$.

Further assume that the force generated per unit cross-sectional area, f, and the constants k_1, k_2, and k_3 are the same for the horse and the armadillo.

(a) Neglecting the mass of the forelegs, and assuming that only the forelegs generate force on the ground, show that the acceleration, a, of each animal is proportional to:

$$a \sim \frac{l}{h}\frac{f}{L}. \tag{8.27}$$

(b) Show that the top speed of the animal is proportional to Lh/l.

(c) Assuming that the acceleration derived in (a) is constant while the animal accelerates to top speed, derive an expression for the ratio

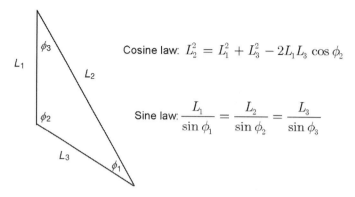

Figure 8.34

For Problem 8.11.

$\Delta t_{\mathrm{horse}}/\Delta t_{\mathrm{armadillo}}$, where Δt is the time taken to reach top speed. Using data in the text, plus $L_{\mathrm{horse}}/L_{\mathrm{armadillo}} = 8$, numerically compute the ratio $\Delta t_{\mathrm{horse}}/\Delta t_{\mathrm{armadillo}}$.

(d) Does the value in (c) seem reasonable? If not, what are the shortcomings of the analysis leading to this answer?

8.11 Compute the magnitude of the weight W that can be supported in the hand when the forearm is angled downward at 30° to the horizontal. You may make the same assumptions as were made in the text for the situation in which the forearm was horizontal. Specifically, you may use values given in the text for f, and for L_{H}, L_{F}, and muscle area (from Table 8.1). You may also assume that the distance from the center of rotation of the elbow to the weight in the hand is $L_{\mathrm{tot}} = 31$ cm, and you can neglect the weight of the forearm. Some handy formulae from the trigonometry of triangles are given in Fig. 8.34.

8.12 Consider a person standing with their legs slightly bent, as shown in Fig. 8.35. Each leg supports a constant load P, effectively applied at point A. Posture is primarily maintained by a muscle group on the front of the femur that attaches to the femur and to the tibia just below the knee.

(a) Neglecting the mass of the femur, and considering the knee to be frictionless and to pivot around point B, show that the tension T in the muscle is:

$$T = \frac{Pl \sin \theta}{r} \tag{8.28}$$

where l is the distance from A to B and r is the radius of the knee, equal to half the thickness of the femur. State assumptions that you make.

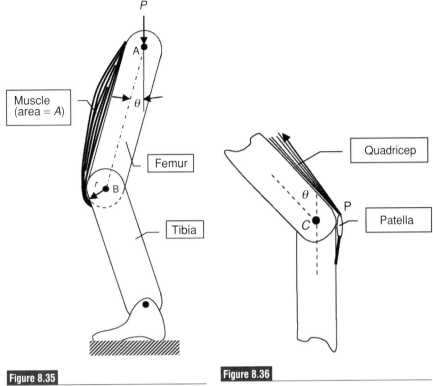

Figure 8.35

For Problem 8.12.

Figure 8.36

For Problem 8.13.

(b) Suppose now that the same person squats a little further, so that θ becomes $\theta + \delta\theta$. This causes a slight stretch in the muscle: initially it has length L, which becomes $L + \delta L$. If L is less than the optimal length, the stretch will produce an increase δT in steady-state muscle tension. It is acceptable to assume

$$\delta T = \beta A \frac{\delta L}{L} \qquad (8.29)$$

where A is the muscle cross-sectional area (constant) and β is a proportionality constant. If the leg is to avoid buckling, show that β must satisfy:

$$\beta = \frac{LlP}{Ar^2} \cos\theta \qquad (8.30)$$

Hint: relate δL to $\delta\theta$, and use the results of (a).

8.13 A schematic diagram of the knee joint is shown in Fig. 8.36, where C denotes the effective center of rotation of the knee and P is the point of insertion of the quadriceps tendon to the patella. The distance from C to P is 10 cm. The quadriceps, responsible for extension of the knee, are known to produce maximum isometric tetanic tension at an effective optimal length of 30 cm. This occurs when the knee flexion angle, θ, is 45°. Obtaining muscle performance data from the appropriate figure in this chapter, determine the range of knee flexion angles for which the quadriceps produce an isometric tetanic tension at least 80% of maximal. Hint: be careful to use radians (rather than degrees) in this question when appropriate.

References

1. A. L. Lehninger. *Biochemistry*, 2nd edn (New York: Worth, 1975).
2. A. J. Vander, J. H. Sherman and D. S. Luciano. *Human Physiology: The Mechanisms of Body Function*, 4th edn (New York: McGraw-Hill, 1985).
3. D. W. Fawcett. *Bloom and Fawcett: A Textbook of Histology* (Philadelphia, PA: W.B. Saunders, 1986).
4. A. Despopolous and S. Silbernagl. *Color Atlas of Physiology* (New York: Georg Thieme Verlag, 1986).
5. J. Desaki and Y. Uehara. The overall morphology of neuromuscular junctions as revealed by scanning electron microscopy. *Journal of Neurocytology*, **10** (1981), 101–110.
6. B. R. Jewell and D. R. Wilkie. The mechanical properties of relaxing muscle. *Journal of Physiology*, **152** (1960), 30–47.
7. A. J. McComas. *Skeletal Muscle Form and Function* (Champaign, IL: Human Kinetics, 1996).
8. A. M. Gordon, A. F. Huxley and F. J. Julian. The variation in isometric tension with sarcomere length in vertebrate muscle fibers. *Journal of Physiology*, **184** (1966), 170–192.
9. A. V. Hill. The heat of shortening and the dynamic constants of muscle. *Proceedings of the Royal Society of London. Series B, Biological Sciences*, **126** (1938), 136–195.
10. T. A. McMahon. *Muscles, Reflexes and Locomotion* (Princeton, NJ: Princeton University Press, 1984).
11. G. F. Inbar and D. Adam. Estimation of muscle active state. *Biological Cybernetics*, **23** (1976), 61–72.
12. Y. C. Fung. *Biomechanics: Mechanical Properties of Living Tissues* (New York: Springer Verlag, 1981).
13. J. Z. Young. *The Life of Mammals* (London: Oxford University Press, 1975).
14. E. S. Hsu, A. G. Patwardhan, K. P. Meade, T. R. Light and W. R. Martin. Cross-sectional geometrical properties and bone mineral contents of the human radius and ulna. *Journal of Biomechanics*, **26** (1993), 1307–1318.

15. W. M. Murray, T. S. Buchanan and S. L. Delp. Scaling of peak moment arms of elbow muscles with upper extremity bone dimensions. *Journal of Biomechanics*, **35** (2002), 19–26.

16. M. Gerbeaux, E. Turpin and G. Lensel-Corbeil. Musculo-articular modeling of the triceps brachii. *Journal of Biomechanics*, **29** (1996), 171–180.

17. C. I. Morse, J. M. Thom, N. D. Reeves, K. M. Birch and M. V. Narici. In vivo physiological cross-sectional area and specific force are reduced in the gastrocnemius of elderly men. *Journal of Applied Physiology*, **99** (2005), 1050–1055.

18. W. M. Murray, T. S. Buchanan and S. L. Delp. The isometric functional capacity of muscles that cross the elbow. *Journal of Biomechanics*, **33** (2000), 943–952.

19. C. S. Klein, C. L. Rice and G. D. Marsh. Normalized force, activation, and coactivation in the arm muscles of young and old men. *Journal of Applied Physiology*, **91** (2001), 1341–1349.

20. M. J. Hubal, H. Gordish-Dressman, P. D. Thompson, T. B. Price, E. P. Hoffman *et al.* Variability in muscle size and strength gain after unilateral resistance training. *Medicine and Science in Sports and Exercise*, **37** (2005), 964–972.

21. J. A. Kent-Braun and A. V. Ng. Specific strength and voluntary muscle activation in young and elderly women and men. *Journal of Applied Physiology*, **87** (1999), 22–29.

22. P. T. Prendergast, F. C. T. van der Helm and G. N. Duda. Analysis of muscle and joint loads. In *Basic Orthopaedic Biomechanics and Mechanobiology*, ed. V. C. Mow and R. Huiskes. (Philadelphia, PA: Lippincott, Williams & Wilkins, 2005), pp. 29–90.

23. L. Wang and T. S. Buchanan. Prediction of joint moments using a neural network model of muscle activations from EMG signals. *IEEE Transactions on Neural Systems and Rehabilitation Engineering*, **10** (2002), 30–37.

24. D. B. Burr and R. B. Martin. The effects of composition, structure and age on the torsional properties of the human radius. *Journal of Biomechanics*, **16** (1983), 603–608.

25. C. T. Rubin and L. E. Lanyon. Dynamic strain similarity in vertebrates; an alternative to allometric limb bone scaling. *Journal of Theoretical Biology*, **107** (1984), 321–327.

26. R. A. Kaufmann, S. H. Kozin, A. Barnes and P. Kalluri. Changes in strain distribution along the radius and ulna with loading and interosseous membrane section. *Journal of Hand Surgery*, **27** (2002), 93–97.

27. J. C. McGinley and S. H. Kozin. Interosseus membrane anatomy and functional mechanics. *Clinical Orthopaedics and Related Research*, (2001), 108–122.

28. L. A. Poitevin. Anatomy and biomechanics of the interosseous membrane: its importance in the longitudinal stability of the forearm. *Hand Clinics*, **17** (2001), 97–110, vii.

29. F. W. Werner and K. N. An. Biomechanics of the elbow and forearm. *Hand Clinics*, **10** (1994), 357–373.

30. G. Finerman (ed.) *Symposium on Sports Medicine: The Knee* (St. Louis, MO: Mosby, 1985).

31. H. Gray. *Anatomy of the Human Body*, 20th edn: thoroughly reviewed and re-edited by W. H. Lewis (Philadelphia, PA: Lea and Febiger, 1918).

32. G. J. Romanes. *Cunningham's Manual of Practical Anatomy*, 15th edn, Vol. I (Oxford: Oxford University Press, 1986).

33. P. G. J. Maquet. *Biomechanics of the Knee*, 2nd edn (New York: Springer Verlag, 1984).

34. H. H. Huberti and W. C. Hayes. Patellofemoral contact pressures. The influence of Q-angle and tendofemoral contact. *Journal of Bone and Joint Surgery (American Volume)*, **66** (1984), 715–724.

35. H. H. Huberti and W. C. Hayes. Determination of patellofemoral contact pressures. In *Symposium on Sports Medicine: The Knee*, ed. G. Finerman. (St. Louis, MO: Mosby, 1985), pp. 45–53.

9 Skeletal biomechanics

The skeletal system is made up of a number of different tissues that are specialized forms of *connective tissue*. The primary skeletal connective tissues are bone, cartilage, ligaments, and tendons. The role of these tissues is mainly mechanical, and therefore they have been well studied by biomedical engineers. Bones are probably the most familiar component of the skeletal system, and we will begin this chapter by considering their function, structure, and biomechanical behavior.

9.1 Introduction to bone

The human skeleton is a remarkable structure made up of 206 bones in the average adult (Fig. 9.1). The rigid bones connect at several joints, a design that allows the skeleton to provide structural support to our bodies, while maintaining the flexibility required for movement (Table 9.1). Bones serve many functions, both structural and metabolic, including:

- providing structural support for all other components of the body (remember that the body is essentially a "bag of gel" that is supported by the skeleton)
- providing a framework for motion, since muscles need to pull on something rigid to create motion
- offering protection from impacts, falls, and other trauma
- acting as a Ca^{2+} repository (recall that Ca^{2+} is used for a variety of purposes, e.g., as a messenger molecule in cells throughout the body)
- housing the marrow, the tissue that produces blood cells and stem cells.

In the following sections, we focus our discussion on the biomechanical functions of bone, and to do so we start by describing the composition and structure of bone tissue.

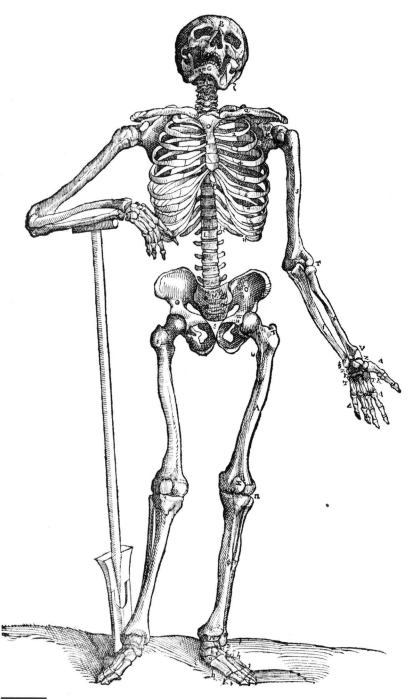

Figure 9.1

The figure on this page is from a woodcut engraving of the bones of the human body first published in 1543 in *De Humani Corporis Fabrica Libri Septem* by Andreas Vesalius. In the text that accompanied this figure, Vesalius enumerated the bones in the body, adding them up to arrive at a total of 304. His addition was incorrect – his figures actually added up to 303. In subsequent publications, he revised his count to 246, 40 more than what we now know is the true number of bones in the adult skeleton. The major bones of the anatomically correct adult skeleton are labeled in the figure on the following page (from Young [1] reproduced with kind permission of Oxford University Press). A modern edition of Vesalius' work is available [2].

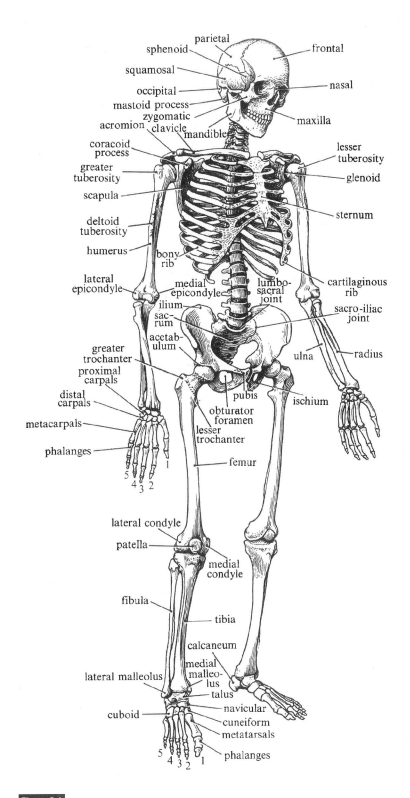

parietal

sphenoid

squamosal

occipital

mastoid process

zygomatic

acromion

clavicle

coracoid process

greater tuberosity

scapula

deltoid tuberosity

humerus

bony rib

lateral epicondyle

medial epicondyle

ilium

sac- rum

acetab- ulum

greater trochanter

proximal carpals

distal carpals

metacarpals

phalanges

5

4 3 2

1

frontal

nasal

maxilla

lesser tuberosity

glenoid

sternum

lumbo- sacral joint

cartilaginous rib

sacro-iliac joint

radius

ulna

obturator foramen

lesser trochanter

pubis

ischium

femur

lateral condyle

patella

medial condyle

fibula

tibia

calcaneum

medial malleo- lus

talus

navicular

cuneiform

metatarsals

phalanges

lateral malleolus

cuboid

5

4 3 2

1

mandible

Figure 9.1

(cont.)

Table 9.1. Distribution of bones in the arms and legs: flexibility is conferred by having more bones (and hence more degrees of freedom) as one moves distally; an exception is the spine, which has 33 *vertebrae*

Arms	Legs
1 bone in upper arm (humerus)	1 bone in thigh (femur)
2 bones in forearm (radius, ulna)	2 bones in lower leg (tibia, fibula)
8 bones in wrist	7 bones in ankle
20 bones in hand	20 bones in foot

Table 9.2. Approximate composition of the extracellular matrix of bone tissue based on consensus values from several sources

Component	Mass (%)
Mineral phase (hydroxyapatite)	70
Organic matrix (osteoid)	
Collagen (mostly type I)	18
Non-collagenous proteins and proteoglycans	2
Water	10

9.2 Composition and structure of bone

Like other connective tissues, bone tissue consists of cells embedded in an extracellular matrix (ECM). Hence, although bone is sometimes considered to be inanimate, it is actually a living, dynamic tissue that is constantly being renewed. In fact, it has been estimated [3] that 10–15% of the bone in the whole body is replaced with new bone every year. Bone is unique, however, in that its matrix contains not only an organic, collagen-based phase but also a mineral phase that provides bone with its characteristic rigidity and strength (Table 9.2). This mineral phase is composed primarily of crystalline *hydroxyapatite* ($[Ca_3(PO_4)_2]_3 \cdot Ca(OH)_2$), with small amounts of other mineral substances.

The process of bone remodeling is accomplished by specialized cells within bone tissue, of which there are three types:

- *osteoblasts*, which are recruited to synthesize bone
- *osteoclasts*, which are recruited to dissolve bone
- *osteocytes* and *lining cells*, which reside permanently in bone tissue.

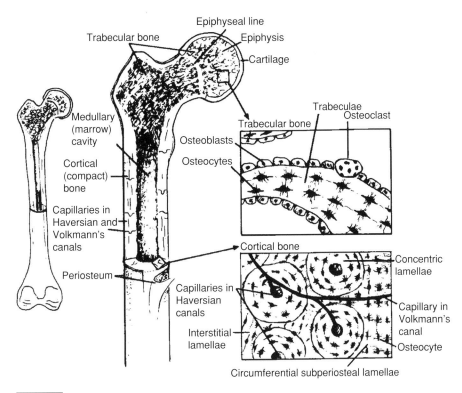

Figure 9.2

Overview of bone architecture (including locations of trabecular and cortical bone). The insets show more details of trabecular and cortical bone structure, from which the reader can appreciate the complex layered architecture of the different bone regions. The *epiphyseal line* is a remnant of the growth plate, a region that separates the growing shaft of the bone from the end of the bone (*epiphysis*) during childhood. This figure is of the femur, but the other long bones demonstrate similar features. From Hayes and Bouxsein [4]. With kind permission of Lippincott Williams & Wilkins.

Osteocytes are osteoblasts that became embedded in the bone matrix as it was being secreted, while lining cells reside on the surface of bone tissue. The lining cells and embedded osteocytes are connected by small canals and so are able to communicate with one another. These cells are believed to play a role in the maintenance of bone tissue by surveying the local microenvironment.

Bone tissue is the basic material from which bones are built. In whole bones, this material is organized such that two types of bone are apparent when one looks at the interior of a bone: *cortical* and *trabecular* bone (Fig. 9.2). The primary difference between these two bone types is their porosity, with trabecular bone being the more porous. We can quantify this difference by first considering the *apparent density* of a bone specimen, defined as its mass divided by its bulk volume. The apparent density of solid cortical bone is approximately 1.8 g/cm^3. We then define the *relative density*, which is the ratio of the apparent density of a specimen to that of solid cortical bone. A bone specimen with a relative density > 0.7 is considered cortical bone, reflecting the presence of minor amounts of porosity in normal cortical bone, as explained below. Trabecular bone is defined as bone with a relative density < 0.7, although typical values are lower, approaching 0.05 (or over 90% porosity) in some cases. The apparent density of trabecular bone ranges from 0.1 to 1.0 g/cm^3 [5].

9.2.1 Cortical bone

Cortical bone (also called compact bone) lines the outer surface of most bones and is found in the shafts of the long bones. Owing to its low porosity (typically $< 10\%$), cortical bone is the stronger and heavier of the two types of bone. The microarchitecture of cortical bone is quite complex and plays an important role in dictating its mechanical properties. Cortical bone tissue is laid down in *lamellae*, or layers, about 5 μm thick. Within a layer, the collagen fibers run parallel to one another; however, the orientation of fibers is different in neighboring layers. This is a structure similar to that seen in another important composite, namely plywood.

The lamellae are arranged in different ways in different parts of the bone (Fig. 9.3). Near the outer and inner surfaces of the bone, the lamellae are circumferentially arranged and parallel to one another; these are known as the *circumferential lamellae*. Between the outer and inner circumferential lamellae, cortical bone tissue is primarily made up of *osteonal bone*, which consists of cylindrically shaped structures formed from concentric lamellae. These cylindrical structures are approximately 200 μm in diameter and 1 cm long, and are typically aligned with the long axis of the bone. This solid bone matrix is permeated by a highly interconnected network of canals and channels. Osteocytes live within this network, where they are supplied by nutrients dissolved in the extracellular fluid. The canals are arranged in a characteristic pattern: a blood vessel runs within a *Haversian canal*, which is surrounded by concentric lamellae. The lamellae are connected to the central canal by microscopic channels called *canaliculi*. A canal plus surrounding lamellae and intervening bone is called an *osteon* (Figs. 9.2 and 9.3). The osteons are packed together to form the bone proper. The space between osteons is called *interstitial lamellar bone*, and the junction between osteons and interstitial bone is the *cement line*. The cement line is a highly mineralized, collagen-free layer of < 1 μm in thickness. Its importance will become clear when we talk below about bone fracture.

9.2.2 Trabecular bone

Trabecular bone (also called cancellous or spongy bone) is found in the vertebrae and the ends of the long bones, such as the femur, tibia, and radius (Fig. 9.4). The

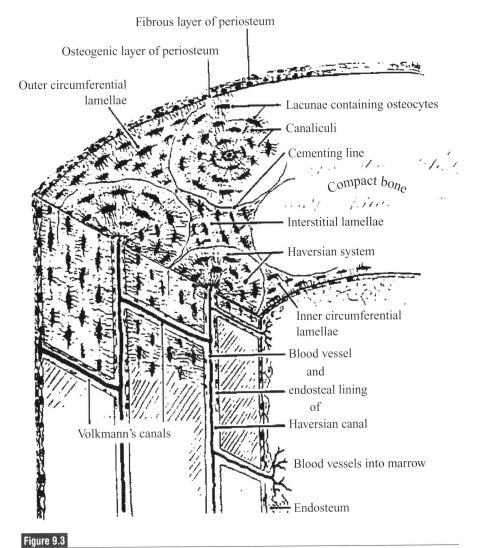

Fibrous layer of periosteum

Osteogenic layer of periosteum

Outer circumferential lamellae

Lacunae containing osteocytes

Canaliculi

Cementing line

Compact bone

Interstitial lamellae

Haversian system

Inner circumferential lamellae

Blood vessel and endosteal lining of Haversian canal

Volkmann's canals

Blood vessels into marrow

Endosteum

Figure 9.3

Microstructure of cortical bone. Although cortical bone appears solid, it is actually permeated with a network of canals, which house blood vessels and nerves, and canaliculi, through which nutrient-rich fluid flows to osteocytes contained within lacunae (holes) within the bone tissue. Shear stresses exerted on osteocytes by the flow of fluid within the canaliculi are believed to play an important role in the biological response of bone to mechanical stimulation, as discussed in Section 2.9.4. From Ham and Cormack [6] with kind permission of Lippincott Williams & Wilkins.

study of trabecular bone is important because age-related fractures primarily occur at trabecular bone sites, such as the proximal femur (hip), the distal radius (wrist), and vertebral bodies (spine).

Like cortical bone, trabecular bone tissue has a lamellar structure, with lamellae running parallel to the trabeculae (Fig. 9.2). In contrast to cortical bone, however, the matrix of trabecular bone is organized as a three-dimensional porous

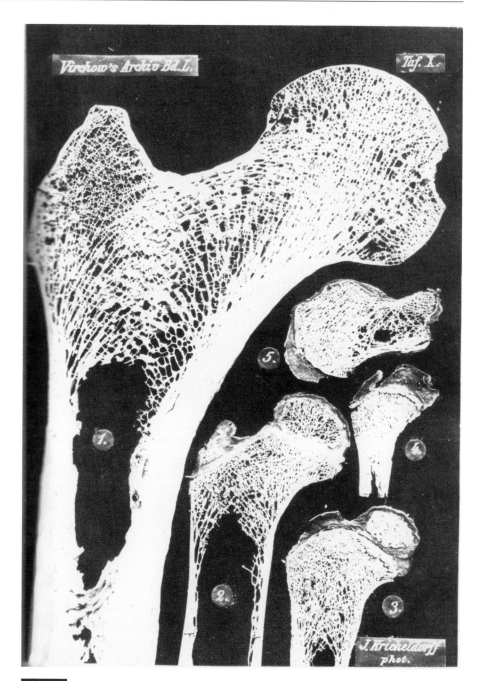

Figure 9.4

Cross-sections showing the open porous structure of trabecular bone. The large sample on the left is the femoral head seen in a longitudinal section. Note the transition from cortical bone in the shaft of the femur to trabecular bone at the head, with only a thin layer of cortical bone on the surface. This photograph is from an 1870 article by the German anatomist Julius Wolff in which he compared the orientation of the trabeculae in the proximal femur to the predicted trajectories of the principal stresses [7].

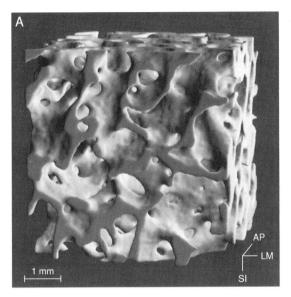
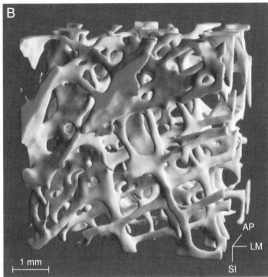

Figure 9.5

Three-dimensional structure of trabecular bone from the iliac crest of the pelvis of a 37-year-old man with no known bone disorders (A) and of a 73-year-old woman with osteoporosis (B). Note the dense, plate-like structure of the young bone, whereas the osteoporotic bone is less dense, and has a more rod-like structure with thinner trabeculae. These images were obtained by microcomputed tomography of biopsy samples. AP, anterior-posterior; LM, lateral-medial; SI, superior-inferior. From Muller *et al.* [8], with permission from Elsevier.

network of interconnected struts called *trabeculae*. The pores between trabeculae are filled with bone marrow. Because of its higher porosity, trabecular bone is somewhat weaker than "solid" cortical bone, but it is also much lighter. (The bone in birds, for example, is primarily trabecular.) The porosity of the trabecular network varies with anatomical location, biomechanical function, and age. In dense trabecular bone, such as is found in the pelvis or femoral condyle at the knee, the trabeculae are organized as a network of nearly closed cells formed by parallel plates and interconnecting rods (Fig. 9.5A). In low-density trabecular bone, such as is found in the vertebrae or in people with osteoporosis, the trabeculae are organized as a network of open cells formed by interconnected rods (Fig. 9.5B). In healthy bone, trabeculae are on average about 200 μm thick; in osteoporosis – a condition characterized by loss of bone mass and increased fracture risk – there are fewer trabeculae and they become thinner (Fig. 9.5B).

9.3   Biomechanical properties of cortical and trabecular bone

The biomechanical properties of bone are determined by its composition and structure at multiple length scales. On a length scale of the order of micrometers, bone

Table 9.3. Summary of mechanical properties of human cortical bone based on data from Reilly and Burstein, *Journal of Biomechanics*, **8**(1975), 393–405.

Parameter	Value
Modulus (GPa)	
Longitudinal	17.0
Transverse	11.5
Shear	3.3
Poisson's ratio	0.3–0.6
Ultimate strength: longitudinal (MPa)	
Tension	133
Compression	193
Shear	68
Ultimate strength: transverse (MPa)	
Tension	51
Compression	133

tissue behaves like a fiber-reinforced composite, with the hydroxyapatite mineral providing stiffness and the collagen fibers providing tensile strength and ductility to prevent brittle cracking. It is generally thought that the bone tissue making up an individual trabecular strut is slightly (20–30%) less stiff than cortical bone tissue owing to subtle differences in mineralization and lamellar microstructural organization and orientation [9]. The biomechanical properties of cortical and trabecular bone diverge significantly, however, when the structural organization of the tissues at length scales of millimeters is taken into account, as discussed below.

9.3.1 Cortical bone mechanics

The longitudinal alignment of the osteons and the orientation of the lamellae give cortical bone strongly anisotropic properties. Specifically, stiffness and strength along the axis of cortical bone (longitudinal direction) are greater than properties transverse to the bone's axis (Table 9.3). Further, cortical bone is stronger in compression than in tension. This makes sense when we realize that this "optimizes" cortical bone for typical loading conditions, in which the bone is usually loaded along its axis in compression. The stiffness and strength of cortical bone depend

Table 9.4. Summary of mean compressive properties of human trabecular bone from different anatomic locations. Values in parentheses are standard deviations. Femur specimens were pooled from both the proximal and distal femur. The specimens from the tibia, distal femur, and spine were tested in the longitudinal (inferior-superior) direction. The proximal femur specimens were oriented along the neck of the femur. Adapted from Keaveny [11]. Copyright 2001 from *Bone Mechanics Handbook* by Cowin. Reproduced by permission of Routledge/Taylor & Francis Group, LLC.

Anatomic site	Relative density	Modulus (MPa)	Ultimate stress (MPa)	Ultimate strain (%)
Proximal tibia	0.16 (0.056)	445 (257)	5.33 (2.93)	2.02 (0.43)
Femur	0.28 (0.089)	389 (270)	7.36 (4.00)	Not reported
Lumbar spine	0.094 (0.022)	291 (113)	2.23 (0.95)	1.45 (0.33)

not only on the orientation but also on the rate at which the bone is strained. This is a characteristic of viscoelastic materials, which we will examine in more detail when we consider soft skeletal tissues. For cortical and trabecular bone, the strain rate dependency is weak: the stiffness and strength are proportional to the strain rate to the power 0.06 for strain rates in the physiological range of $0.001–0.03 \ s^{-1}$ [10]. Nonetheless, the result is that bone is stiffer and stronger at higher strain rates. This is desirable, since this is a way to compensate for the higher loads and stresses imposed by vigorous activity or superphysiological loading.

9.3.2   Trabecular bone mechanics

The mechanical properties of a specimen of trabecular bone, which is made up of many trabecular struts and has dimensions > 1 mm, are called the *apparent* mechanical properties. The apparent properties depend on the material properties of bone tissue matrix, the amount of tissue, and the structural organization of the trabeculae (see also the discussion of structural versus material properties in Section 9.9). In particular, the apparent properties are strongly dependent on the relative density and architecture. Since the relative density and architecture of trabecular bone vary with anatomic site, age, and disease, there is significant variation in the stiffness and strength of trabecular bone (Table 9.4).

A typical compressive stress–strain curve for trabecular bone of different relative densities is shown in Fig. 9.6. At small strains, the bone behaves as a linearly elastic material. The linear elastic response ends once the trabecular rods and plates begin to collapse under the increasing stress. The subsequent large anelastic region corresponds to the progressive collapse of the trabeculae. Once the collapsing trabeculae contact one another (thereby eliminating the pores within the bone), the

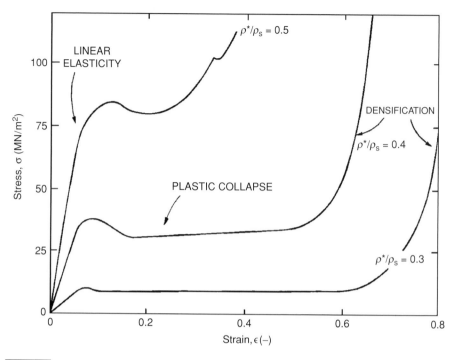

Figure 9.6

Compressive stress–strain curves for trabecular bone of different relative densities. Relative density is the ratio of the apparent density (ρ^* in this figure) to the density of solid cortical bone (ρ_s). The reader will appreciate from this figure that the biomechanical behavior of trabecular bone is extremely complex, as described more fully in the text. From Gibson and Ashby [12], based on data reported in [13]. Reprinted with permission of John Wiley & Sons, Inc.

stress rises steeply. The collapse of trabeculae allows trabecular bone to undergo large compressive strains at approximately constant stress, so that large amounts of energy can be absorbed from impacts without high stresses being generated. Thus, trabecular bone functions similar to foam packaging, providing high energy absorption with minimal density and weight. As we will see below, the analogy with engineering foams is a useful one for understanding several aspects of the mechanical behavior of trabecular bone.

9.3.3 Trabecular bone mechanics: density dependence

The strong dependence of the mechanical properties of trabecular bone on relative density is evident when Young's modulus or compressive strength is plotted against apparent (or relative) density, as is shown in Fig. 9.7. The data points in these plots come from a variety of studies and include bone samples of unspecified architecture that were loaded in tension or compression and in unspecified directions over a

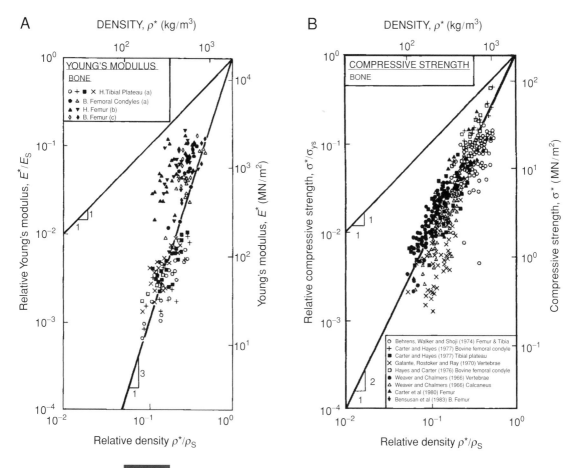

Relationship between relative density and Young's modulus (A) or compressive strength (B) of trabecular bone. The solid lines are not fits; rather, they indicate linear relationships of the specified slope between the log of the relevant mechanical property and the log of the relative density. The lines are drawn such that the intercept is defined assuming a "solid" or non-porous piece of trabecular bone would have the same properties as cortical bone, namely $\rho_s = 1800$ kg/m^3, $E_s = 17$ GPa, and yield stress $\sigma_{ys} = 193$ MPa. From Gibson and Ashby [12] (see the reference for the original sources of the data). Reproduced with permission of the authors and publisher.

wide range of strain rates. It is remarkable that, despite the variability in specimen origin and test conditions, there is clearly a strong dependence of the modulus and strength of trabecular bone on apparent density alone, even over a range of mechanical properties spanning three orders of magnitude.

Some of the data in Fig. 9.7 are from a study by Carter and Hayes [10] in which they tested bovine and human trabecular bone specimens with a wide range of apparent densities. Based on their data, they concluded that the compressive elastic modulus, E (in GPa), was correlated with the cube of the apparent density,

ρ (in g/cm^3) by

$$E = 3790 \, \dot{\varepsilon}^{0.06} \rho^3, \tag{9.1}$$

where $\dot{\varepsilon}$ is in units of s^{-1}. Further, the compressive ultimate strength, σ_{ult} (in MPa), was correlated with the square of the apparent density by

$$\sigma_{ult} = 68 \, \dot{\varepsilon}^{0.06} \rho^2. \tag{9.2}$$

Later retrospective statistical analyses of data from several studies on *human* trabecular bone alone [14] confirmed the quadratic relationship of Equation (9.2), but found that a quadratic relationship for the modulus was slightly more accurate than the cubic relationship of Equation (9.1), such that

$$E \propto \dot{\varepsilon}^{0.06} \rho^2 \text{ for human trabecular bone.} \tag{9.3}$$

The quadratic dependence of stiffness and strength on bone density has important implications: even a modest loss of bone mass, such as occurs in osteoporosis, can result in a significant reduction in mechanical properties and therefore increased fracture risk. The non-linear dependency of mechanical properties on density is also clear from the data summarized in Table 9.4. Notice that the *coefficients of variation* (CV)[1] for the moduli and ultimate compressive strengths are quite large (typically $> 50\%$). Some of this variability between specimens from the same location is likely to result from variation in the densities of the specimens. However, the variability in the specimen densities (CV $\sim 30\%$) was much lower than the variability in the mechanical properties, suggesting the properties are very sensitive to the density, as predicted by Equations (9.2) and (9.3).

The observant reader will notice that some of the data in Table 9.4 and Fig. 9.7 are not consistent with the mechanical properties being proportional to the square of the density. For example, the femoral samples in Table 9.4 are denser than the tibial samples, but not as stiff. Clearly, bone properties are not *solely* determined by the apparent density but depend on other factors, such as the trabecular architecture. Also note that there is little variability in the ultimate strains in Table 9.4 (CV $\sim 20\%$), suggesting a weak dependence of failure strain on apparent density. Several other studies have confirmed that the yield and ultimate strain of trabecular bone are relatively constant over a wide range of relative densities and architectures (see Keaveny [11] for a summary). The unit cell or foam model, to which we now turn, also predicts that yield strain is independent of relative density (see Problem 9.2 [p. 436]).

---

[1] The coefficient of variation is the ratio of the standard deviation to the mean. It is a measure of the variability within the population.

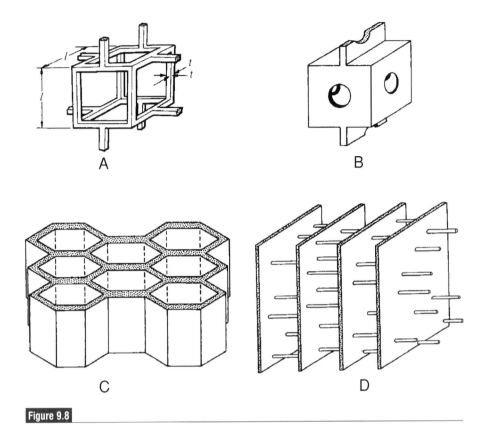

Figure 9.8

Idealized models for the microstructure of trabecular bone: (A) low-density equiaxed structure; (B) higher density equiaxed structure; (C) oriented prismatic structure; (D) oriented parallel plate structure. From Gibson and Ashby [12]. Reproduced with permission of the authors and publisher.

9.3.4 Trabecular bone mechanics: unit cell models

The structure of trabecular bone and its mechanical behavior is similar in many respects to that of engineering foams and other cellular materials, including wood, and, as we discussed in Ch. 2, the actin cytoskeleton. Accordingly, we can use the unit cell approach of Section 2.6.3, where we estimated the Young's modulus and shear modulus of the actin cytoskeleton, to estimate the mechanical properties of trabecular bone based on its relative density. Again, the details of the following derivations are detailed in Gibson and Ashby [12].

Because of the variability of trabecular bone architecture, several idealized models have been proposed to represent trabecular bone microstructure (Fig. 9.8). The model in Fig. 9.8A is appropriate for low-density bone without preferred orientation of the trabeculae (equiaxed, open cell rod-like structure). The bone in certain regions of the femoral head has this microstructure. Recall we used the same model in Section 2.6.3 to represent the filamentous actin cytoskeleton and derived a relationship between the relative modulus, E^*/E_s, and relative density, ρ^*/ρ_s, given in equation 2.47 and repeated here:

$$\frac{E^*}{E_s} = C_1 \left(\frac{\rho^*}{\rho_s} \right)^2.$$ (9.4)

In this case, E^* and ρ^* are the apparent properties of a trabecular specimen, and E_s and ρ_s are the modulus and density of the bone tissue matrix itself, with the properties of cortical bone tissue often used for normalization ($E_s = 17$ GPa; $\rho_s = 1.8$ g/cm^3). Comparison of the predicted relationship with the relationship observed empirically (Equation [9.3]) indicates that the equiaxed, open-cell model is reasonable for human trabecular bone. Analysis of the other models in Fig. 9.8 suggests that the exponent of the density in the modulus–density relationship should lie between 1 (linear dependence) and 3 (cubic dependence), depending on the structure. For example, for the prismatic structure (Fig. 9.8C), the modulus is linearly dependent on the density in the "longitudinal" direction (along the prism axis), but has cubic dependence in the transverse direction (refer to Gibson and Ashby [12] for the derivation). Most of the moduli in Fig. 9.7 do fall between the lines with slopes 1 and 3, broadly consistent with the model predictions.

What about the model predictions for the dependency of compressive strength on density? For the low-density, rod-like model (Fig. 9.8A), failure occurs primarily by elastic buckling of the vertical struts. Once the trabeculae buckle and collapse, the bone is deformed permanently (Fig. 9.6). The compressive strength at which buckling occurs is the *yield strength*, which is the stress at which the material no longer behaves as a linear elastic material. Typically the yield strength is less than the ultimate strength, which is the maximum stress that can be supported before catastrophic failure. Although trabecular bone continues to support loads as it collapses plastically and densifies (Fig. 9.6), we are often interested in the yield strength because it represents the limit of elastic or non-permanent deformation. For the low-density, rod-like model, it can be shown that the compressive yield strength is proportional to the square of the density (see Problem 9.3 [p. 436])

$$\frac{\sigma^*}{\sigma_{ys}} = C_2 \left(\frac{\rho^*}{\rho_s} \right)^2$$ (9.5)

where σ^* is the apparent compressive yield strength of the trabecular bone and σ_{ys} is the compressive yield strength of the tissue matrix itself, often assumed to be that of cortical bone ($\sigma_{ys} = 193$ MPa). Analysis of the other models in Fig. 9.8 suggests that the exponent of the density in the compressive strength–density relationship should lie between 1 (linear dependence) and 2 (square dependence), depending on the structure. Again, the model predictions are generally consistent with experimental evidence (Fig. 9.7 and Equation [9.2]).

These cellular solid models, while simple and idealized, provide useful insight into the mechanical behavior of trabecular bone. For example, the models clearly demonstrate the dependence of the mechanical properties on both the density and architecture of the bone, and they are useful for predicting the general effects of changes in either density or structure on the mechanical integrity of bone. They also provide insight into the mechanisms by which trabecular bone deforms and ultimately fails. However, the model predictions agree with the experimental data only in a broad sense, suggesting there are additional material and structural factors that determine the properties of trabecular bone. Identification of these factors so as to gain improved understanding of the relationship between the mechanical properties of bone and its "quality" (i.e., structure, composition, and density) will help with diagnosis in patients with osteoporosis at risk for fracture and with efforts to engineer replacement bone tissue. Accordingly, this is an active area of research utilizing advanced imaging methods, sophisticated approaches to quantify bone quality, and high-resolution finite element analyses (see, for example, van Rietbergen and Huiskes [15]).

9.4 Bone fracture and failure mechanics

The interest in bone from a biomechanical perspective is largely motivated by the fact that bones fracture, with serious medical consequences and significant health-care costs. An improved understanding of the mechanisms underlying bone fracture and the ability to predict when fractures will occur may help in the development of new strategies for fracture prevention and treatment.

Bones can fail in a variety of ways. If a bone is loaded monotonically to a stress level that exceeds the failure or ultimate stress of the bone tissue, then fracture will occur. This type of failure can occur with trauma or impact, such as a fall on the hip. Bone can also fail at stress levels well below the yield stress through the growth of preexisting cracks. Bone is filled with micro-cracks (Fig. 9.9). In normal, healthy bone, these micro-cracks are repaired by bone cells. In fact it has been suggested that micro-fractures might be required to initiate remodeling and normal turnover of the skeleton [17]. With age and some bone-related diseases, however, the repair mechanisms may not keep up with the accumulation of damage, leading to increased fracture risk. Growth of micro-cracks, if unchecked, can lead to failure and fracture.

We consider in this section two modes of fracture by which cracks can grow, leading to catastrophic failure. The first is *fast fracture* in which a crack rapidly propagates across the material. In the context of bone, this is most commonly associated with trauma or impact but can occur at stress levels below the yield stress. The second fracture mechanism is *fatigue*, caused by the repeated, cyclic application

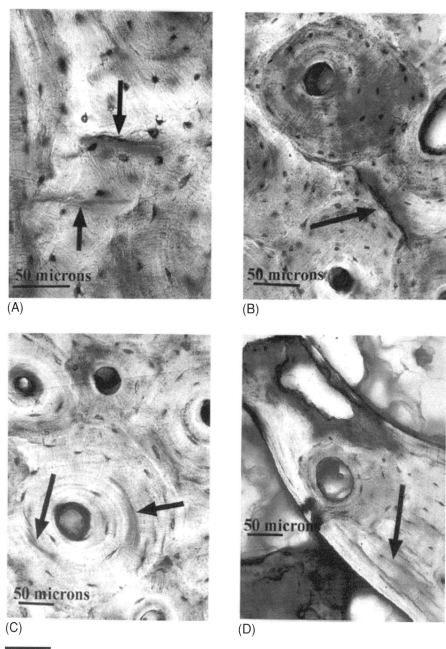

Figure 9.9

Micro-cracks in human metatarsal bones (arrows). Four types of cracks were observed. (A) cracks within the interstitial lamellar bone matrix, not in contact with osteons; (B) Cracks in the interstitial bone ending at cement lines; (C) Cracks within osteons; and (D) cracks within trabecular struts. The metatarsal bones are in the distal end of the foot and connect with the bones in the toes. From Donahue *et al.* [16], with permission from Elsevier.

of stresses that eventually lead to failure. This is the type of failure that occurs in stress fractures suffered by military recruits on long marches, ballet dancers who repeatedly load their foot bones, basketball players who jump a lot, and recreational athletes who run a marathon without training properly. Much of our discussion involves linear fracture mechanics and is based on Ashby and Jones [18].

9.4.1 Fast fracture

Consider an inflated balloon that is pricked with a pin. This causes the fabric of the balloon to rupture (i.e., to undergo fast fracture). This yielding phenomenon occurs even through the stress in the balloon wall prior to the pin damaging the balloon was less than the yield stress. Why is this?

This phenomenon can be best understood by thinking about the system from an energy viewpoint. It is clear that the pin creates a defect in the balloon wall. The question is then what controls whether this defect grows or not. The answer is that the defect will grow so long as there is sufficient energy in the system to allow the defect to propagate. In the case of the balloon, the air pressure in the balloon provides the energy by doing work on the material of the balloon wall. This work can be used to change the elastic energy stored in the balloon fabric, or it can be used to create a tear surface in the balloon fabric. If we denote:

$\delta U^{el} \equiv$ the increase in elastic energy of material (in this case, the rubber in the wall of the balloon),

$\delta W \equiv$ the incremental work done on the material by external loads (in this case, by pressure in the balloon), and

$\delta U^{cr} \equiv$ the energy needed to increase the crack size,

then the crack will propagate if and only if

$$\delta W \geqslant \delta U^{el} + \delta U^{cr}. \tag{9.6}$$

If this inequality is not satisfied, then there is not sufficient energy available to "grow" the crack. This corresponds, for example, to the case of a partly inflated balloon that is pricked with a pin.

To use Equation (9.6), we must be able to quantify each of the terms. Consider first δU^{cr}. Suppose that we have a sample, chosen for simplicity to be rectangular with thickness t, containing a preexisting crack of length a (Fig. 9.10). The sample is subjected to a tensile stress σ. We define the *toughness*, G_c, to be the energy absorbed by the material per unit area of new crack surface created.[2] To enlarge the crack from length a to δa therefore requires $G_c t \delta a$ units of energy.

---

[2] G_c is a material property that can be looked up in handbooks. It has units of energy per unit area, e.g., J/m^2 in the SI system. Note that the crack surface is not the area enclosed by the crack that is visible in Fig. 9.10; it is the area of the actual crack *interface*.

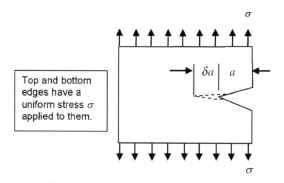

σ

Top and bottom edges have a uniform stress σ applied to them.

δa a

σ

Figure 9.10

Schematic drawing of a rectangular plate subjected to a uniaxial tension resulting in a uniaxial stress σ. The plate contains a crack of length a, and is of thickness t (into the plane of the page). If the crack enlarges from a to $a + \delta a$, then the crack surface area increases by $\delta a t$.

Now we consider the terms δW and δU^{el}. They depend on the specifics of the problem (i.e., they are not material properties). We will assume that the edges of the sample are clamped so they cannot move. (This can be accomplished by subjecting the plate to the required stress, σ, allowing it to reach equilibrium, and then clamping the edges.) After clamping, the work done by the externally imposed stress on the sample, δW, is zero. (Recall that for work to be done, a displacement is required.) In problems in which the edges of the sample are not clamped then the external work will be non-zero. For the special case of $\delta W = 0$, the condition for fast fracture (Equation [9.6]) becomes:

$$\delta U^{\mathrm{el}} + G_c t\, \delta a \leqslant 0 \tag{9.7}$$

Estimating δU^{el} is slightly trickier. First, we note that δU^{el} will be < 0 if crack formation *releases* elastic energy. For the simplest case of uniaxial tension and perfectly elastic behavior, the energy stored in the sample per unit volume is given by $\sigma^2/2E$, where E is Young's modulus for the material and σ is the stress resulting from the applied tension (Box 9.1).

The crack causes a local stress relaxation in the material around the crack (Fig. 9.11). Of course, the stress decays continuously from zero at the crack interface to its unrelieved value far from the crack, but we will simplify this situation by saying that the stress is identically zero in some finite region surrounding the crack, and unrelieved outside this region. For a crack of length a, the "stress relief" region will have surface area proportional to $\frac{1}{2}\pi a^2$. This corresponds to a sample volume of $\frac{1}{2}\pi a^2 t$, and a stress energy relief of

$$U^{\mathrm{el}} \approx -\frac{\sigma^2}{4E}\pi a^2 t. \tag{9.8}$$

Box 9.1 What is the energy stored in the plate per unit volume?

Consider a small element of material acted upon by a uniaxial stress, σ, and undergoing a deformation in the same direction, producing a strain ε. If the undeformed surface area of the face that the traction is applied to is A_0 and the undeformed length of the material in the direction of traction is Λ_0, then the volume of the element is $V_0 = A_0\Lambda_0$. The work done on the element is equal to the energy stored, so we may write

$$\frac{U^{\text{el}}}{V_0} = \frac{\text{Energy stored}}{\text{Unit volume}} = \frac{\int \mathbf{F} \cdot d\mathbf{r}}{A_0\Lambda_0} = \frac{1}{A_0\Lambda_0} \int \sigma A_0 \, d\Lambda \quad (9.9)$$

where Λ is the length of the element in the direction that the traction is applied. We substitute $\varepsilon = \Lambda/\Lambda_0$ into the above equation to write

$$\frac{U^{\text{el}}}{V_0} = \int \sigma \, d\varepsilon = \frac{\sigma^2}{2E} \quad (9.10)$$

where we have assumed linear elastic behavior ($\sigma = E\varepsilon$) and that there was no elastic energy stored in the reference state ($\varepsilon = 0$).

As the crack grows by an amount δa, then U^{el} will change by an amount

$$\delta U^{\text{el}} \approx -\frac{\sigma^2}{2E} \pi a t \, \delta a. \quad (9.11)$$

Therefore, the crack growth condition (9.7) becomes

$$-\frac{\sigma^2}{2E} \pi a t \, \delta a + G_{\text{c}} t \, \delta a \leqslant 0. \quad (9.12)$$

Rearrangement allows us to state that the crack will grow if

$$\frac{\pi a \sigma^2}{2} \geqslant E G_{\text{c}}. \quad (9.13)$$

It turns out that this simplified analysis is wrong by a factor of two (because of the unrealistic assumption made about how the stress is relieved around the crack). A more detailed derivation shows that the critical condition for crack propagation is

$$\pi a \sigma^2 \geqslant E G_{\text{c}}. \quad (9.14)$$

Note that the right-hand side of Equation (9.14) involves only material properties and is itself therefore a material property. We define a new property called the *fracture toughness*, or *critical stress intensity factor*, K_{c}, by

$$K_{\text{c}} = \sqrt{E G_{\text{c}}} \quad (9.15)$$

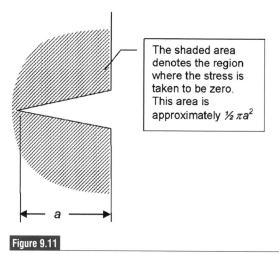

The shaded area denotes the region where the stress is taken to be zero. This area is approximately $\frac{1}{2}\pi a^2$

Figure 9.11

Magnified view of the crack tip in the specimen shown in Fig. 9.10. The shaded region is the zone where it is assumed that the stress has been relieved to zero.

(which has units of MN/m$^{3/2}$). We also define the *stress intensity factor*

$$K = \sqrt{\pi a}\ \sigma \qquad (9.16)$$

Therefore, the crack propagation condition can be rewritten as

$$K \geqslant K_c \qquad (9.17)$$

Note that the stress intensity factor K depends on both the imposed stress σ and the crack size a. If the combination is too large, fast fracture occurs.

All of the above analysis assumes fully linear elastic material behavior. This is not usually the case in practice. The crack tip acts to concentrate stresses, so σ can exceed σ_{yield} locally (Fig. 9.12). This causes local plastic flow, thereby absorbing more energy and increasing the effective G_c. In the case of bone, classical plastic behavior is not observed. Instead, energy is absorbed by a slightly different mechanism. This involves osteons being pulled out of the surrounding interstitial bone along the cement line, which greatly increases the toughness of bone, since significant energy is expended in osteon pull-out (Fig. 9.13). A complication is that osteon pull-out depends on crack-propagation speed: fast-propagating cracks shear off osteons, rather than pulling them out (Fig. 9.14). This is an example of a ductile/brittle transition and results in a rather complex relationship between fracture toughness K_c and the speed of crack propagation, or rate of strain (Fig. 9.15). The implication is that a fast-moving crack will propagate more easily for a given σ value than will a slow-moving crack. Note also that the value of K_c depends on factors such as the orientation of the bone, the age of the person (and therefore

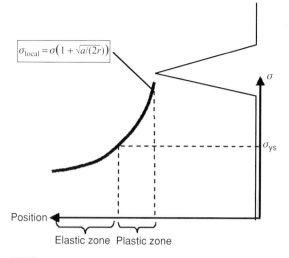

$$\sigma_{local} = \sigma\left(1 + \sqrt{a/(2r)}\right)$$

Elastic zone   Plastic zone

Figure 9.12

Plot of local stress, σ_{local}, versus position in the vicinity of the crack tip. It can be seen that there is a region where the stress exceeds the yield stress, σ_{ys}, leading to plastic behavior in the neighborhood of the crack tip.

Figure 9.13

Bone fracture showing osteon pull-out. This mode of bone fracture dissipates significant amounts of energy, as osteons are pulled out from surrounding interstitial bone. Compare this with a fracture associated with osteon shearing, as shown in Fig. 9.14. From Piekarski [19]. Copyright 1984 from *Natural and Living Biomaterials* by Hastings. Reproduced by permission of Routledge/Taylor & Francis Group, LLC.

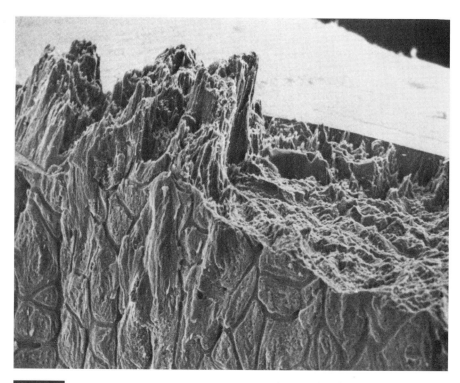

Bone fracture showing osteon cleavage. A fracture moving from left to right was originally slowly propagating, causing pull-out. At the midway point of the photograph it developed into a fast-moving crack. From Piekarski [19]. Copyright 1984 from *Natural and Living Biomaterials* by Hastings. Reproduced by permission of Routledge/Taylor & Francis Group, LLC.

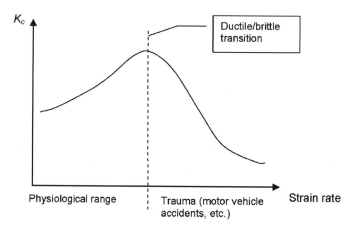

Schematic of relationship between fracture toughness, K_c, and the rate of strain for bone. At low rates of strain in the physiological range, osteons are pulled out from the surrounding matrix. It takes more energy to pull out an osteon rapidly than slowly; therefore, in this range, bone toughens with increasing strain rate. Above a critical rate, osteons begin to fracture, which requires less energy than being pulled out from the matrix. As more and more osteons fracture with increasing strain rate, the fracture toughness of the bone decreases.

Table 9.5. Fracture toughness (K_c) and toughness (G_c) of cortical bone loaded in the longitudinal and transverse directions. These values are for mode I loading, as shown in Fig. 9.10. The fracture toughness values for human cortical bone are comparable to those of a tough ceramic, but an order of magnitude less those for aluminium and steel. Adapted from Martin *et al.* [20] with kind permission of Springer Verlag.

Direction of fracture	Bone type	K_c (MN/m$^{3/2}$)	G_c (J/m^2)
Longitudinal	Bovine femur	3.21	1388–2557
	Bovine tibia	2.8–6.3	630–2238
	Human femur	2.2–5.7	350–900
Transverse	Bovine femur	5.49	3100–5500
	Bovine tibia	2.2–4.6	780–1120
	Human tibia	2.2–5.7	350–900

the relative density), and the water content of the bone. Some representative data are shown in Table 9.5.

9.4.2 Fatigue fracture

Cracks in bones can propagate at stress levels lower than those that cause fast fracture, if the stress is applied in a repeated (cyclic) loading and unloading pattern. Consider again the sample shown in Fig. 9.10 but now suppose that the applied stress σ varies periodically between $\sigma_{\max}$ and $\sigma_{\min}$ instead of being constant (Fig. 9.16). Since the stress intensity K is equal to $\sigma\sqrt{\pi a}$, this implies that K will vary cyclically with σ. In addition, each time the stress goes through one cycle, the crack length, a, increases by a small amount. This means that the extreme values of K will also increase by a small amount each cycle in such a way that the difference between the maximum and minimum value of K increases over time (Fig. 9.16B).

For many materials, it is found empirically that cracks grow in a manner shown in Fig. 9.17. After a rapid initiation phase, they grow at a steady rate (on a log–log plot), before suddenly causing fracture when they reach a critical size. The rate of growth during the "steady growth" phase can be expressed by Paris' law

$$\frac{da}{dN} = C(\Delta K)^m \tag{9.18}$$

where N is the number of cycles, and C and m are material constants that are empirically determined. Note that da/dN is the rate of crack enlargement. Because

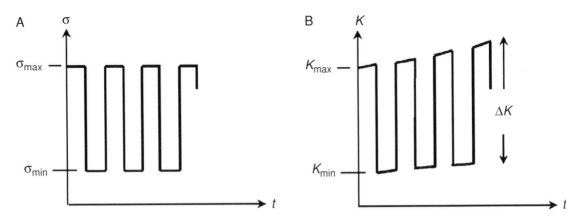

A σ

σ_{max}

σ_{min}

t

B K

K_{max}

K_{min}

ΔK

t

Figure 9.16

Variation of imposed stress, σ (A), and resulting stress intensity factor, K (B), with time, t. The change in K over one cycle, $\Delta K = K_{max} - K_{min}$, is given by $\Delta K = \Delta\sigma\sqrt{\pi a}$, where $\Delta\sigma = \sigma_{max} - \sigma_{min}$. Notice that as the crack length, a, increases, so too does ΔK.

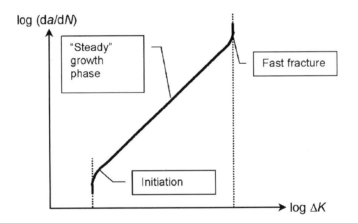

log (d*a*/d*N*)

"Steady" growth phase

Fast fracture

Initiation

log ΔK

Figure 9.17

Crack growth characteristics during cyclical fatigue, where da/dN is the rate of growth (linear enlargement per cycle), and ΔK is the variation in the stress intensity factor over one loading cycle. The central linear portion is described by Paris' law (Equation (9.18)).

most of the crack's "lifetime" is spent in this steady growth phase, to a good approximation we can ignore the other phases. Therefore, if we are told that a crack has an initial length a_0, we can integrate Equation (9.18) to determine the number of cycles required for the crack to grow to some final size, a_f. Denoting this number of cycles by N_f, we can rearrange Equation (9.18) to write:

$$N_f = \int_0^{N_f} dN = \int_{a_0}^{a_f} \frac{da}{C(\Delta K)^m}. \tag{9.19}$$

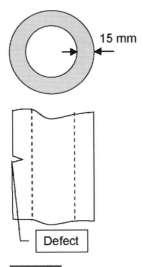

15 mm

Figure 9.18

Schematic view of tibia with a small defect.

Substituting $\Delta K = \Delta\sigma\sqrt{\pi a}$ into Equation (9.19), integrating, and rearranging gives:

$$N_{\mathrm{f}} = \frac{1}{C(\Delta\sigma)^m \pi^{m/2}} \left. \frac{a^{1-m/2}}{1-m/2}\right|_{a_0}^{a_{\mathrm{f}}} \quad \text{for } m \neq 2 \tag{9.20}$$

where we have assumed that the stress variation over one cycle, $\Delta\sigma$, is constant from cycle to cycle.

The concepts of fast versus fatigue fracture are considered in the following example. During a rather rough hockey game you are slashed in the tibia, creating a 0.01 mm defect. The cortical bone thickness in the tibia at the slash location is 15 mm, and we can assume that the tibia has the cross-section of a hollow cylinder at this point (see Fig. 9.18). Tomorrow you plan to run a marathon. Will your tibia fracture during the marathon?

To answer this question, we need some background data. We will assume that the stress distribution in the tibia during running is of the form shown in Fig. 9.19, with peak-to-trough amplitude of 11 MPa. Let us first check to see if fast fracture will occur (hopefully not). To evaluate this possibility, we need to know the maximum stress that the tibia is exposed to. We are not given this information, but from the shape of the curve in Fig. 9.19, it is $\leqslant 11$ MPa, so let us take 11 MPa as a conservative estimate of $\sigma_{\max}$. Then we can compute the maximum value of the stress intensity factor, $K_{\max}$:

$$K_{\max} = \sigma_{\max}\sqrt{\pi a} = (11 \times 10^6 \,\mathrm{Pa}) \sqrt{\pi(0.01 \times 10^{-3} \,\mathrm{m})} = 0.060 \,\mathrm{MN/m^{3/2}}.$$
$$\tag{9.21}$$

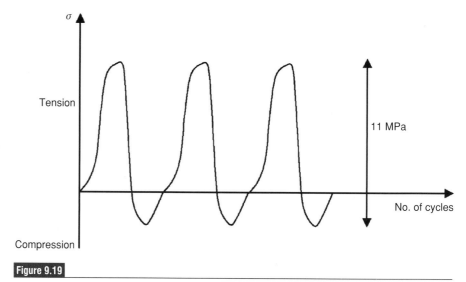

Figure 9.19

Assumed stress distribution in the tibia at the defect location during running for the example described in the text.

From Table 9.5, this is less than the critical value, K_c, of 2.2–6.3 MN/m$^{3/2}$, so fast fracture will not occur. Thought of in another way, if we conservatively estimate K_c as 2.2 MN/m$^{3/2}$, the critical crack length at which fast fracture will occur is 12.7 mm, which is obviously longer than the initial crack.

Now let us look at the issue of fatigue fracture. We use Equation (9.20) with values of $m = 2.5$ and $C = 2.5 \times 10^{-6}$ m(MN/m$^{3/2}$)$^{-2.5}$, which are suitable for cortical bone. Inserting these and using $a_0 = 0.01 \times 10^{-3}$ m and the crack length at which fast fractures will occur (12.7 mm $= 12.7 \times 10^{-3}$ m) as a_f, we get

$$N_f = \frac{1}{2.5 \times 10^{-6}(11)^{2.5}\pi^{1.25}} \frac{(12.7 \times 10^{-3})^{-1/4} - (0.01 \times 10^{-3})^{-1/4}}{-1/4} = 14\,112$$

(9.22)

where we are careful to use consistent units! If you have a stride length of 2 m while running (right heel strike to right heel strike), then this number of cycles corresponds to 28.2 km, which is less than the 42 km length of a marathon. Consequently, it seems risky to try running the race (not to mention the intense pain that you would experience in your tibia while running).

Of course, this example is oversimplified. Our analysis ignores the microstructure of bone, which is similar in many ways to engineered fiber-reinforced composites. In composite materials, the presence of lamellae or fibers limits damage. In bone, the osteons and collagen fibers play a similar role (we have already discussed how osteons limit crack propagation in fast fracture). In fatigue, cracks tend to propagate to the interface between the osteon and surrounding matrix (i.e., to the cement line), where they are redirected to travel along fibers. Because the fibers in bone are oriented longitudinally, the cracks do not progress across the bone, preventing (in most cases) catastrophic failure. Standard engineering analyses of fracture also do not account for the ability of bone to limit damage accumulation by remodeling to repair itself.

Because of the barriers to crack growth inherent in bone tissue, Paris' law does not do a great job of predicting crack growth in bone. This is primarily because the crack growth rate decreases rapidly with increasing length as cracks are arrested, thus limiting the time spent in the steady growth phase for which Paris' law is applicable. A modified Paris law for bone has been developed to account for the effects of the short distance between osteons by adding a second term that describes the rate of crack growth when cracks are small [21]. Details on other approaches to model bone fracture are presented in an excellent book on skeletal tissue mechanics by Martin *et al.* [20].

9.5 Functional adaptation and mechanobiology

Bone is a dynamic tissue in which multiple cell types respond to their local microenvironment to regulate the growth, maintenance, and repair of the tissue. The cues that regulate these processes are numerous, and certainly include biochemical and hormonal signals. As we discussed in Section 2.9.4, bone cells are also exquisitely sensitive to mechanical stimuli, such as stretch, pressure, and flow. Bone cells respond to local mechanical factors by making new bone tissue and resorbing old tissue. Rather than being a random, haphazard process, however, the ultimate structure of bone is regulated in a way that meets the local mechanical demands of the skeleton.

As described in Section 1.1, this concept – that bone senses and adapts to its mechanical environment (i.e., functional adaptation) – is commonly referred to as *Wolff's law*, named after Julius Wolff, a nineteenth century orthopedic surgeon. It is an interesting historical note that while Wolff is often given credit for this concept, the idea of functional adaptation was not original, having been observed and proposed by many before him. It was Wolff, however, who "took charge of the subject" [22]. As stated by Keith [23]:

. . . the idea which underlies Wolff's law was abroad in men's minds before his time, but he was the first to give that idea a definite expression. He was the first to devote 30 years of constant work and observation to prove that the shape and structure of growing bones and adult bones depend on the stress and strains to which they are subjected.

Parameter	Non-playing humerus	Playing humerus
D_o (cm)	2.195	2.45
D_i (cm)	1.10	0.975
A (cm^2)	2.83	3.97
I_z (cm^4)	1.07	1.72

Figure 9.20

Comparison of the dimensions of the humerus in the non-playing arm of an active tennis player with those of the playing arm. Measurements are taken from the long shaft of the bone. D_o and D_i are outer and inner diameters, respectively; A is cross-sectional bone area; I_z is moment of inertia. Based on data from Jones *et al.* [24].

Wolff's law is now widely recognized by biomechanicians not as a mathematical law, but rather as the concept of functional adaptation that encompasses ideas that emerged from the work of several investigators in the nineteenth century. Three core ideas of functional adaptation that continue to intrigue researchers today are [22]:

- bone structure is optimized for strength with respect to its weight[3]
- trabeculae are aligned with the principal directions of stress[4]
- bone adapts its structure through the responsiveness of bone cells to local mechanical stimuli.

An example of structural optimization and bone mechanobiology is the adaptations that occur in response to exercise. A nice example of this comes from a study by Jones *et al.* [24] in which the dimensions of the humerus (upper arm bone) of professional male and female tennis players were measured from X-ray films. The dimensions of the arm the players used to swing the racquet (the playing arm) were compared with those of the non-playing arm (paired control). Their findings are summarized in Fig. 9.20. Note that the mechanically loaded bone in the playing arm has a greater outer diameter and smaller inner diameter than the non-playing arm bone, resulting in a thicker cortex.

Let us consider the implications of these geometric changes to the mechanical properties of the whole humerus. To do so, we idealize the humerus as a hollow

---

[3] We will reconsider this point in Section 9.6 on the design of bone.

[4] Cowin has noted that principal stresses in a homogeneous isotropic elastic material do not match those in a discontinuous structure like trabecular bone, and therefore the mathematically perfect correspondence between the two, as proposed by Wolff, is not valid. However, experimental studies have shown that trabeculae do align with directions of continuum level principal stresses, i.e., the directions that the principal stresses would be if the bone were homogeneous [20], confirming the general observations of Wolff and others.

circular cylinder with a moment of inertia about the z axis, $I_z = \dfrac{\pi}{64}(D_o^4 - D_i^4)$, and subject it to a pure bending moment, M. As described in Section 8.4.2, we can estimate the strain from bending based on Euler–Bernoulli beam theory as:

$$\varepsilon_x = \pm\frac{My}{EI_z} \tag{9.23}$$

where ε_x is the strain in the longitudinal direction at a point in the bone that is a distance y from the neutral axis (the midline for our idealized bone) and E is the Young's modulus of cortical bone. To compare the consequences of the change in geometry, we will determine the change in moment that can be supported by the playing humerus versus the non-playing humerus, maintaining the same surface strain (i.e., ε_x at $y = D_o/2$) in the two cases. Rearranging Equation (9.23), taking the ratio of the moment for the playing arm (M_p) to that of the non-playing arm (M_{np}), and substituting the expression for the moment of inertia, we get

$$\frac{M_p}{M_{np}} = \frac{I_{z,p}}{I_{z,np}} \frac{D_{o,np}}{D_{o,p}} = \frac{\left(D_{o,p}^4 - D_{i,p}^4\right)}{\left(D_{o,np}^4 - D_{i,np}^4\right)} \frac{D_{o,np}}{D_{o,p}} \tag{9.24}$$

where the subscripts p and np indicate playing arm and non-playing arm, respectively. Clearly, the ratio of the moments is strongly dependent on the relative geometries of the two bones.

Substituting the dimensions measured by Jones *et al.* [24], we find that a 45% greater moment can be applied to the playing arm while maintaining the same surface strain as in the non-playing arm. Thus, the playing arm humerus has adapted its geometry in response to the extra forces imposed on it. The question of how the skeleton does this is intriguing but unanswered. Measurements of peak strains on the periosteal (outer) surface of a variety of bones in a variety of animals (including humans) are remarkably consistent, ranging between 2000 and 3000 microstrain (0.2–0.3% strain) [25]. Thus it appears that the skeleton adapts to control strain. But the mechanisms by which the bone cells responsible for remodeling the tissue sense and respond to strain are largely unknown.

9.6 The design of bone

It is clear that there are a wide range of design requirements for bone, including:

- strength to resist axial, bending, torsion, and impact loading, but with greater strength in some directions than others depending on the predominant modes of loading

Table 9.6. Incidence of fracture of various bones in males aged 20–45 years. Bone nomenclature is given in Fig. 9.1. Adapted from Currey [26] with permission of the American Society for Bone and Mineral Research.

Bone or structure	Incidence per 10 000 at risk
Skull	370
Scapula	22
Vertebrae	57
Pelvis	18
Ribs	192
Clavicle	134
Humerus	118
Radius/ulna	369
Carpal bones	978
Hip	43
Femur	0.4
Patella	28
Tibia/fibula	244
Ankle	243
Foot	378

- stiffness under axial, bending, and torsion loading, but again stiffer in some directions than others depending on the predominant modes of loading
- low weight
- high fatigue fracture resistance
- able to absorb energy during impact
- able to repair itself
- able to optimize its structure dynamically.

These requirements are clearly mutually inconsistent; for example, our bones could be much stronger, but then they would be too heavy. Because of this, bone is not optimally designed for any one function or to meet any single requirement. Instead, it is optimized to perform acceptably well in many categories over the range of normally encountered physiological conditions. This is interesting, because one might assume bones are designed primarily to resist fracture considering the serious medical consequences associated with fractured bones. Currey [26] considered to what extent bones are designed to resist fracture by examining the incidences of fracture of different bones in a large number of young people (Table 9.6). The propensity to fracture was quite different in different bones in the body, leading him to make two interesting observations. First, the fracture incidence of flat bones that have strong muscle attachments (e.g., scapula, pelvis, and vertebrae) is much lower than that of other bones. Currey hypothesized that this may be an example of bones that are designed to be very stiff to resist flexing when pulled on by the muscle attachments, which they do at the expense of increased mass. The result is that such bones are stronger than required for normal activities (and therefore clearly not optimized only to resist fracture).

The second observation from the fracture incidence data was that the more distal bones in the upper and lower limbs (e.g., foot or hand bones) fractured much more frequently than did the proximal limb bones (e.g., humerus or femur). Currey suggested that this is an example of minimizing the mass of the distal bones at the expense of increased fracture risk. Light distal bones have lower mass moments of inertia about the joint center (recall the mass moment of inertia increases as the *square* of the distance to the center of rotation) and therefore require less muscle force to accelerate. This is a significant advantage in terms of maximizing speed of movement with minimal energy expenditure, particularly for fast moving animals like horses. Thus, the strength of the distal bones is compromised by the need to make them as light as possible.

These examples support the notion that bone adapts to meet the functional requirements of multiple normal, everyday activities. Therefore, most bones are not adapted primarily to resist fracture but adapt to the best compromise for a wide range of different design constraints. Currey also presented the counter-example of the skull bone. The top of the human skull, under normal conditions, is loaded trivially by muscle forces[5] and so should be very thin if it adapts to loading as long bones do. The skull is not thin, however, and so does not appear to be adapted to the functional loads that occur during ordinary living. Instead, the relatively thick skull is adapted to the extreme loads that occur with a potentially fatal blow to the skull and is an example of a bone that is designed primarily to resist fracture.

9.7 Introduction to soft connective tissues

The skeletal system not only contains bones but also soft connective tissues, most notably ligaments, tendons, and articular cartilage. These tissues are critical for normal movement, and in the following sections we will consider their structure, function, and biomechanical behavior.

---

[5] This was tested by Richard Hillam, a graduate student at the University of Bristol, who for his thesis work had strain gauges implanted on his skull and tibia and then measured strains for a wide range of activities, including smiling, eating a hamburger, walking, and jumping. The strains in the skull did not exceed 200 microstrain, even when heading a heavy ball [26].

Table 9.7. Biochemical constituents of soft connective tissues. The values for ligament refer to ligaments of the extremities; elastic ligaments (e.g., in the spine) have substantially more elastin (see the description of ligaments in Section 9.9.1). Minor non-collagenous proteins are not listed and make up the remainder of the dry weight.

Tissue	% of dry weight			Weight % water in wet sample
	Collagen	Elastin	Proteoglycans	
Tendon	75–85	< 3	1–2	65–70
Ligament (extremity)	75–80	< 5	1–3	55–65
Articular Cartilage	50–75	Trace	20–30	60–80
Fibrocartilage	65–75	Trace	1–3	60–70

Ligaments, tendon, and cartilage are composed primarily of an ECM consisting of collagen and proteoglycans, interspersed with water and cells (see Section 2.4. for more information about proteoglycans and ECM). The relative proportions of these constituents and their biochemical and structural organization differs between connective tissues (Table 9.7), imparting unique biomechanical properties that are optimized for the function of the specific tissue. Several other soft fibrous tissues in the body are based on collagen, including skin, fat, blood vessel walls, heart valves, and the connective tissues that hold organs in place. While we will not consider their biomechanical behavior specifically, many of the principles that govern the relationship between the structure and function of connective tissues of the skeletal system also apply to other soft tissues.

It is clear from Table 9.7 that collagen is the most abundant component of the ECM of skeletal connective tissues. As collagen plays an important role in defining the biomechanical properties of skeletal tissues, we start our discussion of connective tissue mechanics by considering the structure of collagen.

9.8 Structure of collagen

As discussed in Section 2.4, there are several types of collagen, with type I collagen being the most common in ligaments, tendons, and bone, and type II being the primary collagen in cartilage. Collagen is organized in a structural hierarchy that allows it to provide tensile stiffness and strength to tissues. The basic unit of collagen is *tropocollagen* or the *"collagen triple helix."* Tropocollagen contains

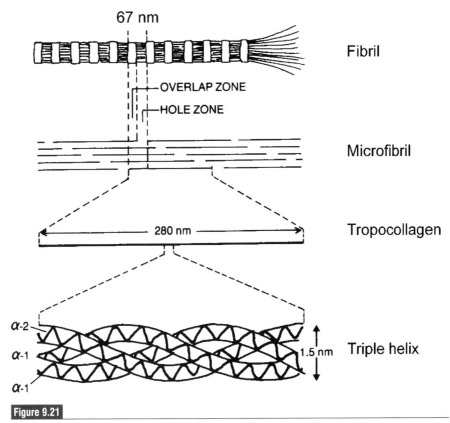

67 nm

Fibril

OVERLAP ZONE

HOLE ZONE

Microfibril

280 nm

Tropocollagen

α-2

α-1

α-1

1.5 nm

Triple helix

Figure 9.21

The structural hierarchy of collagen, showing how progressively larger structures are constructed by collections of smaller units. Collagen fibers (not shown) are formed by the aggregation of several fibrils. Adapted from Nordin and Frankel [27] with permission of Lippincott Williams & Wilkins.

three polypeptide chains (α chains) wound around each other to form a single helical structure approximately 1.5 nm in diameter and 280–300 nm in length (Fig. 9.21). The three collagen subunits are closely packed and stabilized by hydrogen bonding. Tropocollagen molecules are then covalently cross-linked to neighboring tropocollagens to form long strands, with neighboring molecules staggered at approximately one quarter of their length ($\sim$67 nm, called the *D-period*). Collagen microfibrils are formed by the aggregation of five tropocollagens. Microfibrils group to form subfibrils, which, in turn, group to form collagen fibrils that are 50–500 nm in diameter. Collagen fibers, which are 50–300 μm in diameter and visible by light microscopy, are made up of aggregations of fibrils. In connective tissues, the fibers have waviness or crimping (Fig. 9.22) that, as we will soon find out, plays an important role in determining the biomechanical properties of the tissue.

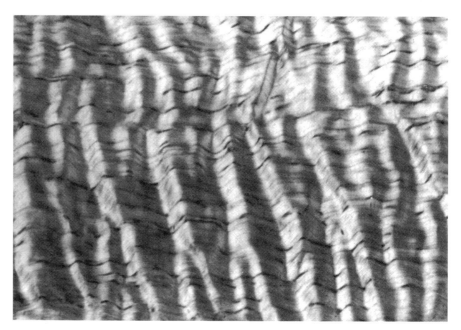

9.9 Structure of ligament, tendon, and cartilage

9.9.1 Ligament

Ligaments in the skeletal system attach one bone to another across a joint and therefore help to guide joint movement and maintain joint congruency. Two ligaments you may be familiar with are the anterior cruciate ligament (ACL) and medial collateral ligament (MCL) of the knee. Ligaments are also found around other joints and along the spine. Outside the skeletal system, ligaments support internal organs and structures including the uterus, bladder, liver, breasts, and diaphragm. Ligaments are even found around teeth, connecting the teeth to the jaw bone.

Ligaments are primarily composed of collagen (mostly type I and some type III), elastin fibers, and proteoglycans. The collagen in the middle section of the ligament is arranged in fibers (Fig. 9.23) that aggregate to form *fascicles*. The fibers are generally oriented along the long axis of the ligament in order to resist tensile forces, although some fibers align off-axis. Close to the points where the ligament

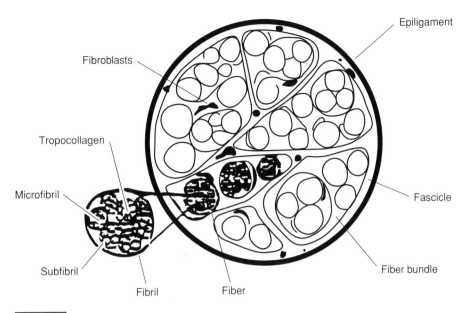

Figure 9.23

Schematic cross-section of the middle section of a ligament, showing how collagen fibers aggregate into fascicles, which in turn aggregate to form the ligament. From Nigg and Herzog [28]. Copyright John Wiley & Sons Limited. Reproduced with permission.

inserts into bone, the composition and structure of the ligament changes so that it is stiffer, which may help to reduce the risk of tearing at the bone–ligament interface. Ligaments also contain elastin fibers, although only in small amounts in the ligaments of the extremities. Some ligaments, such as the ligamentum flavum that connects the vertebrae in the spine, are highly elastic, containing twice as many elastin fibers as collagen fibers [29]. As discussed in Section 2.4, elastin is more compliant and more extensible than collagen. In ligament, it acts like a rubber band to return the ligament to its original length after a stress is removed. The matrix of ligaments also contains a small amount of proteoglycans, which aid in the regulation of water movement within the tissue.

Almost two thirds of the wet weight of a ligament is water. Water in ligaments may provide lubrication and assist in providing nutrients to cells within the ligaments. Additionally, the presence of water and its interaction with the proteoglycans influences the viscoelastic behavior of the tissue, as explained later in this chapter. Ligament cells (fibroblasts) vary in shape, size, and density throughout the tissue, but they are relatively sparse and tend to orient with the long axis of the ligament. Presumably, fibroblasts monitor their environment and in response to local stimuli produce proteins that help to maintain and repair the ligament tissue.

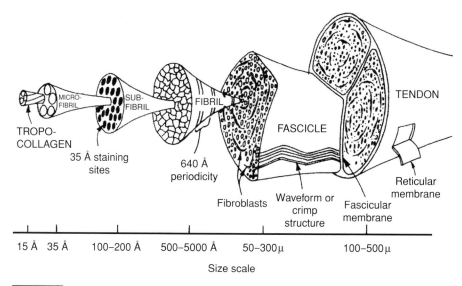

MICRO-
FIBRIL

SUB-
FIBRIL

FIBRIL

TENDON

TROPO-
COLLAGEN

FASCICLE

35 Å staining
sites

640 Å
periodicity

Reticular
membrane

Fibroblasts

Waveform or
crimp
structure

Fascicular
membrane

15 Å 35 Å   100–200 Å   500–5000 Å   50–300 μ   100–500 μ

Size scale

Figure 9.24

Hierarchy of tendon structure and associated size scales. (1Å = 0.1 nm = 0.0001 μm) Copyright 1978 from "The multicomposite structure of tendon", *Connective Tissue Research* **6**:11–23 by Kastelic *et al.* [30]. Reproduced by permission of Taylor & Francis Group, LLC.

9.9.2 Tendon

Tendons connect muscles to bones and are responsible for transmitting force between the two. Tendons vary in size and shape throughout the body, depending on the specific functional requirements for the muscle and bone to which they are attached. In some cases, tendons pass through bony features or specialized connective tissue sheaths that guide the tendon across joints to maintain the correct orientation and efficient force transmission.[6] The matrix of tendons is composed primarily of type I collagen (Table 9.7). In the main body of the tendon, the collagen is organized in a structural hierarchy with features similar to ligaments, including fiber crimping. In tendon, collagen fibers aggregate to form fascicles that are surrounded by a connective tissue sheath called the *endotenon* or fascicular membrane, which contains blood, lymphatic vessels, and nerves for the tendon (Fig. 9.24). The fascicles aggregate into fascicle bundles surrounded by *epitenons*, and several bundles are surrounded by the outermost tendon sheath called the *paratenon*. Some

---

[6] The tendons that flex your fingers are one such example. The muscles responsible for flexing your fingers are located in your forearm. They are attached to the bones of your fingers by long tendons that run through a sheath at the wrist and then through sheaths along the fingers that guide the tendons past your knuckle joints.

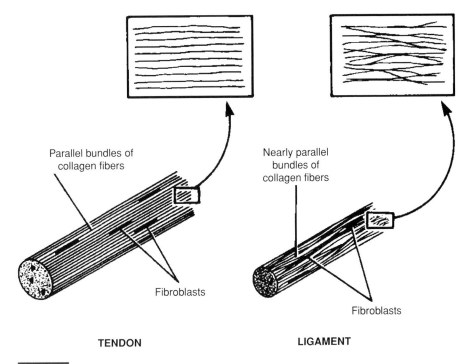

Parallel bundles of
collagen fibers

Nearly parallel
bundles of
collagen fibers

Fibroblasts

Fibroblasts

TENDON

LIGAMENT

Figure 9.25

Alignment of collagen fibers in tendons and ligaments. The different arrangement of collagen fibers in tendons compared with ligaments has important implications for differences in their biomechanical behavior, as explained in Section 9.10. From Nordin and Frankel [27] with permission of Lippincott Williams & Wilkins.

tendons are surrounded by an additional sheath that is filled with synovial fluid, which acts as a lubricant to reduce friction associated with movement of the tendon. In the regions where the tendon connects with muscle (the *myotendinous junction*) or with bone (the *osteotendinous junction*), the tendon composition and structure change to help to reduce the risk of tearing and ensure efficient force transfer. For example, at the osteotendinous junction, the tendon tissue changes to fibrocartilage (a tough form of cartilage; see below) and then to calcified fibrocartilage just prior to inserting into the bone. Because the primary function of tendons is to transmit muscle force, they must be relatively inextensible; therefore, their elastin content is less than that of ligaments. Additionally, the collagen fibers in tendon tend to be aligned in a more parallel fashion than in ligament (Fig. 9.25). Like ligaments, tendons contain proteoglycans and are highly hydrated. Tendon fibroblast cells align longitudinally among the fiber bundles; while they are relatively sparse, they play important roles in the maintenance and repair of the tendon tissue.

9.9.3 Cartilage

The three main types of cartilage in the body are *hyaline cartilage*, *elastic cartilage*, and *fibrocartilage*. These cartilages differ in their proportions of collagen, elastin, and proteoglycans. The most common type of cartilage is hyaline cartilage, which is found in several places in the body, including the larynx, nasal septum, and between the ribs and sternum. It is also found as the growth plate in developing bones. Most commonly, however, hyaline cartilage is found in joints where there is a lot of motion between opposing bones (such as the ankle, knee, hip, shoulder, or elbow joints). These joints are called *synovial joints* or *diarthrodial joints*, and the hyaline cartilage in these regions is called *articular cartilage*. It forms a thin layer on the ends of the long bones, such as the femur or tibia, and on sesamoid bones, such as the patella (kneecap). In diarthrodial joints, articular cartilage transmits and distributes loads from one bone to another. The forces that the thin articular cartilage layer supports are impressive: in the hip, peak pressures as high as 18 MPa have been measured [31].[7] Articular cartilage also provides a smooth surface to allow one bone to move relative to another with minimal friction. *Synovial fluid*, a clear viscous fluid between opposing cartilage layers, acts as a lubricant and helps to reduce the kinetic coefficient of friction for cartilage on cartilage to 0.005 to 0.05 [32]. This is similar to the coefficient of friction for ice on ice (0.03) and over an order of magnitude lower than that for steel on steel (0.5)!

Articular cartilage consists of a porous ECM of type II collagen, proteoglycans, and other non-collagenous proteins interspersed with interstitial water and electrolytes. Mature articular cartilage does not contain nerves, blood vessels, or lymphatics, and therefore, it is nourished and drained only by diffusion or convection from the surrounding synovial fluid. Articular cartilage contains cells, called chondrocytes, which are embedded in the ECM, but are sparse, making up less than 5–10% of the total volume. Articular cartilage is an anisotropic tissue in that its composition and the organization of its components vary through its thickness. Four zones are evident in mature articular cartilage (Fig. 9.26). Near the articulating surface, the collagen is arranged in dense fibrils that run parallel to the articular surface. This region is called the *superficial zone* and makes up 10–20% of the total thickness of the articular cartilage layer. The organization of the fibers tangent to the surface is thought to help to resist shear forces generated by joint movement. This region also is relatively impermeable to fluid flow and therefore helps to retain fluid within the collagen matrix when it is compressed. The retention of fluid and its pressurization with compression of the tissue allows cartilage to support large compressive loads, with more than 90% of the load supported by the pressurized fluid

---

[7] To put this in perspective, the pressure applied by a large adult elephant standing on a hockey puck would be approximately equal to the peak pressure measured in the hip!

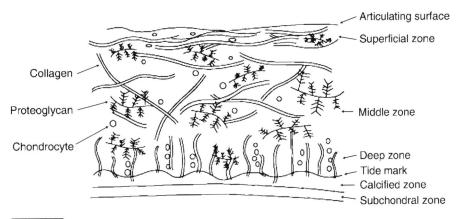

Figure 9.26

Layered microstructure of articular cartilage. The different zones have specialized compositions and structures according to their function, as discussed in the text. In healthy adult humans, the total thickness of articular cartilage in the knee is approximately 1.5 to 3 mm, although there is significant variability with location in a joint and between subjects [33]. Osteoarthritic cartilage can be much thinner because of disruption of collagen fibrils and loss of proteoglycans, a process that starts in the superficial zone but can progress to the deep zone, resulting in complete loss of mechanical strength of the tissue. Modified from Setton *et al.* [34], with permission from the OsteoArthritis Research Society International.

within the cartilage [35]. In the next 40–60% of the thickness, called the *middle zone*, the collagen content is reduced and the fibrils are more randomly organized. Water content is also lower in the middle region compared with the superficial region, but the proportion of proteoglycans (especially aggrecan) is higher. In the *deep zone* (the next 20–30% of the tissue), the collagen fibers run perpendicular to the surface of the underlying bone and the proteoglycan content is the highest. Between the deep zone and the underlying (subchondral) bone is the *calcified zone*. The collagen fibers from the deep zone cross the *tide mark* into the calcified zone, where they are mineralized and provide anchorage between the cartilage and bone.

Two other main types of cartilage are elastic cartilage and fibrocartilage. Elastic cartilage is found in the external ear, auditory canal, and parts of the larynx. It differs from hyaline cartilage in that it has more elastin in its matrix. Fibrocartilage is found in the intervertebral discs, the meniscus of the knee, and at transition regions between bones and ligaments or tendons. Fibrocartilage contains thick layers of dense collagen fibers between layers of hyaline cartilage and therefore has features and properties intermediate between cartilage and dense connective tissues.

9.10 Biomechanical properties of ligament, tendon, and cartilage

Although ligament, tendon, and cartilage are made up of similar components, the relative proportions of those components and their organization differ substantially between the three tissues. Because the biomechanical behavior of a tissue is determined by its composition and structure, the mechanical properties of ligaments, tendons, and cartilage are also considerably different. In this section, we will explore the structure–function relationships for these three tissues. Before doing so, we should emphasize the important distinction between *structural* and *material properties.*

Structural properties refer to the mechanical properties of an object as a whole and so will depend not only on the material of which the object is composed, but also its shape and size. This should be obvious: it is much easier to break a thin wooden pencil than a wooden hockey stick, although both are made of the same material. Structural properties are described by the force–deformation relationship, the stiffness, the ultimate or failure load, and the ultimate elongation. For tendons and ligaments, structural properties are determined by measuring the mechanical response of the entire bone–tendon–muscle or bone–ligament–bone structure to loading. In this case, the structural properties are influenced not only by the properties and geometry of the tissue but also by the mechanical properties of the bone–tissue and muscle–tissue junctions.

Material properties, in contrast, describe the intrinsic mechanical behavior of the tissue constituents only, irrespective of the tissue sample's overall geometry. In the pencil and hockey stick example, the material properties would be those of the wood. Tissue material properties are determined by the biomechanical properties of the tissue's constituents (i.e., collagen, elastin, proteoglycans) and their microstructural organization, orientation, and interaction. Material properties are described by parameters such as the stress–strain relationship, the modulus, and the ultimate stress and strain. Since stress and strain are normalized by the cross-sectional area and length of the sample, respectively, material properties are independent of the geometry of the specimen.

9.10.1 Structural properties

Experimental studies on the structural properties of soft connective tissues in the skeletal system have focussed on ligament and tendon. Because ligaments and tendons are primarily responsible for transmitting tensile loads, experimental studies are performed in tension by clamping the bone–tendon–muscle or bone–ligament–bone complex in a test machine and measuring the force that is generated with increasing displacement of the two ends of the complex. The resulting load–deformation curve can be used to define several structural properties, including the stiffness (the rate of change of force with deformation), the energy absorbed

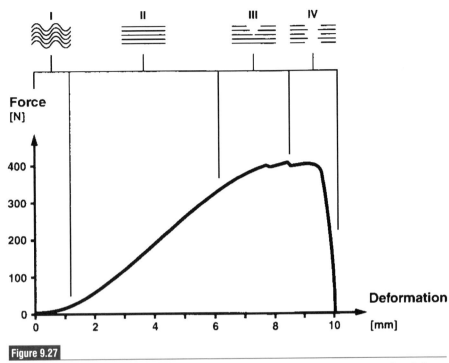

Figure 9.27

A typical force–displacement (deformation) curve for a rabbit ligament. The schematics at the top of the figure indicate the state of the collagen at each stage of the force–displacement curve, as explained in the text. From Nigg and Herzog [28]. Copyright John Wiley & Sons Limited. Reproduced with permission.

(the area under the force–deformation curve), and the ultimate load and ultimate deformation (at failure).

The load–deformation curves of all collagen-based tissues have a characteristic upwards concave shape in which the stiffness varies non-linearly with force (Fig. 9.27). In the initial portion of the curve, called the toe region, the tissue deforms easily without much tensile force. This is followed by a region in which the load and deformation are approximately linearly related; the slope of this region is often used to represent the elastic stiffness of the tissue. The linear region is followed by a region in which sharp falls in tensile force are seen, presumably as a result of sequential microfailure of collagen fibers. Finally, with further loading, the force reduces to zero as the tissue fails completely. For ligaments and tendons, the tensile behavior in the toe region permits initial stretching without much resistance. As the force or displacement increases, however, the ligament or tendon stiffens, providing more resistance to deformation. For ligament, this protects the joint from excessive relative movement between bones, while for tendon, it ensures efficient load transfer from muscles to bones.

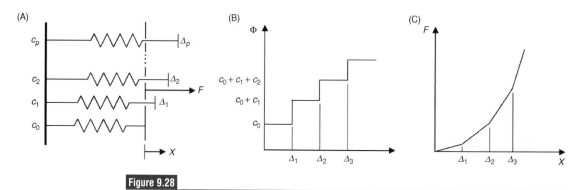

Simple model to explain how collagen crimp and alignment result in non-linear structural behavior of collagenous tissues. (A) Collagen fibers, represented by parallel linear springs with stiffness c_i, are recruited to resist the applied force at different displacements, simulating different degrees of crimp and off-axis orientation. (B) As a result of progressive recruitment, the effective stiffness of the tissue, Φ, increases in a step-wise fashion. (C) The step-wise increase in stiffness results in a non-linear relationship between force, F, and displacement, X. From Frisen *et al.* [36], with permission from Elsevier.

The non-linear force–deformation relationship can be explained by considering the microstructure of the collagen fibers and their organization in connective tissues. As shown in Fig. 9.22, collagen (the principal component of connective tissues that provides tensile strength) is crimped. The crimp allows the tissue to be extended with minimal force; this phenomenon is responsible for the toe region of the load–deformation curve. As the deformation increases, the crimp flattens out and the apparent stiffness increases as the force required to deform the tissue is resisted by the inherent material stiffness of the collagen fibers. A second source of non-linearity in the force–deformation curves of skeletal connective tissues is the alignment of the collagen fibers. During tensile loading of tissues like tendon, little realignment of the fibers occurs because they are already oriented parallel to the direction of the applied load (Fig. 9.25). As a result, the toe region of the load–deformation curve is relatively short. In ligament, many of the collagen fibers are parallel to the long axis, but there are also non-axial fibers that run at a variety of angles with respect to the long axis (including running perpendicular). As a result, the toe region for ligament is longer than that of tendon, since some realignment of collagen fibers must occur during tensile loading to resist deformation.

A simple, but illustrative, model of how the progressive recruitment of individual fibers can result in a non-linear force–displacement curve is shown in Fig. 9.28. In this model, individual collagen fibers are represented by linear elastic springs arranged in parallel. The fibrils have different degrees of crimp and different orientations relative to the loading direction; they are therefore recruited to resist the applied force at different displacements ($\Delta_1, \Delta_2, \ldots, \Delta_P$). The stiffness of the tissue, Φ, at any deformation, X, is given by the slope of the force–deformation

(F–X) curve and is expressed as:

$$\Phi(X) = \mathrm{d}F/\mathrm{d}X. \tag{9.25}$$

Because of the progressive recruitment of collagen fibers, the stiffness is a summation of step functions (Fig. 9.28B):

$$\Phi(X) = \sum_{i=0}^{p} c_i u(X - \Delta_i) \tag{9.26}$$

where c_i are the stiffnesses of the individual springs, and $u(\xi)$ is the Heaviside function, $u(\xi) = 0$ for $\xi < 0$ and $u(\xi) = 1$ for $\xi > 0$. As a result, with increasing deformation, more parallel springs are recruited, resulting in increased stiffness and a non-linear force–displacement response (Fig. 9.28C).

From an engineering design perspective, the collagen arrangements in tendons and ligaments make sense. Tendon must provide high tensile strength and stiffness to transmit muscle forces to bone without substantial deformation. The highly oriented parallel fiber bundles provide the high tensile strength and minimal extensibility but allow flexibility in bending (similar to braided wire ropes). Ligaments must also provide high tensile strength but, depending on the function of the ligament, may experience off-axis loads and therefore require some strength in off-axis directions. Ligaments also need to be slightly extensible to allow for proper joint movement. These features are provided by the nearly parallel arrangement of the collagen fibers in ligaments of the extremities. Some ligaments such as the ligamentum flavum in the spine contain more elastin than collagen; therefore, they deform substantially before the stiffness increases (long toe region). However, the ligamentum flavum is "pre-tensed" when the spine is in its neutral position and so stabilizes the spine to prevent excessive movement between vertebral bodies.

9.10.2 Material properties

The material properties of biological tissues are based on stress–strain relationships of the tissue substance itself. These properties are more difficult to measure than structural properties for several reasons: it is difficult to grip the tissue without damaging it; accurate measurement of tissue cross-sectional area is challenging; strain is best measured without contacting the tissue; and material properties are sensitive to external factors such as the source of the tissue and how the tissue is handled, stored, or prepared prior to testing. Nonetheless, several advances in test methods and a better understanding of the effects of various external factors on the measurements have improved characterization of the tensile, compressive, and time-dependent (viscoelastic) material properties of ligaments, tendons, and cartilage [37].

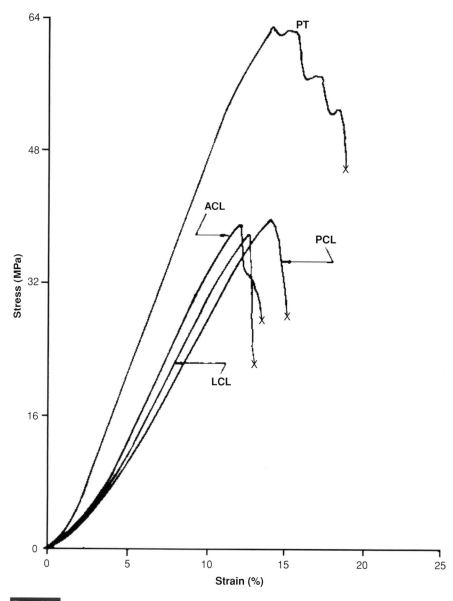

Figure 9.29

Tensile material (stress–strain) curves for the patellar tendon (PT), anterior cruciate ligament (ACL), posterior cruciate ligament (PCL), and lateral collateral ligament (LCL) of the knee of a young human donor. Note that the tendon has a shorter toe region and larger failure stress than the ligament. Modified from Butler *et al.* [38], with permission from Elsevier.

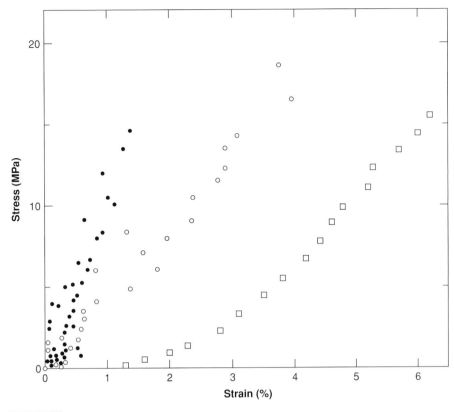

Figure 9.30

Stress-strain curves for tropocollagen (●), collagen fibrils (○), and a tendon (□). Strain in the tropocollagen was estimated from measurements of the elongation of the helical pitch of the collagen triple helix. Strain in the collagen fibril was determined using an X-ray diffraction technique that measured the change in D-period spacing. The moduli of the collagen fibril (430 MPa) and the tendon (400 MPa in the linear portion) are remarkably similar. The modulus of tropocollagen is about three times greater than that of the fibril. From Sasaki and Odajima [39], with permission from Elsevier.

9.10.3 Material properties: tension

The material (stress–strain) behavior of connective tissues under tension is non-linear, in large part because of the crimping and alignment of collagen fibers (Fig. 9.29). The linear region after the toe region is believed to represent the resistance to deformation of the collagen fibers themselves. Consistent with this theory, stress–strain curves of collagen fibrils are linear and have a slope similar to that of the linear region for tendon (Fig. 9.30). It is possible, however, that the non-linear tissue behavior is also partly a result of complex molecular interactions between the various components of connective tissues, which is a phenomenon that is not well understood at present.

Table 9.8. Equilibrium tensile moduli of skeletal connective tissues. Cartilage specimens were harvested from high-weight-bearing areas (HWA) or low-weight-bearing areas (LWA). The higher tensile moduli in the LWA correlated with a higher collagen to proteoglycan ratio in these regions.

Tissue	Species	Location	Tensile modulus (MPa)	Source
Tendon	Human	Achilles tendon	819	Wren [40]
	Human	Patellar tendon	643	Butler [38]
Ligament	Human	Cruciate ligaments	345	Butler [38]
	Human, young	Ligamentum flavum	98	Nachemson [29]
	Human, aged	Ligamentum flavum	20	Nachemson [29]
Articular cartilage	Human	HWA of femoral condyle, superficial zone	5.0	Akizuki [41]
	Human	HWA of femoral condyle, middle zone	3.1	Akizuki [41]
	Human	LWA of femoral condyle, superficial zone	10.1	Akizuki [41]
	Human	LWA of femoral condyle, middle zone	5.4	Akizuki [41]

The tensile material properties of ligament, tendon, and cartilage can be described by the tensile modulus, E, which is equal to the slope of the linear portion of the stress–strain curve. When a soft connective tissue is subjected to an applied force, its length will change with time until it reaches an equilibrium length. This process is called "creep" and is the result of the viscoelastic properties of the tissue. Ideally, the tensile modulus should be measured once the tissue stops deforming and has reached equilibrium. The equilibrium modulus therefore describes the quasi-static behavior of the tissue and is dependent on the intrinsic tensile properties of the tissue matrix (mostly collagen). The transient, time-dependent properties of the tissue must be described with other (viscoelastic) parameters, which we will consider below. Typical equilibrium tensile moduli for skeletal connective tissues are listed in Table 9.8. By comparison with the composition of the tissues shown in Table 9.7, you can see there is a correlation between collagen content and tensile stiffness.

9.10.4 Material properties: compression

Because articular cartilage is loaded in compression, it makes sense to study its compressive properties. The compressive properties of cartilage can be measured using several test configurations, the most popular of which are shown in Fig. 9.31. A typical compression test for cartilage is a confined compression test, in which

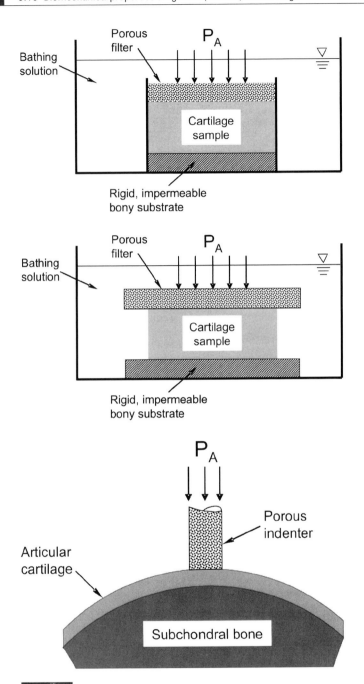

Figure 9.31

Schematics of three test configurations used to measure the compressive material properties of cartilage: confined compression (top), unconfined compression (middle), and indentation (bottom). From Mow and Guo [42] with kind permission of Annual Reviews.

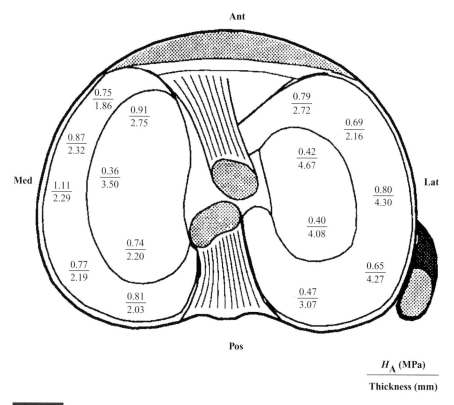

Figure 9.32

A map of aggregate compressive equilibrium moduli (H_A, the number above the line) and cartilage thickness (the number below the line) measured in the tibial plateau of a normal 21-year-old male. Significant regional differences in both parameters are evident. The tibial plateau is the surface of the tibia (shin bone) that forms the base of the knee joint. Ant, anterior; Pos, posterior; Med, medial; Lat, lateral; see text for definition of H_A. From Mow *et al.* [43] with permission of Lippincott Williams & Wilkins and Dr. Shaw Akizuki.

a small cylindrical plug of cartilage tissue is placed in a cylindrical chamber (Fig. 9.31, top). The walls of the chamber, which contact the cartilage plug, prevent the cartilage from expanding laterally, thereby ensuring that deformation occurs only in the direction of loading. The cartilage is loaded under a constant compressive load applied with a porous filter. The pores in the filter allow fluid from the cartilage to flow through the filter as the cartilage is compressed. Once equilibrium is reached (i.e., the fluid stops flowing out of the cartilage), the stress and strain are recorded and the ratio is defined as the aggregate equilibrium modulus (H_A). Like the tensile equilibrium modulus, the aggregate modulus is a measure of the stiffness of the solid matrix and is independent of fluid flow (since it is measured at equilibrium once fluid movement has ceased).

Aggregate compressive equilibrium moduli for human articular cartilage range from 0.1 to 2 MPa and vary significantly with location, even in a single joint (Fig. 9.32). Experiments have shown that H_A is directly related to cartilage proteoglycan content [44], but not collagen content. Hence, proteoglycans, with their ability to bind water electrochemically, are primarily responsible for providing the compressive stiffness of cartilage. In osteoarthritis, a degenerative disease of cartilage, there is often loss of proteoglycans, leading to decreased compressive stiffness of the tissue [34].

9.10.5 Material properties: viscoelasticity

An important characteristic of the material properties of skeletal connective tissues is that they display properties of both elastic and viscous materials and so are known as "viscoelastic." Some of the viscoelastic characteristics displayed by most biological tissues are time-dependent responses, hysteresis, and history-dependent responses.

Time-dependent responses. Such responses include creep, an increase in deformation over time under constant load, and stress relaxation, a decline in stress over time under constant deformation (Fig. 9.33A,B). The material properties of viscoelastic tissues are also time-dependent in that they respond differently to variations in the *rate* at which the material is loaded or strained (Fig. 9.33C).

Hysteresis. During loading and unloading of viscoelastic materials, the force–displacement curves do not follow the same path but instead follow a hysteresis loop, indicating that internal energy is dissipated during the loading–unloading cycle (Fig. 9.33D).

History-dependent responses. If a viscoelastic tissue is subjected to several different stress paths that arrive at the same stress value, σ_0, at time t_1, the corresponding strains, ε, at time t_1 will differ (Fig. 9.34). In other words, the strain at time t_1 depends on the entire sequence or "history" of stresses up to that time.

The mechanisms responsible for these viscoelastic behaviors in soft connective tissues likely include the intrinsic viscoelasticity of the solid phase of the tissue arising from intermolecular friction between the collagen, elastin, and proteoglycan polymeric chains; deformation of these molecules; and other complex intermolecular interactions. For biological tissues that contain significant amounts of water, viscoelasticity also results, in part, from the interactions of the water with the proteins, specifically the frictional drag the water creates when it flows through the porous ECM.

As we demonstrated in Section 2.6.1 for cells, linear viscoelastic behavior can be modeled by using lumped parameter approaches in which the viscoelastic material is represented by linear springs and linear dashpots. Consider the model shown in Fig. 2.35 (p. 56); as we will see below, this "standard linear model" is the simplest

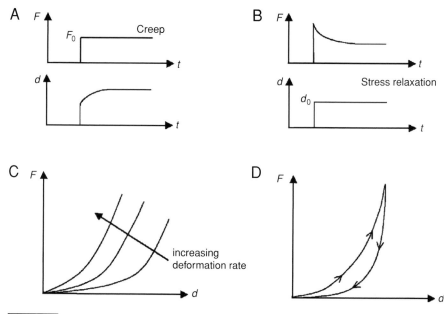

Figure 9.33

Typical force–deformation (*F*–*d*) characteristics of viscoelastic materials. (A) Creep, which occurs following a change in applied load, in this case a step change from zero to a constant F_0. After the instantaneous deformation when the load is applied, a continued gradual increase in deformation is seen until equilibrium is reached. (B) Stress relaxation, which is a gradual decrease in stress under constant deformation (d_0) until equilibrium is reached. (C) Rate dependency of force–displacement relationship, in this case showing that increased force is required to obtain a given deformation if the rate of deformation is increased. (D) Hysteresis, or different loading and unloading paths indicating dissipation of energy. The arrows show the direction of loading and unloading.

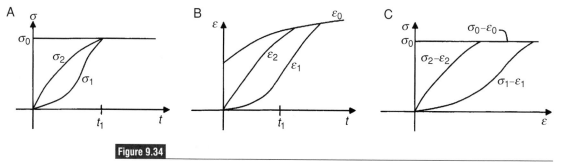

Figure 9.34

The response of viscoelastic materials to loading depends on the "stress history." Although the three different stress paths in (A) arrive at the same stress level at time $t = t_1$, the corresponding strains differ (B), and hence the resulting stress–strain relationships are each different (C). Thus, the strain at $t = t_1$ depends on the stress path (or its history) to that point. From Wineman and Rajagopal [45]. Reproduced with the permission of the authors and publisher.

equivalent lumped parameter representation that predicts many of the characteristics of a viscoelastic solid. As we did for the cell, we can use such models to generate constitutive relationships that describe the behavior of biological tissues under various conditions, including creep and stress relaxation. However, because we are interested in material properties, we need first to reformulate the relationships between force and displacement in terms of stress, strain, and material parameters. We will assume linear material behavior, so that the stress, σ, and strain, ε, are linearly related by Young's modulus, E (compare with Equation [2.7]):

$$\sigma = E\varepsilon. \tag{9.27}$$

Similarly, the stress is related to the time derivative of strain by the viscosity, μ (compare with Equation [2.8]):

$$\sigma = \mu\dot{\varepsilon}. \tag{9.28}$$

Additionally, the spring constants and damping coefficients in the model in Fig. 2.35 are replaced with the corresponding material constants; that is, Young's moduli E_0 and E_1 in place of the spring constants k_0 and k_1, and viscosity μ in place of the damping coefficient η_0. Following the force balance approach outlined in Section 2.6.1, it can be shown that the constitutive equation for the standard linear model in Fig. 2.35 is (see Problem 9.5):

$$\sigma + \frac{\mu}{E_0}\dot{\sigma} = E_1\varepsilon + \mu\left(1 + \frac{E_1}{E_0}\right)\dot{\varepsilon}. \tag{9.29}$$

Equation (9.29) can be rewritten as:

$$\sigma + \tau_\varepsilon\dot{\sigma} = E_R(\varepsilon + \tau_\sigma\dot{\varepsilon}) \tag{9.30}$$

where we have defined the relaxation time for constant strain, τ_ε, as

$$\tau_\varepsilon = \frac{\mu}{E_0}; \tag{9.31}$$

the relaxation time for constant stress, τ_σ, as

$$\tau_\sigma = \frac{\mu}{E_1}\left(1 + \frac{E_1}{E_0}\right); \tag{9.32}$$

and the relaxed elastic modulus E_R as E_1.

As we did for the cell model in Section 2.6.3, Equation (9.30) can be solved for the creep and stress relaxation response by imposing step changes in stress or strain at time $t = 0$ and considering times $t > 0$ with appropriate initial conditions. Consider first the case of a step change in stress at time $t = 0$, in which an initially unloaded specimen ($\sigma(t \leqslant 0) = 0$) is subjected to a constant stress σ_0 at time zero, i.e., $\sigma(t > 0) = \sigma_0$. There will be an immediate elastic response, resulting in a strain at time $t = 0^+$ of $\sigma_0/(E_0 + E_1)$. By solving Equation (9.30) subject to this initial condition, we can then compute the creep response as:

$$\varepsilon(t) = \frac{\sigma_0}{E_R}\left[1 - \left(1 - \frac{\tau_\varepsilon}{\tau_\sigma}\right)e^{-t/\tau_\sigma}\right]. \tag{9.33}$$

A similar approach holds for the response to a step change in strain at $t = 0$ (i.e., $\varepsilon(t > 0) = \varepsilon_0$), giving the stress relaxation response as:

$$\sigma(t) = \varepsilon_0 E_R\left[1 - \left(1 - \frac{\tau_\sigma}{\tau_\varepsilon}\right)e^{-t/\tau_\varepsilon}\right]. \tag{9.34}$$

The creep and stress relaxation responses are often written in terms of a creep function, $J(t)$, and a relaxation function, $G(t)$, which are defined as:

$$J(t) = \frac{\varepsilon(t)}{\sigma_0}, \quad G(t) = \frac{\sigma(t)}{\varepsilon_0}. \tag{9.35}$$

For the standard linear model, the creep and relaxation functions are determined simply from the definitions in Equation (9.35) and the expressions in Equations (9.33) and (9.34). The general shape of the creep and stress relaxation responses for the standard linear model are shown in Fig. 9.33A and Fig. 9.33B, respectively, if force (F) is replaced by stress and displacement (d) is replaced by strain. The standard linear model captures many of the features of viscoelastic materials, including an immediate elastic response, non-linear creep that tends towards non-zero equilibrium strains, and non-linear stress relaxation that tends towards non-zero equilibrium stresses.

The creep and stress relaxation tests provide measures of viscoelastic response to a static load. However, tissue is more typically loaded in a time-varying manner, and hence viscoelastic properties are often measured by dynamically loading the material over a range of frequencies and measuring the stress–strain response. As shown below, this approach allows one to determine the degree to which a material exhibits viscoelastic behavior, ranging from purely elastic to purely viscous.

Here we consider the response of the material to the application of an unsteady sinusoidal strain, $\varepsilon(t)$, of small amplitude, ε_0. Recalling Euler's formula,[8] the periodic applied strain can be written as:

$$\varepsilon(t) = \varepsilon_0 e^{i\omega t} \tag{9.36}$$

---

[8] Euler's formula states that for any number, θ, $e^{i\theta} = \cos\theta + i\sin\theta$. This formula allows us to use a complex exponential function to represent trigonometric functions. In our case, this makes the derivation of the viscoelastic parameter equations more convenient and their interpretation clearer.

where $i = \sqrt{-1}$, ω is the circular frequency (in radians per second[9]) at which the periodic strain is applied and it is understood that we are concerned only with the real part of the strain and corresponding stresses.

For a linear material, we expect that application of a sinusoidal strain will result in stress that also varies sinusoidally. However, because of viscous flow and the time-dependent response of viscoelastic materials, we also expect that the stress will be out of phase with the strain, such that:

$$\sigma(t) = \sigma_0 e^{i(\omega t + \delta)} \tag{9.37}$$

where σ_0 is the stress amplitude and δ is the phase lag. As will be shown below, the degree to which the stress and strain are out of phase is indicative of the degree of viscosity, with the two being in phase ($\delta = 0$) for a purely elastic material and completely out of phase ($\delta = \pi/2$) for a purely viscous material.

The ratio of the stress to the strain is the *complex modulus*, $E^*(\omega)$, defined as:

$$E^*(\omega) = \frac{\sigma}{\varepsilon} = \frac{\sigma_0}{\varepsilon_0} e^{i\delta} = \frac{\sigma_0}{\varepsilon_0}(\cos\delta + i\sin\delta). \tag{9.38}$$

If we define $E'(\omega)$ and $E''(\omega)$, which are called (for reasons that will be clear below) the *storage modulus* and *loss modulus*, respectively, as:

$$E'(\omega) = \frac{\sigma_0}{\varepsilon_0}\cos\delta, \quad E''(\omega) = \frac{\sigma_0}{\varepsilon_0}\sin\delta, \tag{9.39}$$

the complex modulus can then be written as:

$$E^* = E' + iE''. \tag{9.40}$$

The storage modulus, E', is the real part of the complex modulus and is proportional to the portion of the stress that is in phase with the strain. It can be shown that, for a purely elastic material, the only contribution to E^* is from the storage modulus, and that E' is proportional to the elastic energy stored in the tissue due to the applied strain. The imaginary part of the complex modulus, the loss modulus E'', is proportional to the portion of the stress that is 90° out of phase with the strain. For a Newtonian fluid, $E^* = iE''$, and E'' is proportional to the energy dissipated or lost in the material per cycle.

Viscoelastic materials are often characterized by the magnitude of the complex modulus and the phase lag, δ, which are related to E' and E'' by

$$|E^*| = \sqrt{E'^2 + E''^2}, \quad \delta = \tan^{-1}\left(\frac{E''}{E'}\right). \tag{9.41}$$

---

[9] Recall that for any harmonic phenomenon, the circular frequency ω is related to the frequency in Hertz, f, by $\omega = 2\pi f$.

The magnitude of the complex modulus reflects the overall stiffness of the tissue, whereas the phase lag characterizes the ratio of the energy dissipated to the energy stored, or "internal friction." As mentioned above, $\delta = 0°$ for a purely elastic material, whereas $\delta = 90°$ for a purely viscous material. Tendon and ligament are only slightly viscoelastic, with $\delta \sim 3°$ [43,46]. In contrast, the phase lag for bovine articular cartilage tested in pure shear under infinitesimal strain (< 0.001 rad) ranges from 9° to 20° over a frequency range of 0.01 to 20 Hz [43]. When cartilage is tested in pure shear under very small strain, little water movement occurs and therefore the viscoelastic properties measured in a pure shear experiment are the intrinsic viscoelastic properties of the cartilage matrix itself (collagen and proteoglycans) and do not describe viscoelastic effects owing to fluid flow. It is likely that the higher proteoglycan content in cartilage and its interaction with collagen and other matrix molecules is responsible for the more viscous behavior of cartilage compared with ligament and tendon.

It must not be forgotten that the dynamic viscoelastic parameters are functions of the loading frequency, ω, and are dependent on the stress–strain relationship of the tissue. For the standard linear model, the stress–strain relationship is given in Equation (9.29). This equation can be solved for the sinusoidal strain input (Equation [9.36]) to generate expressions for E^* and δ in terms of the time constants and relaxed modulus derived from the material constants and the frequency ω:

$$|E^*| = \left(\frac{1 + \omega^2 \tau_\sigma^2}{1 + \omega^2 \tau_\varepsilon^2} \right)^{1/2} E_R, \quad \delta = \tan^{-1}\left(\frac{\omega(\tau_\sigma - \tau_\varepsilon)}{1 + \omega^2(\tau_\sigma \tau_\varepsilon)} \right). \tag{9.42}$$

Details regarding the derivation of these equations can be found in Fung [47], Haddad [48], or any other textbook on viscoelasticity. Here we consider interpretation of these results only in a qualitative manner. The frequency dependence of the complex modulus and phase lag for the standard linear model are shown in Fig. 9.35. From this plot, it can be seen that the tissue stiffness ($|E^*|$) increases with higher frequencies. This is a typical response for viscoelastic materials (Fig. 9.33C). Notice that the phase lag or internal friction is sizeable only over a fairly narrow range of frequencies, centered on the point where $\omega = (\tau_\varepsilon \tau_\sigma)^{-1/2}$. For some biological tissues, hysteresis caused by viscous losses is actually relatively insensitive to loading frequency over a wide range [49]. To account for the insensitivity of viscous losses to strain rate in some tissues, generalized linear models have been proposed (e.g., Fung [49]). One such model and its implications for modeling hysteresis in soft connective tissues is considered in Problem 9.7.

The models and equations presented above assume that the material being studied behaves linearly. While this assumption may be true for oscillations of small amplitude (i.e., infinitesimal strain), we know that most soft tissues exhibit non-linear stress–strain behavior for the finite deformations that occur during physiological

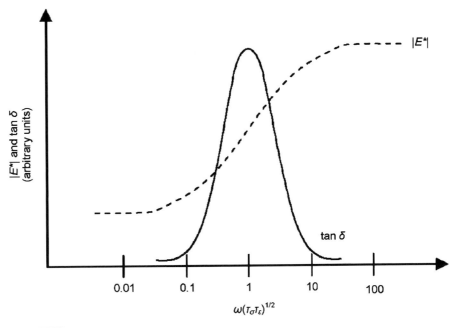

Figure 9.35

Dependency of the magnitude of the complex modulus ($|E^*|$) and the tangent of the phase lag (δ) on the loading frequency, ω, for the standard linear model with relaxation times τ_σ and τ_ε. Adapted from Fung [47], with kind permission of Springer-Verlag.

loading. In order to improve modeling of biological tissues, Fung [47] proposed the quasilinear theory of viscoelasticity, in which the non-linear elastic behavior was accounted for within the framework of linear viscoelasticity. The details of the theory, which utilizes a relaxation function along the lines of that in Equation (9.35), are provided in [47]. This theory has been used successfully to model the viscoelastic properties of many soft tissues, including tendons and ligaments, with better agreement than can be achieved with linear viscoelastic theory (e.g., Woo *et al.* [50]).

9.10.6 Material properties: biphasic mixture theory of cartilage

While viscoelastic models are useful for describing and predicting the mechanical behavior of biological tissues, they are phenomenological and reveal little about the actual physical mechanisms responsible for the observed mechanical behavior. Recognizing that the composition and microstructural organization of cartilage were responsible for much of its viscoelastic behavior, Mow and colleagues proposed the *biphasic theory* of cartilage mechanics to account for the solid and fluid phases of cartilage and their interaction [51]. The biphasic theory idealizes articular cartilage as a mixture of a porous, permeable elastic solid matrix phase (collagen and proteoglycans) swollen with an incompressible interstitial fluid phase (water that can be bound or unbound to the matrix). In the theory, loads are supported by both pressurization of water and by elastic support of the solid matrix. Movement of water through the porous matrix is modeled and governed by frictional drag at the interface between the fluid and solid matrix. Viscoelastic behavior therefore is accounted for through viscous losses owing to solid–fluid interactions (i.e., flow-dependent properties of the tissue). More recent improvements to the theory incorporate viscoelasticity of the matrix component as well (i.e., flow-independent properties of the tissue). The details of the theory and its solution are beyond our scope here but are described in detail in the original paper by Mow *et al.* [51] and are summarized in Mow *et al.* [43]. More recent advances in theories of cartilage mechanics, including matrix anisotropy and a triphasic mixture theory, that accounts for the electrolytes (ions) in the interstitial water and their interaction with the charged proteoglycans, are described by Mow and Guo [42].

9.11 Problems

9.1 Typical compressive stress–strain curves for cortical bone and for trabecular bone of two different densities are shown in Fig. 9.36. Calculate the approximate strain energy density to failure in each case. Strain energy density, U, is a measure of the ability of a material to absorb energy up to fracture and is given by:

$$U = \int_0^{\varepsilon_u} \sigma \, d\varepsilon, \tag{9.43}$$

where ε_u is the ultimate strain at failure. What does your result imply about the function of trabecular versus cortical bone and the consequences of loss of trabecular bone density, as occurs in osteoporosis?

9.2 Recall that several experimental studies have demonstrated that the yield strain of trabecular bone is relatively constant over a wide range of apparent densities. Demonstrate that this is the case using the density dependency relationships in Section 9.3 and Hooke's law, which is valid for trabecular bone virtually up to the yield strain.

9.3 In this question, we will work through the derivation of Equation (9.5), which states that the compressive strength of trabecular bone is proportional to the

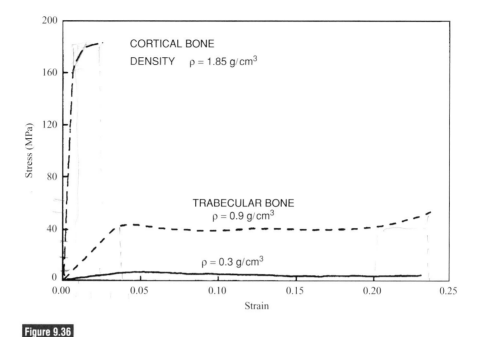

Figure 9.36

For Problem 9.1. Reprinted from Hayes and Gerhart, Biomechanics of bone: applications for assessment of bone strength, *Bone and Mineral Research*, **3**, 259–294 . Copyright 1985, with permission from Elsevier.

square of the relative density, assuming that the microstructure of trabecular bone can be represented by a low-density, rod-like model (Fig. 9.8A).

(a) Begin by showing that the ratio of the apparent density to the tissue density is approximately proportional to $(t/l)^2$.

(b) When low-density trabecular bone is loaded in compression, failure occurs when the vertical struts buckle (Fig. 9.37). The critical load, F_{crit}, at which a strut of length l, Young's modulus E_s, and second areal moment of inertia, I, buckles is given by Euler's formula:

$$F_{crit} \propto \frac{E_s I}{l^2}. \tag{9.44}$$

Derive a proportional relationship between the moment of inertia, I, of a single strut and the dimensions of the strut, assuming it has square cross-section with dimension t, as shown in Fig. 9.8A. In this case we are interested in the moment of inertia of the cross-sectional area; that is with respect to the axis that runs through the middle of a strut perpendicular to its long axis.

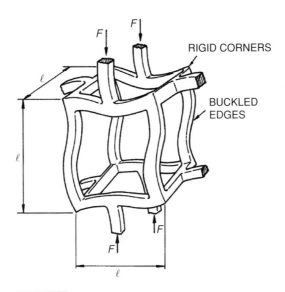

F

F

RIGID CORNERS

BUCKLED
EDGES

ℓ

ℓ

F

F

ℓ

(c) You should now be able to use Euler's formula (Equation [9.44]) and the relationships you derived in (a) and (b) to derive Equation (9.5). To do so, you will need to relate F_{crit} to the compressive strength at collapse, σ^*, and the cross-sectional area of the unit cell, l^2; that is, $F_{\text{crit}} \sim \sigma^* l^2$.

9.4 Recall the example in Section 8.4.2 about the person lifting a weight and the resulting stresses in the bones of the forearm. Assuming there is a 0.01 mm defect in the radius bone 8 cm from the elbow, determine how many times the subject can lift the weight before his/her bone fractures. Does your calculation likely overestimate or underestimate the actual number of lifts? Why?

9.5 Derive the constitutive Equation (9.29) for the standard linear viscoelastic model in Fig. 2.35.

9.6 Show mathematically that the phase lag or internal friction for the standard linear viscoelastic model is a maximum when $\omega = (\tau_\varepsilon \tau_\sigma)^{-1/2}$, where τ_ε and τ_σ are defined in Equations (9.31) and (9.32)

9.7 Consider the load–extension (force–deformation) curves shown in Fig. 9.38. These data are from tensile tests on relaxed papillary muscles from the ventricle of the rabbit heart. Note that the tissue exhibits hysteresis owing to viscous

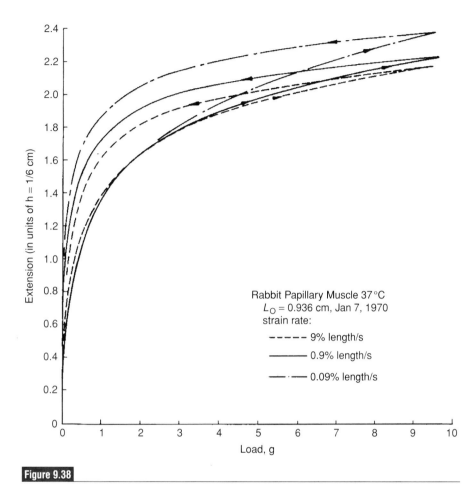

Figure 9.38

For Problem 9.7. From Fung [47] with kind permission of Prof. Y. C. Fung.

losses, as expected of a viscoelastic material. Also notice, however, that the amount of hysteresis is essentially insensitive to the applied strain rate over two orders of magnitude. Is this experimental observation consistent with the predictions made by the standard linear model? If not, how can the model be improved to account for insensitivity to loading frequency?

References

1. J. Z. Young. *The Life of Mammals* (London: Oxford University Press, 1975).
2. A. Vesalius. *Bones and Cartilages. On the Fabric of the Human Body* (San Francisco, CA: Norman, 1998).

3. A. M. Parfitt. The physiologic and clinical significance of bone histomorphometric data. In *Bone Histomorphometry*, ed. R. R. Recker (Boca Raton, FL: CRC Press, 1983), pp. 143–224.

4. W. C. Hayes and M. Bouxsein. Biomechanics of cortical and trabecular bone: implications for assessment of fracture risk. In *Basic Orthopaedic Biomechanics*, 2nd edn, ed. V. C. Mow and W. C. Hayes. (Philadelphia, PA: Lippincott-Raven, 1997), pp. 69–111.

5. R. Huiskes and B. van Rietbergen. Biomechanics of bone. In *Basic Orthopaedic Biomechanics and Mechano-biology*, 3rd edn, ed. V. C. Mow and R. Huiskes. (Philadelphia, PA: Lippincott, Williams & Wilkins, 2005), pp. 123–139.

6. A. Ham and D. Cormack. *Histophysiology of Cartilage, Bone, and Joints* (Toronto: Lippincott, 1979).

7. J. Wolff. Ueber die Innere Architectur der Knochen und Ihre Bedeutung für die Frage Vom Knochenwachsthum. *Archiv für Pathologische Anatomie und Physiologie und für Klinische Medicin*, **50** (1870), 389–450.

8. R. Muller, H. van Campenhout, B. van Damme, G. van der Perre, J. Dequeker, *et al.* Morphometric analysis of human bone biopsies: a quantitative structural comparison of histological sections and micro-computed tomography. *Bone*, **23** (1998), 59–66.

9. X. E. Guo and S. A. Goldstein. Is trabecular bone tissue different from cortical bone tissue? *Forma*, **12** (1997), 185–196.

10. D. R. Carter and W. C. Hayes. The compressive behavior of bone as a two-phase porous structure. *Journal of Bone and Joint Surgery (American Volume)*, **59** (1977), 954–962.

11. T. M. Keaveny. Strength of trabecular bone. In *Bone Mechanics Handbook*, 2nd edn, ed. S. C. Cowin. (New York: CRC Press, 2001), pp. 16.1–16.42.

12. L. J. Gibson and M. F. Ashby. *Cellular Solids: Structure and Properties*, 2nd edn (Cambridge: Cambridge University Press, 1997).

13. W. C. Hayes and D. R. Carter. Postyield behavior of subchondral trabecular bone. *Journal of Biomedical Materials Research*, **10** (1976), 537–544.

14. J. C. Rice, S. C. Cowin and J. A. Bowman. On the dependence of the elasticity and strength of cancellous bone on apparent density. *Journal of Biomechanics*, **21** (1988), 155–168.

15. B. van Rietbergen and R. Huiskes. Elastic constants of cancellous bone. In *Bone Mechanics Handbook*, 2nd edn, ed. S. C. Cowin. (Boca Raton, FL: CRC Press, 2001), pp. 15.1–15.24.

16. S. W. Donahue, N. A. Sharkey, K. A. Modanlou, L. N. Sequeira and R. B. Martin. Bone strain and microcracks at stress fracture sites in human metatarsals. *Bone*, **27** (2000), 827–833.

17. D. B. Burr, R. B. Martin, M. B. Schaffler and E. L. Radin. Bone remodeling in response to in vivo fatigue microdamage. *Journal of Biomechanics*, **18** (1985), 189–200.

18. M. F. Ashby and D. H. R. Jones. *Engineering Materials: An Introduction to Their Properties and Applications* (Oxford: Pergamon Press, 1980).

19. K. Piekarski. Fractography of bone. In *Natural and Living Biomaterials*, ed. G. W. Hastings and P. Ducheyne. (Boca Raton, FL: CRC Press, 1984), pp. 99–117.

20. R. B. Martin, D. B. Burr and N. A. Sharkey. *Skeletal Tissue Mechanics* (New York: Springer Verlag, 1998).

21. D. Taylor and P. J. Prendergast. A model for fatigue crack propagation and remodeling in compact bone. *Proceedings of the Institute of Mechanical Engineers, Series H*, **211** (1997), 369–375.

22. H. Roesler. The history of some fundamental concepts in bone biomechanics. *Journal of Biomechanics*, **20** (1987), 1025–1034.

23. A. Keith. Hunterian Lecture on Wolff's law of bone-transformation. *Lancet*, **16** (1918), 250–252.

24. H. H. Jones, J. D. Priest, W. C. Hayes, C. C. Tichenor and D. A. Nagel. Humeral hypertrophy in response to exercise. *Journal of Bone and Joint Surgery (American Volume)*, **59** (1977), 204–208.

25. C. T. Rubin and L. E. Lanyon. Dynamic strain similarity in vertebrates; an alternative to allometric limb bone scaling. *Journal of Theoretical Biology*, **107** (1984), 321–327.

26. J. D. Currey. How well are bones designed to resist fracture? *Journal of Bone and Mineral Research*, **18** (2003), 591–598.

27. M. Nordin and V. H. Frankel. *Basic Biomechanics of the Musculoskeletal System* (Malvern, PA: Lea & Febiger, 1989).

28. B. M. Nigg and W. Herzog. *Biomechanics of the Musculo-Skeletal System* (Chichester, UK: John Wiley, 1999).

29. A. L. Nachemson and J. H. Evans. Some mechanical properties of the third human lumbar interlaminar ligament (ligamentum flavum). *Journal of Biomechanics*, **1** (1968), 211–220.

30. J. Kastelic, A. Galeski and E. Baer. The multicomposite structure of tendon. *Connective Tissue Research*, **6** (1978), 11–23.

31. W. A. Hodge, R. S. Fijan, K. L. Carlson, R. G. Burgess, W. H. Harris *et al.* Contact pressures in the human hip joint measured in vivo. *Proceedings of the National Academy of Sciences USA*, **83** (1986), 2879–2883.

32. G. A. Ateshian and V. C. Mow. Friction, lubrication, and wear of articular cartilage and diarthrodial joints. In *Basic Orthopaedic Biomechanics and Mechano-biology*, 3rd edn, ed. V. C. Mow and R. Huiskes. (Philadelphia, PA: Lippincott Williams & Wilkins, 2005), pp. 447–494.

33. F. Eckstein, M. Reiser, K. H. Englmeier and R. Putz. In vivo morphometry and functional analysis of human articular cartilage with quantitative magnetic resonance imaging: from image to data, from data to theory. *Anatomy and Embryology*, **203** (2001), 147–173.

34. L. A. Setton, D. M. Elliott and V. C. Mow. Altered mechanics of cartilage with osteoarthritis: human osteoarthritis and an experimental model of joint degeneration. *Osteoarthritis and Cartilage*, **7** (1999), 2–14.

35. M. A. Soltz and G. A. Ateshian. Experimental verification and theoretical prediction of cartilage interstitial fluid pressurization at an impermeable contact interface in confined compression. *Journal of Biomechanics*, **31** (1998), 927–934.

36. M. Frisen, M. Magi, L. Sonnerup and A. Viidik. Rheological analysis of soft collagenous tissue. part I: theoretical considerations. *Journal of Biomechanics*, **2** (1969), 13–20.

37. S. L. Y. Woo, T. Q. Lee, S. D. Abramowitch and T. W. Gilbert. Structure and function of ligaments and tendons. In *Basic Orthopaedic Biomechanics and Mechano-biology*, 3rd edn, ed. V. C. Mow and R. Huiskes. (Philadelphia, PA: Lippincott, Williams & Wilkins, 2005), pp. 301–342.

38. D. L. Butler, M. D. Kay and D. C. Stouffer. Comparison of material properties in fascicle-bone units from human patellar tendon and knee ligaments. *Journal of Biomechanics*, **19** (1986), 425–432.

39. N. Sasaki and S. Odajima. Elongation mechanism of collagen fibrils and force–strain relations of tendon at each level of structural hierarchy. *Journal of Biomechanics*, **29** (1996), 1131–1136.

40. T. A. L. Wren, S. A. Yerby, G. S. Beaupre and D. R. Carter. Mechanical properties of the human Achilles tendon. *Clinical Biomechanics*, **16** (2001), 245–251.

41. S. Akizuki, V. C. Mow, F. Muller, J. C. Pita, D. S. Howell *et al.* Tensile properties of human knee joint cartilage: I. Influence of ionic conditions, weight bearing, and fibrillation on the tensile modulus. *Journal of Orthopaedic Research*, **4** (1986), 379–392.

42. V. C. Mow and X. E. Guo. Mechano-electrochemical properties of articular cartilage: their inhomogeneities and anisotropies. *Annual Review of Biomedical Engineering*, **4** (2002), 175–209.

43. V. C. Mow, W. Y. Gu and F. H. Chen. Structure and function of articular cartilage and meniscus. In *Basic Orthopaedic Biomechanics and Mechano-biology*, 3rd edn, ed. V. C. Mow and R. Huiskes. (Philadelphia, PA: Lippincott, Williams & Wilkins, 2005), pp. 181–258.

44. R. L. Sah, A. S. Yang, A. C. Chen, J. J. Hant, R. B. Halili *et al.* Physical properties of rabbit articular cartilage after transection of the anterior cruciate ligament. *Journal of Orthopaedic Research*, **15** (1997), 197–203.

45. A. S. Wineman and K. R. Rajagopal. *Mechancial Response of Polymers: An Introduction* (Cambridge: Cambridge University Press, 2000).

46. P. Netti, A. D'Amore, D. Ronca, L. Ambrosio and L. Nicolais. Structure–mechanical properties relationship of natural tendons and ligaments. *Journal of Materials Science: Materials in Medicine*, **7** (1996), 525–530.

47. Y. C. Fung. *Biomechanics: Mechanical Properties of Living Tissues* (New York: Springer Verlag, 1981).

48. Y. M. Haddad. *Viscoelasticity of Engineering Materials* (London: Chapman & Hall, 1995).

49. Y. C. Fung. *A First Course in Continuum Mechanics* (London: Prentice-Hall International, 1994).

50. S. L. Woo, G. A. Johnson and B. A. Smith. Mathematical modeling of ligaments and tendons. *Journal of Biomechanical Engineering*, **115** (1993), 468–473.

51. V. C. Mow, S. C. Kuei, W. M. Lai and C. G. Armstrong. Biphasic creep and stress relaxation of articular cartilage in compression? Theory and experiments. *Journal of Biomechanical Engineering*, **102** (1980), 73–84.

10 Terrestrial locomotion

In this chapter we examine terrestrial locomotion (walking and running) from both descriptive and quantitative points of view. The study of locomotion has long been an important part of the field of biomechanics. Although not strictly "locomotion," we also consider the dynamics and kinematics of jumping.

10.1 Jumping

10.1.1 Standing jump

We begin by considering the standing vertical jump, since it is the easiest jump to analyze. A standing jump can be divided into two phases, defined as follows (Fig. 10.1).

1. **Push-off**: starting from a crouch, the jumper pushes against the ground with his/her legs until the feet leave the ground.
2. **Airborne**: from the moment that the feet leave the ground until the maximum elevation of the center of gravity is reached.

Analysis of the standing jump is most easily carried out if we neglect changes in posture and body orientation while jumping, and consider only the elevation of the center of gravity. We therefore define:

c = distance the center of gravity is lowered, measured with respect to the elevation of the center of gravity at the instant the feet leave the ground.

h = maximum elevation of the center of gravity, measured with respect to the same reference datum.

z = vertical location of the center of gravity, measured positive upwards from the floor.

v = vertical velocity of the center of gravity, measured positive upwards.

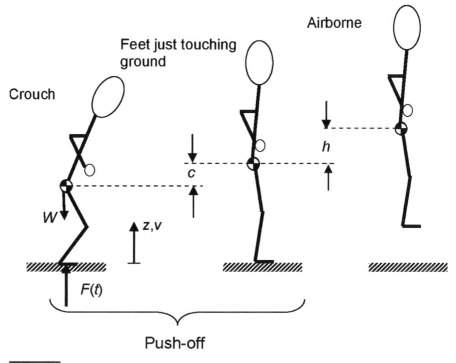

Figure 10.1

Phases of standing vertical jump: starting from a crouch, pushing off until the feet just touch the ground, and airborne. The left panel also shows the forces acting during the push-off phase. The subject's weight is W, and the force exerted by the legs on the ground is $F(t)$. Note the definitions of z and v as positive upwards. Here the center of mass is shown as remaining in a fixed anatomical position, but this is not generally true.

During the *push-off* phase, we recognize that a force $F(t)$ is exerted by the legs on the ground, with a corresponding reaction force exerted by the ground on the legs (Fig. 10.1). Application of Newton's second law to the jumper yields

$$+ \uparrow \sum F_z = ma_z \tag{10.1}$$

$$F(t) - W = \frac{W}{g}\frac{dv}{dt}. \tag{10.2}$$

Integrating Equation (10.2) with respect to time from 0 to t yields

$$v(t) = \frac{g}{W} \int_0^t F(t')\,dt' - gt + constant. \tag{10.3}$$

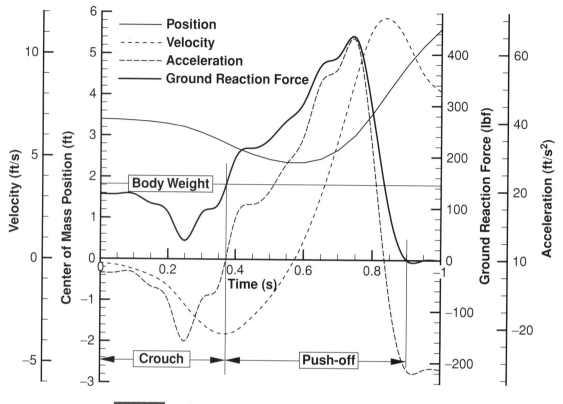

Figure 10.2

Kinematic data observed during the crouch and push-off phases of a vertical jump. The position of the center of mass was measured as a function of time, from which the velocity and acceleration of the center of mass were derived by numerical differentiation. The ground reaction force was computed from a force balance on the jumper, whose mass was 144 lbm. The crouch phase of the jump ends when the acceleration of the center of mass becomes positive; the push-off phase ends when the feet leave the ground and the ground reaction force becomes zero. The ground reaction force corresponding to the body weight is shown for reference. Curves derived from data tabulated by Miller and Nelson [1].

We choose $t = 0$ to be the beginning of the push-off phase, when $v = 0$. The constant in Equation (10.3) is then zero, so that

$$v(t) = \frac{g}{W} \int_0^t F(t')\,dt' - gt. \tag{10.4}$$

To proceed further with the analysis we need to know something about the force history, $F(t)$.[1] An example dataset for the position of a jumper's center of mass as a function of time during the crouch and push-off phases of a standing jump is shown in Fig. 10.2. From these data, we can use Equation (10.2) to derive the

---

[1] Actually, we need to know something about the total vertical impulse due to the ground reaction force, $\int_0^t F(t')\,dt'$. Of course, if we know the behavior of F versus time, then we can calculate the impulse.

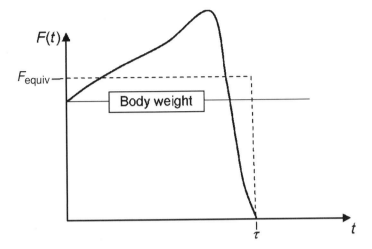

Figure 10.3

Ground reaction force history, $F(t)$, during the push-off phase of the vertical jump. The actual ground reaction force history has a shape similar to that shown by the heavy solid line (see Fig. 10.2). The equivalent idealized history is shown by the dashed line.

ground reaction force, which is also shown in Fig. 10.2. The shape of this force curve makes intuitive sense: during crouching, the reaction force is less than body weight, but it then increases to greater than body weight as the jumper accelerates upwards. Therefore, during the push-off phase, the ground reaction force history has the general shape sketched in Fig. 10.3, where τ is the duration of the push-off phase. To simplify the analysis we replace the actual reaction force history by a simplified history, shown by the dotted line in Fig. 10.3. The magnitude of the applied force in the simplified history, F_{equiv}, is chosen so that the impulses from the actual and simplified forces are equal: that is, so that

$$\int_0^\tau F(t')\,dt' = F_{equiv}\tau. \tag{10.5}$$

Thus, F_{equiv} represents the average force exerted by the legs during the push-off phase. Making the approximation that $F(t)$ equals F_{equiv} in Equation (10.3), we obtain

$$\frac{dz}{dt} = v(t) = gt\left[\frac{F_{equiv}}{W} - 1\right]. \tag{10.6}$$

This expression can be integrated to yield

$$z(t) - z_0 = \frac{1}{2}gt^2\left[\frac{F_{equiv}}{W} - 1\right] \tag{10.7}$$

where z_0 is the elevation of the center of gravity at the beginning of the push-off phase. Eliminating time between Equations (10.6) and (10.7) gives

$$z(t) - z_0 = \frac{v(t)^2}{2g\left[\dfrac{F_{equiv}}{W} - 1\right]}.$$ (10.8)

At the end of pushoff (at time $t = \tau$), $z(t) - z_0$ equals c and $v(t)$ equals $v_{pushoff}$. Substitution of these values into Equation (10.8) produces

$$\frac{v_{pushoff}^2}{2g} = c\left[\frac{F_{equiv}}{W} - 1\right].$$ (10.9)

During the *airborne* phase, the body is effectively a projectile subject to constant gravitational acceleration. The height attained by the center of gravity (h) is, therefore, $v_{pushoff}^2/2g$. Combining this with Equation (10.9) yields

$$h = c\left[\frac{F_{equiv}}{W} - 1\right].$$ (10.10)

Equation (10.10) shows that greater elevation can be achieved by increasing crouch depth c or by increasing average push-off force F_{equiv}. A typical value of F_{equiv} is $2W$, so $h \sim c$.

In order to illustrate the above analysis, consider the world record standing jump.[2] According to *The Guinness Book of World Records, 14th edn* (1976), the world record was held by Johan Christian Evandt (Norway, 1962) who cleared a bar 5 feet 9½ inches (1.76 m) off the ground! What average push-off force, F_{equiv} was exerted during this jump? To answer this question, it is necessary to make several assumptions. We will take the jumper's height to be 6 feet, so that his center of gravity was 36 inches off the ground at the end of push-off. We will also assume that during the jump he oriented his body so his center of gravity just cleared the bar, so the total net elevation of his center of gravity was

$$h = 69.5 - 36 \text{ inches} = 33.5 \text{ inches}.$$ (10.11)

A typical crouch distance for a 6 foot tall person is 20 inches. Using these values, we obtain $F_{equiv}/W = 2.7$, which is about 35% more than a normal person.

10.1.2 Running jumps

Pole vault

The first running jump that we will analyze is the pole vault. This jump essentially involves an interconversion of energy by the vaulter from kinetic energy (speed

---

[2] The highest standing jump of all time is hard to determine. Ray Ewry (USA) set the Olympic record of 1.655 m in the 1900 Games. The event was discontinued as an Olympic sport shortly after that. Sturle Kalstad (Norway) holds the Norwegian standing jump record of 1.82 m, set in 1983. This seems to be the highest standing jump ever recorded, as far as the authors can determine.

down the runway) to potential energy (elevation over bar). During the interconversion, significant amounts of energy are stored in the elastic deformation of the pole, as examination of Fig. 10.4 shows.

Supposing that the energy conversion process is 100% efficient, we have

$$\frac{1}{2}mv^2 = mgh \tag{10.12}$$

where m is the mass of the pole vaulter, v is the vaulter's speed while running down the runway, and h is the net maximum elevation of the vaulter's center of gravity. Thus, we expect that $h = v^2/2g$. Setting $v = 10$ m/s gives a value of $h = 5.1$ m. If the center of gravity starts 1 m off the ground at the beginning of the vault, then the maximum elevation of the center of gravity should be approximately 6.1 m. This is quite close to the men's indoor world record, which is 6.15 m (set in February 1993 by Sergey Bubka, Ukraine).

Although the above analysis gives approximately the right answer, and thus captures the essence of the problem, it is clearly simplistic and neglects several factors:

- We have neglected the mass of the pole, which is approximately 4 kg. This likely has little effect on the height that the vaulter can obtain.
- We have neglected the force exerted by the vaulter's arms on the pole while in the air, specifically the vertical component of this force. This will tend to increase the vault height.
- We have neglected the force applied by the push-off leg on the runway. This force can be used to increase upward momentum, for example by supporting an upward swing of the free leg. It will increase vault height.
- The pole is not 100% efficient in storing energy. This reduced efficiency will decrease vault height.
- The vaulter's center of gravity can actually pass under the bar if the vaulter does all the right contortions (Fig. 10.5). This increases the height of the bar that can be cleared.

Running high jump

The major difference between the standing and running high jump is that there is an opportunity in the running high jump to convert kinetic energy into potential energy. This means that the elevation attained in the running high jump is invariably greater than that attained during the standing jump. However, it is clear that the *stance leg*, responsible for converting kinetic into potential energy, is much less efficient than the pole vaulting pole, otherwise the world record for the running high jump would be about 20 feet (6 m).

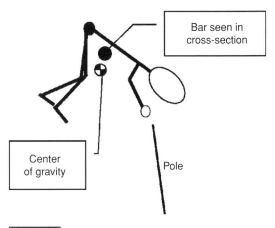

Figure 10.5

Vaulter clearing the bar in the pole vault. Notice that contortions can cause the vaulter's center of gravity to pass under the bar.

There are two general running high jump strategies. *Speed jumpers* try to convert as much forward kinetic energy into potential energy as possible, as discussed above. *Power jumpers* rely much more on an active push-off from the stance leg and are less efficient at energy conversion. Clearly, technique is the major determinant of success in high jumping, and quantitative analysis of this is beyond the scope of this book.

10.2 Description of walking and running

Walking and running are complex activities about which entire books have been written. In this section, we will describe the biomechanics of these activities and do some simple quantitative analyses of them. In Section 10.3 we will quantitatively analyze the biomechanics of walking in more detail.

10.2.1 Walking

We consider humans walking at constant speed on level terrain. With respect to Fig. 10.6, we make the following definitions.

Stance phase: This is the portion of the walking cycle when a given leg is in contact with the ground. The leg (or legs) involved in the stance phase is called the stance leg.

Swing phase: This is the portion of the walking cycle when a given leg is swinging free. The leg involved in the swing phase is called the swing leg.

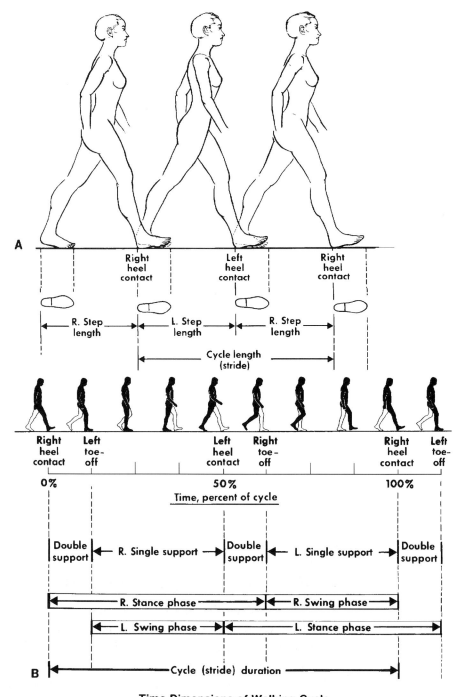

Time Dimensions of Walking Cycle

Figure 10.6

Definitions used in describing walking. From Inman *et al.* [3]. With kind permission of Lippincott Williams & Wilkins.

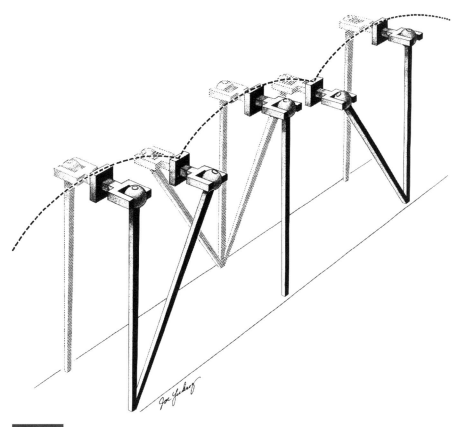

Heel strike: The instant when the swing leg contacts the ground is the heel strike.

Toe off: The instant the swing leg leaves the ground is the toe off.

By its very definition, during a portion of the walking cycle there are short periods when both feet are touching the ground and there are two stance legs. However, for most of the cycle, there is one stance leg and one swing leg.

Walking is a complex superposition of five separate motions: pelvic rotation, pelvic list (tilt), knee flexion in the stance leg, ankle flexion, and toe flexion on the stance foot. The effect of these motions is depicted in Figs. 10.7–10.12. Figure 10.7 shows the path of the center of mass that results from walking without bending of the knees or rotation of the pelvis. As can be verified directly, this results in a great

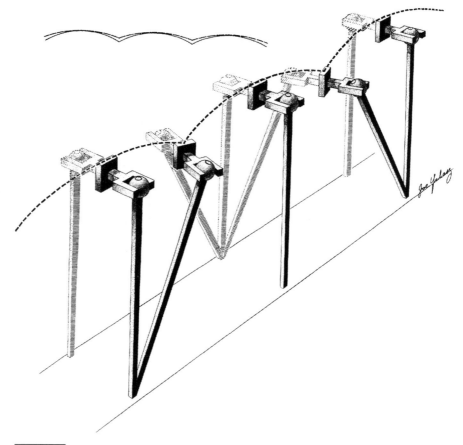

Figure 10.8

Effects of pelvic rotation during walking. When the pelvis rotates in a horizontal plane during each step, the center of mass does not fall as far during the phase of double weight-bearing as is shown in Fig. 10.7. The solid line at the top represents the curve shown in Fig. 10.7, while the dashed line represents the dashed curve from the lower part of this figure, i.e., including pelvis rotation. From Saunders *et al.* [4]. Copyright The Journal of Bone and Joint Surgery, Inc.

deal of up and down motion of the torso and is rather uncomfortable. Figure 10.8 shows the effect of pelvic rotation, which acts to decrease the vertical excursions of the center of mass during walking. In addition, the pelvis lists (tilts) from the horizontal, such that the pelvis drops slightly over the swing leg (Fig. 10.9), which also acts to reduce vertical excursions of the center of mass. However, to accomplish pelvic list, it is necessary to bend the swing leg, otherwise the foot of the swing leg will strike the ground. Fig. 10.10 shows the slight flexion of the knee that occurs on the stance leg just after heel strike. To these pelvic and knee motions must be added the flexion of the ankle and toe, which act to smooth the path of the knee and thus also of the center of mass (Figs. 10.11 and 10.12). The net result

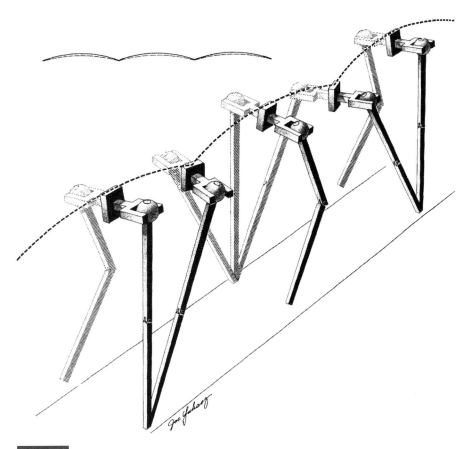

of all these motions is that the center of mass does *not* move forward in a straight line during walking, but instead it oscillates left to right and up and down. When viewed straight ahead, this produces a characteristic "8" pattern, the exact shape of which is a function of walking speed (Figs. 10.13 and 10.14).The resulting path of motion of the center of gravity is much "smoother" than that depicted in Fig. 10.7.

Walking motion is rather complex. However, its essential feature is simply stated: walking motion is designed to *minimize energy expenditure*. To see how this comes about, it is first necessary to understand how energy is used during walking. There are four main modes of energy use.

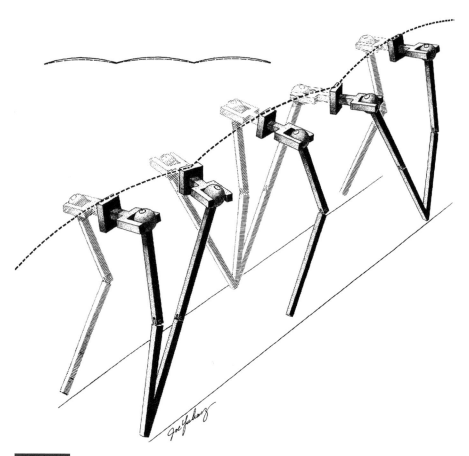

Figure 10.10

Effects of knee flexion during walking. Except at very low walking speed, the knee undergoes approximately 15⁰ of flexion immediately after heel strike and continues to remain flexed until the center of mass has passed over the weight-bearing leg. This has two effects: initially, it absorbs part of the impact of the body at heel strike and later it decreases the amount that the center of mass is elevated as it passes over the weight-bearing leg. The solid line at the top represents the curve shown in Fig. 10.9. From Saunders *et al.* [4]. Copyright The Journal of Bone and Joint Surgery, Inc.

1. The center of gravity moves up and down (about 5 cm per step at normal speed). This can be verified by observing the top of a person's head from the side as they walk on level ground.
2. The center of gravity moves from side to side. Again, this can be verified by direct observation.
3. The center of gravity accelerates and decelerates in the direction of walking during a step cycle. This can be verified by carrying a large flat pan partially full of water. The water invariably begins to slosh back and forth, indicating acceleration and deceleration on the part of the person carrying the pan.
4. The legs accelerate and decelerate.

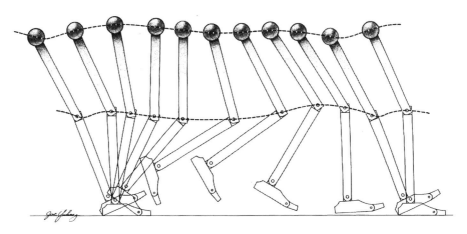

Figure 10.11

Knee and hip pathways during walking at moderate speed. Note that there is a slight increase of knee elevation immediately after heel strike, but for the remainder of stance phase the pathway is relatively straight and shows only a slight variation from the horizontal. From Inman *et al.* [3]. With kind permission of Lippincott Williams & Wilkins.

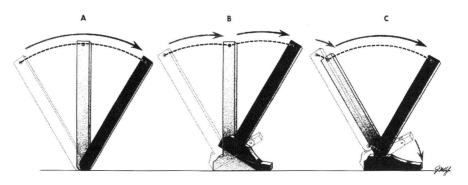

Figure 10.12

The effects of foot and ankle motion on the pathway of the knee during walking. (A) Arc described by the knee when there is no foot. (B) The effect of a foot with a rigid ankle on knee motion, which now comprises two intersecting arcs. However, the knee pathway does not fall abruptly at the end of stance and is more similar to the normal pathway. (C) The effect of a foot with a flail (hyper-flexible) ankle. From Inman *et al.* [3]. With kind permission of Lippincott Williams & Wilkins.

Direct observation will show that the variation in potential energy as a result of the center of gravity moving up and down is largely out of phase with the variation in kinetic energy from the acceleration and deceleration of the center of gravity in the forward direction. This suggests that there is an interconversion of energy between potential and kinetic energy. In practice, after toe off, the weight moves from behind the stance leg over and *up* to a maximum elevation directly over the stance foot, then continues on and moves *down* until heel strike. As the center of gravity moves up, forward speed decreases, and as the center of gravity

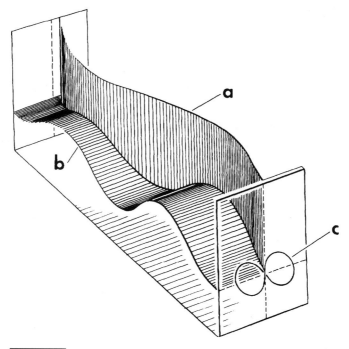

Figure 10.13

Displacements of the center of mass in three dimensions during a single stride cycle. Actual displacements have been greatly exaggerated. From Inman *et al.* [3]. With kind permission of Lippincott Williams & Wilkins.

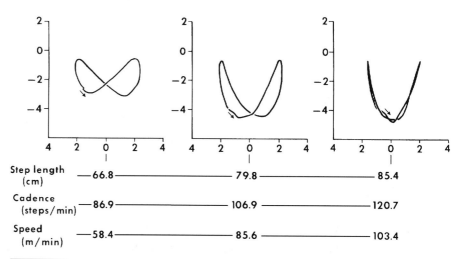

Figure 10.14

Paths (in plane c of Fig. 10.13) showing pelvis motion during walking. The pelvic displacement pattern changes with the speed of walking. From Inman *et al.* [3]. With kind permission of Lippincott Williams & Wilkins.

falls, forward speed increases.[3] This behavior is confirmed by measurements taken during normal walking, shown in Fig. 10.15. The upper curve in Fig. 10.15A shows the temporal variation of kinetic energy associated with forward motion, while the middle trace shows the variation associated with potential energy and kinetic energy associated with vertical motion. The lower curve is the sum of the two upper curves, and clearly displays much less variation than either of the individual curves. Figure 10.15C quantifies this effect by defining an energy recovery, which is seen to be approximately 65% for walking (speeds up to approximately 10 km/h).

Step frequency

A second way in which the body conserves energy during walking is by "tuning" step frequency. In general, the motion of the swing leg will require energy input, as the leg is accelerated after toe off and then decelerated just prior to heel strike. However, there is one condition under which zero energy input is (ideally) required for this motion. This is *ballistic walking*, in which the swing leg behaves as a freely swinging pendulum, and the support leg acts like a rigid "inverted" pendulum. Although this is clearly a simplified model of walking, it incorporates enough of the essential physics to be of interest.

It is clear that the step frequency in this model is determined by the time taken for the swing leg to move from toe off to heel strike. Recall that the frequency of a pendulum, and hence the step frequency, ω, depends on

$l =$ the pendulum (leg) length, having dimensions of L (length)

$g =$ gravitational acceleration, having dimensions of LT^{-2} (length per time^2).

From dimensional arguments, we deduce that ω should be proportional to $(g/l)^{1/2}$. Hence, on earth we expect to observe ω proportional to $(g/l)^{1/2}$ over a wide range of values for ω. This prediction can be contrasted with observations summarized in Fig. 10.16. It can be seen that the simple ballistic walking (resonant) theory approximates the trend of the data.

10.2.2 Running

Interestingly, in running, energy interconversion does not work well. This is primarily because of large variations in potential energy (when both feet have left the ground), which cannot be equalized by changes in forward speed. The net result is a low energy recovery percentage, as shown in Fig. 10.15.

---

[3] Although difficult to observe during normal walking, these effects become obvious if you walk with knees locked.

A

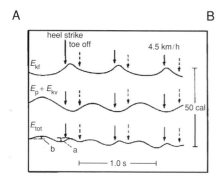

B

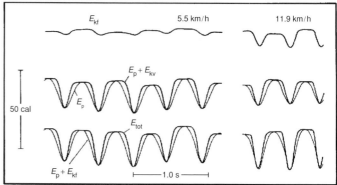

C

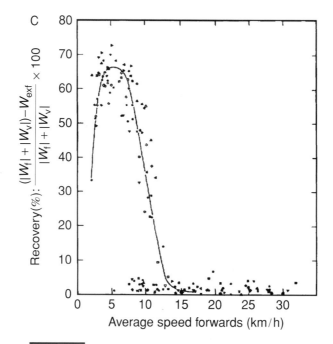

Figure 10.15

Energy interconversion during walking and running, demonstrated by analysis of force-plate records to compute changes in the kinetic and potential energy of the center of mass of the body. (A) Walking at normal speed (4.5 km/h). The upper curve shows the kinetic energy associated with forward motion, $E_{kf} = \frac{1}{2}mv^2$, where v is the forward component of velocity. The middle tracing is the sum of the gravitational potential energy, E_p, and the (small) kinetic energy associated with vertical motion, $E_{fv} = \frac{1}{2}mv_v^2$, where v_v is the vertical component of velocity. The bottom trace shows total energy, $E_{tot} = E_{kf} + E_p + E_{kv}$. Arrows identify heel strike (solid) and toe off of the opposite foot (broken). (B) Corresponding traces for running (5.5 and 11.9 km/h). Unlike walking, E_{tot} experiences large variations. (C) "Recovery" of mechanical energy in walking (open symbols) and running (closed symbols) versus speed. Here, W_f is the sum of the positive increments of the curve E_{kf} in one step, W_v is the sum of the positive increments of E_p, and W_{ext} is the sum of the positive increments of E_{tot} (increments a plus b in panel [A]). In this figure, an increment is defined as the change from a local minimum to a local maximum. From McMahon, T. A. *Muscles, Reflexes, and Locomotion*, © 1984 by T. A. McMahon [5], based on Cavagna *et al.* [6]. Reprinted by permission of Princeton University Press.

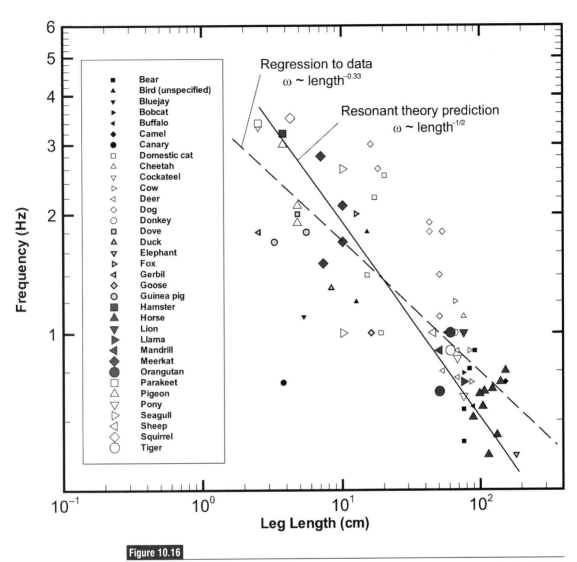

Figure 10.16

Step frequency as a function of leg length, as gathered by the students of the biomechanical engineering class of 1980 (Professor D. F. James, University of Toronto). There is a fair bit of scatter in this data; what happens when your class gathers data and plots them in this way?

10.3 Gait analysis

To this point, our discussion of walking has been qualitative or semi-quantitative. However, it is possible to analyze human walking quantitatively using the techniques of gait analysis. This can yield extensive data on a single subject, including (but not limited to) time-resolved limb locations, velocities, and accelerations, as well as time-resolved forces acting on and within limbs.

Why gather such detailed data on walking? There are a variety of reasons:

- Data on walking can be used as a diagnostic tool for patients having musculoskeletal and/or neurological control problems. Data gathered can also be used as an indicator of disease severity in such patients, although this requires a large database of normals and individuals with confirmed disease.
- Data on walking can be used to evaluate the effects of treatment, such as surgery, drugs, and/or assistive devices, on patients having gait pathologies. In the case of assistive devices, such data can be used to optimize device usage and adjustment.
- Data on walking can be used as a tool for long-term clinical monitoring of surgical patients (e.g., for patients who have received total artificial knees). Such data can form part of an assessment about which type of surgery, knee design, etc. is optimal.

In all of the above applications, a quantitative assessment of walking is more objective, more documentable, and potentially more sensitive than a qualitative analysis. Gait analysis has four main components.

1. **Electromyography.** This measures the electrical signals that cause muscles to contract. Since the force exerted at joints depends on muscle contraction, among other things, knowledge of muscle activity is useful for assessing joint dynamics. Electromyographic data are obtained using electrodes either on the surface (similar to those used for measuring the electrocardiogram) or implanted in the muscle. We will not consider this technique in this chapter.

2. **Anthropometry.** Anthropometry is defined to be the measurement of the human body. In the context of gait analysis, we are primarily interested in the lengths, masses, locations of the centers of mass, and mass moments of inertia of the lower limb segments. Information on sites of attachment of muscles, muscle cross-sectional areas, line of actions of muscles, and so on are also of interest. Below, we will briefly discuss how such data are acquired.

3. **Kinematics.** Kinematics is the study of motion, without regard to the forces and moments responsible for the motion. In a kinematic analysis, we are typically concerned with limb segment position and angle, velocity and angular velocity, and acceleration and angular acceleration. There are a large number of techniques for gathering kinematic data in gait analysis, some of which will be discussed below.

4. **Kinetics.** Kinetics is the study of the forces that produce motion. In the context of gait analysis, the important forces are a consequence of muscle contraction, reaction forces acting on the foot from the ground, and reaction forces acting at joints.

As will be seen, the combination of kinetics, kinematics, and anthropometry, as well as basic biomechanical modeling, allows a great deal of quantitative information about walking to be gathered.

10.3.1 Kinematics

When describing the kinematics of walking, we typically focus on limb segments, such as the foot, the shank (lower leg from knee to ankle), the thigh (upper leg from the hip to the knee), etc. A complete description of the three-dimensional motion of all limb segments is a formidable task because of the large number of degrees of freedom of the system. For this reason, it is essential that we adopt a standardized system of notation. Winter [7] proposes the system shown in Fig. 10.17A, which we will adopt throughout this section. The direction of progression (the anterior–posterior direction) is denoted x. Angles in the x–y plane are measured counter-clockwise with respect to the x axis; and angles in the y–z plane are measured counter-clockwise with respect to the y axis. Note that this convention for measuring angles implies a similar convention for the measurement of angular velocities and accelerations.

There are a number of ways of gathering kinematic data. We will only touch on a few here.

Goniometers. These are devices for measuring joint angles. For example, the knee flexion angle (angle between shank and thigh) is commonly measured with a goniometer. Most goniometers are potentiometers whose resistance varies with angle. They are cheap and easy to use but can be difficult to align with the axis of rotation.

Accelerometers. These devices are mounted onto the surface of the body at convenient points and measure the acceleration of that location. Standard accelerometers can only measure one component of the acceleration vector, so interpretation of complex three-dimensional motions can be difficult.

Imaging. Anatomical markers are placed on key locations, and images of the subject are acquired during walking. To resolve motion in two dimensions, this approach is fairly straightforward; however, in three dimensions it becomes much more complex. In the latter case, at least two cameras must be used, and care must be taken to register the images and ensure that markers are correctly identified.

Here we will consider only a simple two-dimensional analysis of gait, focussing on the motion of the knee joint. We will assume that markers have been placed on the leg as shown in Fig. 10.18. Note that the placement of markers on the shank and thigh means that the vectors $\mathbf{x}_{21}$ and $\mathbf{x}_{43}$ are essentially co-linear with the long

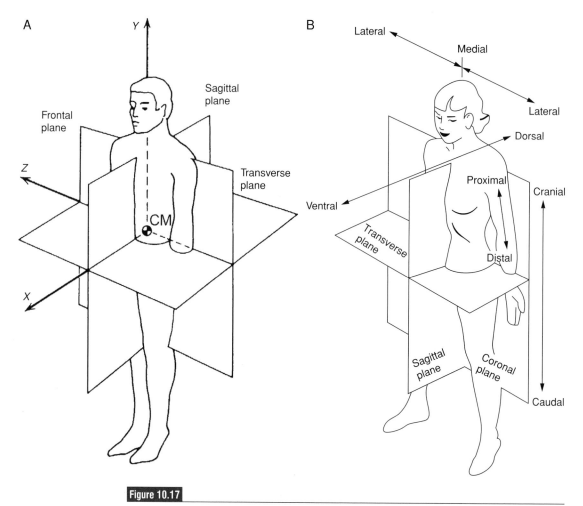

Figure 10.17

(A) A standardized (x, y, z) coordinate system used in gait analysis, as proposed by Winter. (B) Definitions of standard terms used in anatomical descriptions. CM = center of mass. (A) Modified from Winter [7]. Reprinted with permission of John Wiley & Sons, Inc. (B) Modified from NASA [8].

axes of the tibia and femur, respectively. Here the convention is that vector $\mathbf{x_{ij}}$ starts at marker i and points to marker j, so that in the sagittal plane (constant z), we may write:

$$\mathbf{x_{ij}} = (x_j - x_i, \, y_j - y_i). \tag{10.13}$$

Furthermore, the angle θ_{ij} is defined with respect to the x axis, so that from the definition of the dot product, we may write:

$$\theta_{ij} = \arccos\left[\frac{\hat{\mathbf{e}}_x \bullet \mathbf{x_{ij}}}{\|\mathbf{x_{ij}}\|}\right] \tag{10.14}$$

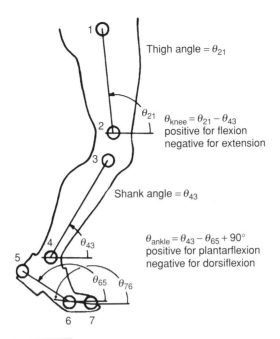

Thigh angle $= \theta_{21}$

$\theta_{knee} = \theta_{21} - \theta_{43}$
positive for flexion
negative for extension

Shank angle $= \theta_{43}$

$\theta_{ankle} = \theta_{43} - \theta_{65} + 90°$
positive for plantarflexion
negative for dorsiflexion

Figure 10.18

Marker locations and limb and joint angle definitions for gait analysis in the sagittal plane. This figure also shows the sign conventions for limb segment angles, consistent with the requirement that all angles be measured counter-clockwise with respect to the x (horizontal) axis, as discussed in the text. Modified from Winter [9]. Reprinted with permission of John Wiley & Sons, Inc.

where $\hat{\mathbf{e}}_x$ is the unit vector in the x direction. Finally, we define the knee flexion angle to be $\theta_{knee} = \theta_{21} - \theta_{43}$, as indicated in Fig. 10.18. Note that this angle is zero when the knee is straight and is positive when the knee is in flexion.

Table 10.1 shows data gathered in the sagittal plane from gait analysis on a normal subject. We can compute the knee flexion angle from noting that

$$\mathbf{x_{21}} \bullet \mathbf{x_{34}} = \|\mathbf{x_{12}}\| \, \|\mathbf{x_{34}}\| \cos(\theta_{knee}). \tag{10.15}$$

For example, for frame 6, we have the following data: $\mathbf{x_{21}} = (-229.2, 409.7)$ mm, $\mathbf{x_{43}} = (-146.5, 377.7)$ mm, which when substituted into Equation (10.15) yields a knee flexion angle of $\theta_{knee} = 8.04°$.

As can be seen, the analysis is fairly straightforward so long as it is restricted to two-dimensional motion in the sagittal plane. The analysis of three-dimensional motion is significantly more complex, and we will only touch on it here. First, a third marker on each limb segment is required, placed so that it is not co-linear with the first two markers. One way to proceed is as follows. Each limb segment is described in terms of its own local coordinate system, defined by the three markers. More specifically, the local y axis is defined to point along the long axis of the limb segment; the local x axis lies in the plane defined by the three markers

Table 10.1. Typical kinematic data gathered from gait analysis of normal subject. Marker locations are as shown in Fig. 10.18. and are measured in the sagittal plane from a fixed reference location following the convention in Fig. 10.17. The frequency of measurement was 50 Hz, with each frame corresponding to one measurement. Data courtesy of Dr. Jennifer A. Moore, University of Toronto.

| Frame | Marker locations (mm) | | | |
	(x_1, y_1)	(x_2, y_2)	(x_3, y_3)	(x_4, y_4)
5	(−294.1, 975.9)	(−68.5, 557.3)	(−73.8, 492.6)	(72.3, 116.8)
6	(−261.4, 970.1)	(−32.2, 560.4)	(−38.1, 492.1)	(108.4, 114.4)
7	(−227.2, 965.0)	(5.0, 563.2)	(−2.6, 491.1)	(137.3, 111.2)
8	(−192.1, 960.9)	(42.4, 565.6)	(32.0, 489.4)	(160.0, 107.7)

and is orthogonal to the local y axis; and the local z axis is orthogonal to the other two axes. The goal is to quantify the relative orientation of limb segments in a time-resolved manner.

Typical data from such a three-dimensional kinematic analysis on the motion of the knee expressed in terms of rotation (Euler) angles are shown in Fig. 10.19. Here the plotted quantities are the rotation angles between the local coordinate systems of the shank and the thigh. This method of presenting the data suffers from the drawback that three-dimensional Euler rotation matrices do not commute. This means that the magnitude of the computed Euler angles depends on the details of the orientation of the limb segment axes and the order in which the angles were extracted. Nonetheless, by using a standardized protocol, useful information can still be obtained. The reader will note that of the three angles, that associated with extension/flexion has the greatest amplitude, which accords with our intuition. This angle is offset from zero by several degrees, which is likely an artefact of how the markers were placed on the leg. Adduction/abduction refers to the rotation of the shank with respect to the thigh about a local axis approximately aligned with the x axis (see Fig. 10.17A). Note that adduction is inward (medial) rotation of the shank; abduction is outward (lateral) shank rotation. Most of the variation seen in this quantity is likely indicative of a small artefact associated with minor misalignment of the marker system. Finally, inward/outward rotation refers to rotation of the shank about the long axis of the femur. Such rotation is physiological, owing to the shape of femoral condyles, which causes the knee to rotate as it is flexed/extended.

A different way of presenting the same data is through the use of helical screw angles. The reader will recall that the three-dimensional motion of any rigid body can be expressed as a translation plus a rotation about an instantaneous axis of rotation. A complete description of the motion of the shank with respect to the thigh therefore requires the time-resolved specification of four degrees of freedom: the

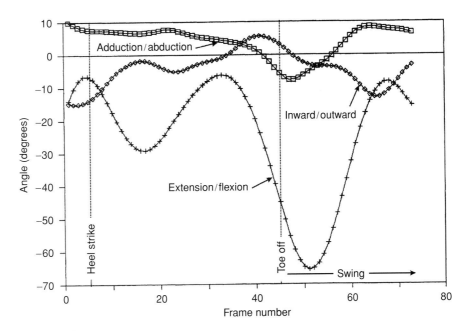

Figure 10.19

Euler angles defining three-dimensional relative motion of the shank with respect to the thigh (knee motion) for a normal subject walking. Note that this figure extends over slightly more than one gait cycle (including the swing phase), with heel strike at between frames 5 and 6, and toe off at approximately frame 45. For extension/flexion, positive values indicate extension; for adduction/abduction, positive values indicate adduction (i.e., inward or medial rotation of the shank with respect to the thigh); for inward/outward rotation, positive values indicate inward rotation. The time interval between frames was 0.02 s. Data courtesy of Dr. Jennifer A. Moore, University of Toronto. These data are from the same dataset as shown in Table 10.1 and Fig. 10.20.

orientation (in three dimensions) of the unit vector aligned with the instantaneous axis of rotation (two degrees of freedom), the angular velocity about the instantaneous axis (one degree of freedom), and the translation along the instantaneous axis of rotation (one degree of freedom). An analogy can be drawn to the motion of a screw; hence the names "helical screw angles" or "helical axis of motion." It is convenient to present the orientation of the instantaneous axis of rotation in terms of the direction cosines of this axis; that is, the values obtained by taking the dot product between the unit vector aligned with the axis of rotation and the unit vectors in each of the x, y, and z directions.[4] Fig. 10.20 shows time-resolved helical axis direction cosines for the same dataset shown in Fig. 10.19. Values of the z direction cosine close to one show that the instantaneous axis is nearly aligned with the z axis, particularly during the swing phase, in agreement with the data shown in Fig. 10.19.

---

[4] Note that any two of the direction cosines are sufficient to define the third, since the sum of the squares of the direction cosines must equal one.

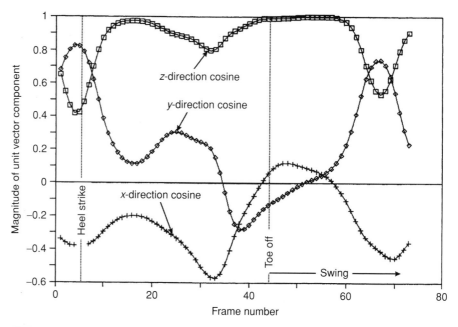

Helical axis of motion cosines defining three-dimensional relative motion of shank and thigh for a normal subject walking. The time interval between frames was 0.02 s. Data courtesy of Dr. Jennifer A. Moore, University of Toronto. These data are from the same dataset as shown in Table 10.1 and Fig. 10.19.

10.3.2 Anthropometry

Anthropometric data on limb segment lengths is relatively easy to obtain, so long as consistent definitions of limb segments are made (see Fig. 10.21 and Table 10.2 for such definitions). Data on the mass, density, location of the center of gravity, and radius of gyration of individual segments are more difficult to obtain. There are indirect ways of obtaining this information (e.g., Winter [7] and Miller and Nelson [1]), but these are prone to inaccuracies. The most accurate method of gathering such data is direct measurements on cadaver limb segments, but this suffers from the drawback of not being patient-specific.

A sense of the inaccuracy in such measurements can be obtained from comparing reported data for body segment masses given from two different sources, shown here in Table 10.2 and Table 10.3. For a subject with mass 160 lbm, the correlations in Table 10.3 give the mass of the upper arms as 10.3 lbm, while the data in Table 10.2 give 9.0 lbm. Considering the variation between one subject and another, this spread between the two correlations is probably within the interindividual differences. More detailed anthropometric data, including population distributions,

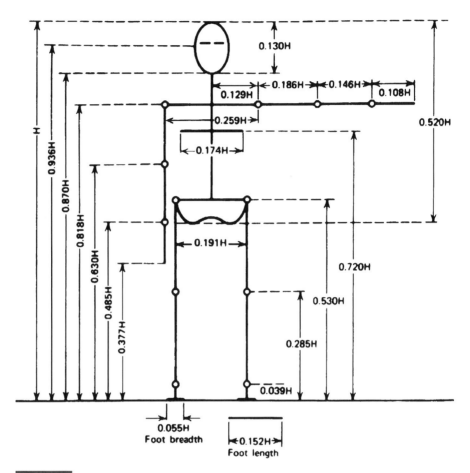

Figure 10.21

Anthropometric data on body segment lengths. All body segment lengths are expressed relative to the total height H. From Winter [7]. Reprinted with permission of John Wiley & Sons, Inc.

are available from a variety of other sources [10,11], including a remarkably complete electronic resource [8]. We will use the data in Table 10.2 for all future calculations. Readers are cautioned that the y and z axes are interchanged in the standard coordinate system for gait analysis versus the coordinate system for anthropometry used by NASA [8].

In using Table 10.2, the reader is reminded that the radius of gyration, k, is related to the mass moment of inertia, I, by the relationship

$$I = mk^2 \tag{10.16}$$

where m is the mass of the limb segment. Further, the mass moment of inertia depends on the center of rotation. The mass moment of inertia for rotation about an arbitrary point G, I_G, is related to the mass moment of inertia for rotation about the

Table 10.2. Anthropometric data on segment mass, location of center of mass, radius of gyration and density. Segment masses are expressed relative to total body mass; locations of the centers of mass are expressed relative to the limb segment length (see Fig. 10.21 for definitions of limb segments); the radii of gyration are expressed relative to the segment length, about three points, namely the center of gravity (C of G) for the segment, the proximal end of the segment, and the distal end of the segment. Note that the radii of gyration are for rotation about the z axis (see Fig. 10.17). Collected from various sources in Winter [7]. Reprinted with permission of John Wiley & Sons, Inc.

Segment	Definition of segment length	Segment weight/total body weight	Center of mass/ segment length		Radius of gyration/ segment length			Density (g/cm³)
			Proximal	Distal	C of G	Proximal	Distal	
Hand	Wrist axis/knuckle II middle finger	0.006	0.506	0.494	0.297	0.587	0.577	1.16
Forearm	Elbow axis/ulnar styloid	0.016	0.430	0.570	0.303	0.526	0.647	1.13
Upper arm	Glenohumeral axis/elbow axis	0.028	0.436	0.564	0.322	0.542	0.645	1.07
Forearm and hand	Elbow axis/ulnar styloid	0.022	0.682	0.318	0.468	0.827	0.565	1.14
Total arm	Glenohumeral joint/ulnar styloid	0.050	0.530	0.470	0.368	0.645	0.596	1.11
Foot	Lateral malleolus/head metatarsal II	0.0145	0.50	0.50	0.475	0.690	0.690	1.10
Shank	Femoral condyles/medial malleolus	0.0465	0.433	0.567	0.302	0.528	0.643	1.09
Thigh	Greater trochanter/ femoral condyles	0.100	0.433	0.567	0.323	0.540	0.653	1.05
Foot and shank	Femoral condyles/medial malleolus	0.061	0.606	0.394	0.416	0.735	0.572	1.09
Total leg	Greater trochanter/medial malleolus	0.161	0.447	0.553	0.326	0.560	0.650	1.06
Head and neck	C7-T1 and 1st rib/ear canal	0.081	1.000	–	0.495	1.116	–	1.11
Shoulder mass	Sternoclavicular joint/glenohumeral axis		0.712	0.288				1.04
Thorax	C7-T1/T12-L1 and diaphragm[a]	0.216	0.82	0.18				0.92
Abdomen	T12-L1/L4-L5[a]	0.139	0.44	0.56				
Pelvis	L4-L5/greater trochanter[a]	0.142	0.105	0.895				
Thorax and abdomen	C7-T1/L4-L5[a]	0.355	0.63	0.37				
Abdomen and pelvis	T12-L1/greater trochanter[a]	0.281	0.27	0.73				1.01
Trunk	Greater trochanter/glenohumeral joint[a]	0.497	0.50	0.50				1.03
Trunk, head, neck	Greater trochanter/glenohumeral joint[a]	0.578	0.66	0.34	0.503	0.830	0.607	
Head, arm, trunk	Greater trochanter/glenohumeral joint[a]	0.678	0.626	0.374	0.496	0.798	0.621	
Head, arm, trunk	Greater trochanter/middle rib	0.678	1.142		0.903	1.456		

[a] These segments are presented relative to the length between the Greater Trochanter and the Glenohumeral Joint.

Table 10.3. Correlations expressing the masses of different body segments in terms of overall body mass (M, in lbm). Compare with the first column of Table 10.2. From Miller and Nelson [1]. With kind permission of Lippincott Williams & Wilkins.

Segment	Segment mass (lbm)
Head	0.028 M + 6.354
Trunk	0.552 M − 6.417
Upper arms	0.059 M + 0.862
Forearms	0.026 M + 0.85
Hands	0.009 M + 0.53
Upper legs	0.239 M − 4.844
Lower legs	0.067 M + 2.846
Feet	0.016 M + 1.826

center of mass, I_0, by the parallel axis theorem, which states that $I_G = I_0 + mx_G^2$, where x_G is the distance between point G and the center of mass. Hence, the radius of gyration about G, k_G, is given by

$$k_G^2 = k_0^2 + x_G^2 \qquad (10.17)$$

where k_0 is the radius of gyration about the center of mass. It is thus seen that any two of the radii of gyration listed in each line of Table 10.2 can be derived from the third radius of gyration as well as from knowledge of the location of the center of mass of the limb segment.

10.3.3 Kinetics

In the context of gait analysis, kinetics is concerned with the measurement of forces generated during walking. One of the most important of these is the ground reaction force (i.e., the force exerted by the ground on the foot during walking). This is measured using a force plate, a floor-mounted plate instrumented with force transducers. One common force plate design is shown in Fig. 10.22, where the plate is supported on an instrumented central column. The horizontal and vertical reaction forces and the reaction moment acting on the column are measured. From these data it is possible to determine F_y and F_x, the vertical and horizontal reaction forces acting on the foot, as well as the center of pressure (i.e., the effective location through which F_y and F_x act).

Typical data for a normal subject walking are shown in Fig. 10.23. Characteristic is the double-humped profile of the vertical reaction force, which can be explained as follows. The vertical reaction force increases from zero at the moment of heel

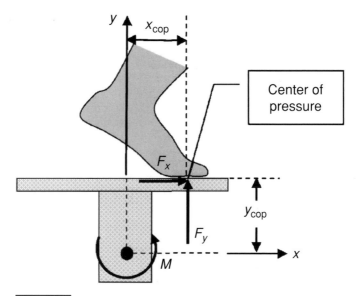

Figure 10.22

Schematic diagram of central support type force plate. F_y and F_x are the vertical and horizontal reaction forces acting on the foot, respectively; x_{cop} and y_{cop} are the x and y locations of the center of pressure, respectively, measured with respect to some convenient reference point. Modified from Winter [7]. Reprinted with permission of John Wiley & Sons, Inc.

strike to a maximum value as the stance leg is dynamically loaded. It then dips slightly below body weight as the stance leg flexes slightly at mid-stride, increases to greater than body weight as the stance leg extends at push-off, and then decreases to zero as the foot leaves the floor.

It is also of interest to note the profile of the horizontal reaction force. Immediately after heel strike it is negative (i.e., the ground is pushing "backwards" on the foot). As weight moves forward on the foot, the magnitude of this force decreases slightly, until the ground is pushing "forwards" on the foot just prior to toe off.

Link segment model

It is possible to combine information about limb segment kinematics, external reaction forces (kinetics), and anthropometry to gain insight into the forces acting at joints during the walking process. To illustrate this process, we consider a link segment analysis of forces at the knee. In the link segment model, the following two assumptions are made:

- joints are considered to be hinge (pin) joints with a single center of rotation at which a set of reaction forces and a reaction moment act
- the anthropometric properties of each limb segment (i.e., location of the center of mass and mass moment of inertia) are fixed during the walking process.

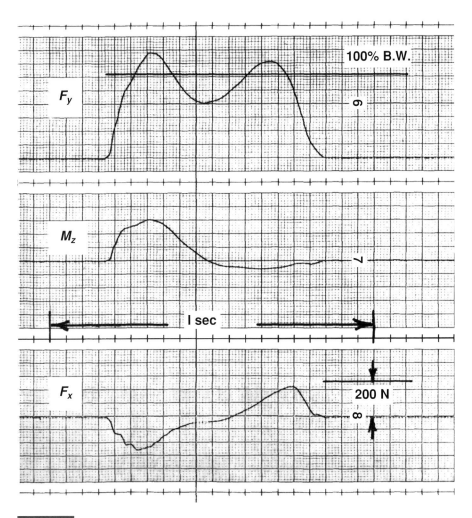

Figure 10.23

Typical ground reaction force and moment data measured with a force plate for a normal subject during walking on level ground. Horizontal axis is time; 100% B. W. is total body weight. From Winter [7]. Reprinted with permission of John Wiley & Sons, Inc.

Consider the forces acting on the shank and foot, as shown in Fig. 10.24. It is clear that the location of the effective center of gravity of the shank plus foot combination, as well as the mass moment of this combination, will depend on the angle that the foot makes with respect to the shank. However, this quantity changes throughout the gait cycle, and inclusion of this effect greatly complicates the analysis. For this reason, we will ignore this effect and simply approximate the shank plus foot as a single limb segment having the properties listed in Table 10.2. Consistent with this approximation, we will treat the center of gravity of the combined segment

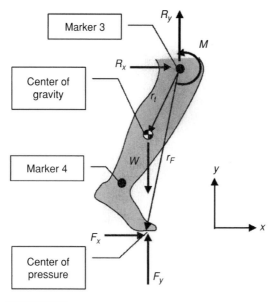

Marker 3

Center of gravity

Marker 4

Center of pressure

Figure 10.24

Forces acting on the shank and foot during walking. F_y and F_x are reaction forces acting on the foot from the floor; R_x, R_y, and M are the reaction forces and moment acting at the knee joint on the shank. $\mathbf{W}$ is the weight of the shank plus foot, given by $\mathbf{W} = -m_t g \hat{\mathbf{e}}_\mathbf{y}$.

as lying along the axis of the shank. Under these assumptions, and with reference to Fig. 10.24, we can apply Newton's second law in the horizontal and vertical directions to obtain:

$$R_x + F_x = m_t a_x \qquad (10.18)$$

and

$$R_y + F_y - m_t g = m_t a_y, \qquad (10.19)$$

where m_t is the mass of the shank plus foot; a_x, a_y are the linear accelerations of the center of mass of the combined segment (shank plus foot) in the x and y directions, respectively; g is the gravitational acceleration constant; and all other quantities are as shown in Fig. 10.24.

In order to illustrate the use of these equations, we will use the data shown in Table 10.4 (corresponding to frames 5–8 in Table 10.1). In particular, we wish to compute the reaction forces and moments acting at the knee for frame 7 of this dataset. Inspection of Equations (10.18) and (10.19) indicates that this requires that we know a_x and a_y, which we can obtain by numerically evaluating $d^2 x_{CG}/dt^2$ and $d^2 y_{CG}/dt^2$, where x_{CG} and y_{CG} are the locations of the center of gravity of the composite segment (shank plus foot). Using a standard centered difference

Table 10.4. Typical kinetic (force plate) data gathered from gait analysis of a normal subject. This data was gathered at the same time as that in Table 10.1, and frame numbers are identical for the two tables. F_x and F_y are the horizontal and vertical reaction forces on the foot, following the sign conventions in Fig. 10.17. x_{cop} and y_{cop} are the center of pressure of the reaction forces. The frequency of measurement was 50 Hz, and the mass of the subject was 65 kg. Data courtesy of Dr. Jennifer A. Moore, University of Toronto.

Frame	F_x (N)	F_y (N)	x_{cop} (mm)	y_{cop} (mm)
5	−0.8	−0.9	129.3	0.0
6	−0.8	6.2	128.5	0.0
7	−1.8	94.0	131.2	0.0
8	−16.6	238.7	141.9	0.0

operator in time to approximate the second derivative d^2/dt^2, we may write:

$$a_x^7 = \frac{x_{CG}^6 - 2x_{CG}^7 + x_{CG}^8}{\Delta t^2} \tag{10.20}$$

and

$$a_y^7 = \frac{y_{CG}^6 - 2y_{CG}^7 + y_{CG}^8}{\Delta t^2} \tag{10.21}$$

where the superscripts indicate the frame number and Δt is the time increment between frames, equal to 0.02 s in the sample dataset (50 Hz sampling rate).

From Table 10.2, it is seen that the location of the center of mass for the shank plus foot segment is at 0.606 of the shank length from the proximal end of the shank. Therefore, we may write:

$$\mathbf{x}_{CG} = \mathbf{x}_3 - 0.606\, \mathbf{x}_{43} \tag{10.22}$$

where it is assumed that markers 3 and 4 were placed on the femoral condyles and medial malleolus, respectively. Use of the data in Table 10.1 allows the values of $\mathbf{x}_{CG} = (x_{CG}, y_{CG})$ shown in Table 10.5 to be computed. Direct substitution into Equations (10.20) and (10.21) then yields $a_x = -10.25$ m/s^2 and $a_y = -1.25$ m/s^2 for frame 7. From Table 10.2, it is seen that the mass of the shank plus foot is 0.061 times total body mass (65 kg), or 3.97 kg. Substitution of numerical values into Equations (10.18) and (10.19) gives $R_x = -38.9$ N and $R_y = -60.0$ N. A similar procedure can be carried out for each frame of data, yielding a time-resolved description of the joint reaction forces (Fig. 10.25).

It remains to carry out a moment balance to compute the joint reaction moment, M (Fig. 10.24). In performing this moment balance, the reader should keep in mind

Table 10.5. Kinematic data derived from values in Table 10.1.
See text for definition of symbols and derivation of the data.

Frame	x_{43} (mm)	x_{CG} (mm)	θ_{43} (rad)
5	(−146.1, 375.8)	(14.7, 264.9)	1.942
6	(−146.5, 377.7)	(50.7, 263.2)	1.941
7	(−139.9, 379.9)	(82.2, 260.9)	1.924
8	(−128.0, 381.7)	(109.6, 258.1)	1.894

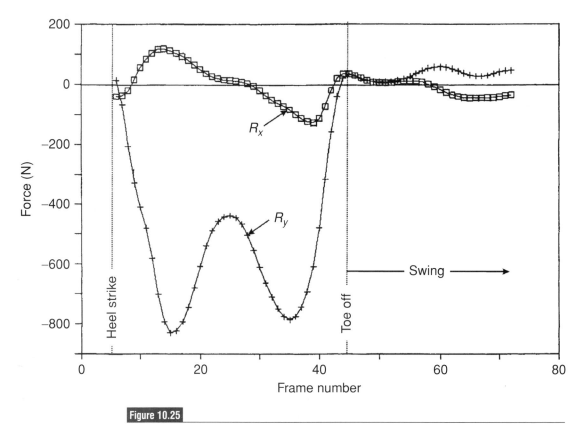

Figure 10.25

Time-resolved knee joint reaction forces derived from link segment model. Data courtesy of Dr. Jennifer A. Moore, University of Toronto. These data are from the same dataset shown in Table 10.1, Table 10.2, and Fig. 10.19.

the fact that we are treating the knee joint as a simple hinge joint, which is not strictly true. Nonetheless, we will make this approximation, taking the pivot point as the point of intersection of the long axes of the femur and tibia. This point is located approximately on the mid-surface of the intercondylar groove, and the computed moment should be interpreted as an effective moment about this point. For the purposes of this example, we will make a further simplification, namely that the location of marker 3 is coincident with the effective center of rotation. In practice this is not the case, and a correction should be made for this relatively small difference in location.

By carrying out a moment balance about the proximal end of the shank (marker 3) we obtain

$$M - gm_t(\mathbf{r_t} \times \hat{\mathbf{e}}_y) \bullet \hat{\mathbf{e}}_z + (\mathbf{r_F} \times \mathbf{F}) \bullet \hat{\mathbf{e}}_z = I_t \alpha \qquad (10.23)$$

where α is the angular acceleration of the shank plus foot; $\hat{\mathbf{e}}_y$ and $\hat{\mathbf{e}}_z$ are the unit vectors in the y and z directions, respectively; I_t is the mass moment of inertia of the shank plus foot about the proximal end of the shank; and $\mathbf{F}$ is the vector sum of F_x and F_y. The interpretation of each of the terms on the left-hand side of Equation (10.23) is as follows: the first term is the joint reaction moment; the second term is the torque due to the weight of the shank plus foot acting through the combined center of gravity; and the last term is the moment due to the reaction force at the floor. The cross products in the second and third terms are dotted with the unit vector $\hat{\mathbf{e}}_z$ so as to obtain a scalar equation.

In order to employ this formula, we must know the angular acceleration of the shank plus foot, α; the cross products $\mathbf{r_t} \times \hat{\mathbf{e}}_y$ and $\mathbf{r_F} \times \mathbf{F}$; and the mass moment of inertia I_t. The computation of each of these quantities is illustrated below.

Calculating α. By using Equation (10.14) it is possible to compute the angle θ_{43} that the shank makes with the horizontal in each frame; this is shown in column four of Table 10.5. Since α is simply equal to $\mathrm{d}^2\theta_{43}/\mathrm{d}t^2$, we can numerically compute α by using the standard centered difference approximation to the second derivative as:

$$\alpha^7 = \frac{\theta_{43}^6 - 2\theta_{43}^7 + \theta_{43}^8}{\Delta t^2} \qquad (10.24)$$

where once again the superscripts indicate the frame number. Using the values in Table 10.5 yields $\alpha = -32.5$ rad/s^2 for frame 7.

Calculating $\mathbf{r_t} \times \hat{\mathbf{e}}_y$. $\mathbf{r_t}$ is equal to $\mathbf{x_{CG}} - \mathbf{x_3}$. Direct substitution of values from Table 10.1 and Table 10.5 gives $\mathbf{r_t} = (84.8 - 230.2)$ mm. Using $\hat{\mathbf{e}}_y = (0, 1, 0)$ and the definition of the cross product then gives:

$$\mathbf{r_t} \times \hat{\mathbf{e}}_y = \begin{vmatrix} \hat{\mathbf{e}}_x & \hat{\mathbf{e}}_y & \hat{\mathbf{e}}_z \\ 84.8 \text{ mm} & -230.2 \text{ mm} & 0 \\ 0 & 1 & 0 \end{vmatrix} \qquad (10.25)$$

which equals $84.8 \, \hat{\mathbf{e}}_z$ mm.

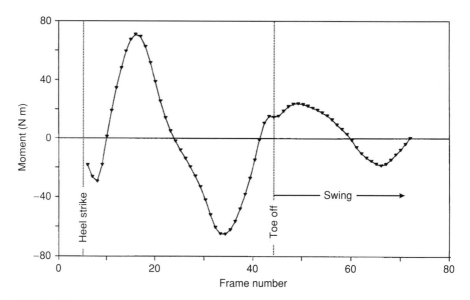

Figure 10.26

Time-resolved knee reaction moment derived from a two-dimensional link segment model. Data courtesy of Dr. Jennifer A. Moore, University of Toronto. These data are from the same dataset shown in Table 10.1, Table 10.2, and Fig. 10.19.

Calculating $\mathbf{r_F} \times \mathbf{F}$. $\mathbf{r_F}$ is equal to $\mathbf{x_{cop}} - \mathbf{x_3}$, where $\mathbf{x_{cop}}$ is the location of the center of pressure. Direct substitution of values from Table 10.1 and Table 10.5 yields $\mathbf{r_p} = (133.8 - 491.1)$ mm. The definition of the cross product then gives:

$$
\mathbf{r_F} \times \mathbf{F} = \begin{vmatrix} \hat{\mathbf{e}}_x & \hat{\mathbf{e}}_y & \hat{\mathbf{e}}_z \\ 133.8\,\text{mm} & -491.1\,\text{mm} & 0 \\ -1.8\,\text{N} & 94.0\,\text{N} & 0 \end{vmatrix}
\tag{10.26}
$$

which equals $11.69\,\hat{\mathbf{e}}_z$ N m.

Calculating $\mathbf{I_t}$. Table 10.2 lists the radius of gyration about the proximal end of the shank plus foot as $0.735\,\|\mathbf{x_{43}}\|$. Computing $\mathbf{x_{43}}$ from data listed in Table 10.5 as 404.8 mm, we obtain the radius of gyration about the proximal end of the shank plus foot to be 297.5 mm. Thus, I_t is $m_t\,(297.5\,\text{mm})^2 = 0.351\,\text{kg m}^2$.

Substitution of the above numerical values into Equation (10.23) yields $M = -19.81$ N m for frame 7. A similar procedure can be carried out for each frame of data, yielding a time-resolved description of the joint reaction moment (see Fig. 10.26).[5]

---

[5] The alert reader will notice that the moment that we calculated $(-19.8\,\text{N m})$ does not exactly match that shown in Fig. 10.26 for frame 7. This is because the calculation used to obtain the figure does not make some of the simplifying assumptions that we have used (e.g., assuming that the center of rotation of the knee is coincident with the femoral condyles).

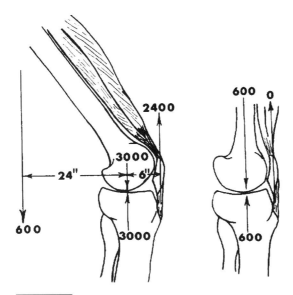

Figure 10.27

Illustration of dependence of bone-on-bone forces on joint configuration. The joint reaction force is 600 N in both cases. On the right, muscle contractile forces are negligible, and the bone-on-bone force is 600 N. On the left the 600 N force applied by the femur on the tibia acts 24 inches to the left of the assumed center of rotation. To balance this moment, the quadriceps must develop 2400 N of tension, assumed to act at 6 inches to the right of the assumed center of rotation. This causes the bone-on-bone force to be 3000 N. From Winter [7]. Reprinted with permission of John Wiley & Sons, Inc.

Joint reaction forces versus bone-on-bone forces

It is important to understand the interpretation of the forces and moment that we have computed in the above analysis. They are the effective net force and moment that act at the presumed center of rotation. Due to muscle contraction, these forces can differ substantially from the so-called bone-on-bone reaction forces (i.e., the force exerted by the femur on the tibia through the intervening cartilage). As a simple example of this effect, consider Fig. 10.27, showing two possible geometric configurations of the knee. Under conditions of no knee flexion (right), there is effectively no tension in the quadriceps and other muscles inserting close to the knee. Therefore, if a vertical joint reaction force of 600 N is computed, the bone-on-bone force is also approximately 600 N. However, if the knee is flexed (left), then the same joint reaction force of 600 N translates to a bone-on-bone force of 3000 N because of contraction of the quadriceps. Such factors, in combination with the dynamic loads placed on the knee during walking and running, can lead to oscillating bone-on-bone forces of several times body weight. This ultimately leads to wear of the cartilage in the joint.

10.4 Problems

10.1 A jumper executes a standing jump from a platform that is moving upwards with constant speed V_p.

    (a) Derive a formula for the maximum elevation of the jumper's center of gravity in terms of the crouch depth, c, the equivalent force to weight ratio, F_{equiv}/W, the platform speed, V_p, and other relevant parameters. The elevation of the center of gravity is to be measured with respect to a stationary frame of reference (i.e., one not attached to the platform).

    (b) If the crouch depth is 18 inches, the ratio F_{equiv}/W is 2, and the platform speed is 5 ft/s, compute the elevation of the center of gravity.

10.2 Your 160 lbm friend agrees to have his standing jump analyzed. Standing on a force plate, he crouches to lower his center of gravity, then executes a jump. The force plate measurement gives a reading that can be described by the equation

$$F(t) = 480 \sin\left(\frac{\pi t}{\tau}\right) + 160\left(1 - t/\tau\right) \qquad (10.27)$$

where $F(t)$ is in lbf. Here the push-off duration τ is 180 ms. How high will your friend's center of gravity be elevated at the peak of his jump?

10.3 A 75 kg stunt man executes a standing jump with the aid of a harness and support wire. In addition to the constant 1100 N force his legs exert during the push-off phase, the wire has a lift mechanism that applies a force given by $550e^{-s/L}$ (in Newtons), where L is 0.4 m and s is the distance traveled. The lift mechanism is engaged at the bottom of the crouch (where $s = 0$) and a safety catch detaches the wire when the stuntman leaves the ground.

    (a) For a crouch depth of 0.4 m, compute the maximal elevation of his center of gravity. Hint: it may be easiest to work from first principles, recalling that $a = v\,dv/dz = \frac{1}{2}\,d(v^2)/dz$.

    (b) If the safety catch fails to disengage, show that his center of gravity is elevated 0.462 m at the top of the jump. Hint: use an approach similar to that for part (a) for the airborne phase.

10.4 Consider the standing high jump, but this time the jumper is on the moon, where the local gravitational field is one sixth that on earth: $g_{moon} = g/6$.

    (a) Using an analysis similar to that developed in Section 10.1.1, derive a formula for the height that a person's center of gravity can be elevated in the standing high jump on the moon,

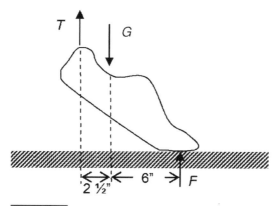

Figure 10.28

For Problem 10.5. G is the force exerted on the foot by the lower leg, which includes the weight, W.

(b)   If J. C. Evandt were to repeat his record-breaking jump on the moon, what bar height could he clear? To compute this height, you may use the same data and assumptions as were used in the text.

10.5   A 150 lbm person is able to jump 22 inches (elevation of center of gravity) if they first crouch so as to lower their center of gravity by 15 inches. What average tension T is present in their Achilles tendon during the push-off phase? See Fig. 10.28 for nomenclature.

10.6   Penelope Polevaulter, having read this book, realizes that her optimal pole-vaulting strategy is to run as fast as possible and push off strongly during take off.
   (a)   If her weight is W, her approach speed is V_a, and she pushes off with an effective force F_0 over an effective crouch distance c, estimate the net elevation of her center of gravity, h. Make and state relevant assumptions.
   (b)   Assume her approach speed is 9 m/s and she is able to push off with an effective force F_0 of two times her body weight with an effective crouch distance of 25 cm. If her center of gravity starts 30 cm from the ground, estimate the height above the floor which her center of gravity can clear.

10.7   Derive an expression to estimate the distance L attainable in the long jump, in terms of the approach velocity V. Neglect air drag and assume that planting the foot at the beginning of the jump does not generate a vertical force but rather produces the optimal angle for take off. Find L for $V = 10$ m/s. (Note: you will have to determine the optimum angle.)

For Problem 10.8. Reproduced with permission of CP (Paul Chiasson).

10.8   Elvis Stojko (former Canadian and world figure skating champion) is going to execute a triple axel jump (three full rotations in the air). He spends 0.95 s in the air for this jump, his mass is 70 kg, and his average radius of gyration about the vertical axis while airborne is 18 cm (Fig. 10.29). Assume that he remains vertical throughout the jump (even though the picture shows this is not quite true). Before starting his set-up for the jump, he is skating in a straight line. The set-up lasts 0.35 s and at the end of the set-up his skates leave the ice. What average moment does the ice exert on him during the set-up? Notice that there is no moment acting on him while he is in the air, so his angular acceleration while airborne is zero.

10.9   A subject of mass 65 kg has her gait analyzed. Suppose that the x component of the force measured by a force plate takes the shape shown in Fig. 10.30 (compare with Fig. 10.23).

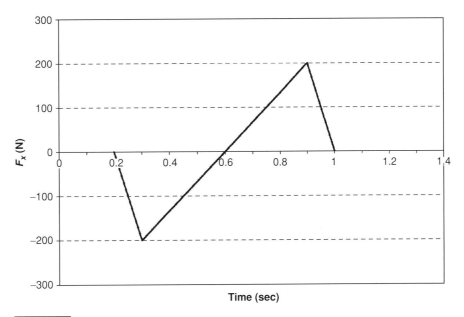

Figure 10.30
For Problem 10.9.

(a) If the forward velocity of the walker is 2 m/s at heel strike, what is it 0.4 s after heel strike?

(b) Estimate the corresponding change in height for the subject's center of gravity (i.e., from heel strike to 0.4 s later). For purposes of this question you should base your analysis on the walking model described in Section 10.2.1 and should make and state suitable simplifying assumptions.

10.10 Consider the force plate data from a gait experiment shown in Fig. 10.31. By considering the motion of the *total leg*, compute the reaction forces at the hip at the instant shown by the heavy vertical line. The total leg includes the thigh, shank and foot (see Table 10.2). For purposes of this question, consider the greater trochanter as the effective center of rotation for the hip joint. Note that you do not need to compute a reaction moment, only forces, and that the vertical scales on the two force traces are not the same. Use the following data: subject's mass 60 kg and height 1.7 m; acceleration of the center of gravity for the total leg is $a_x = -0.25$ m/s^2 and $a_y = -0.75$ m/s^2; length of the total leg segment is 0.530 times height (see Fig. 10.21).

10.11 Handball is a game in which a small rubber ball is hit with the palm of the hand. Suppose a 70 kg handball player who is 1.8 m tall swings his

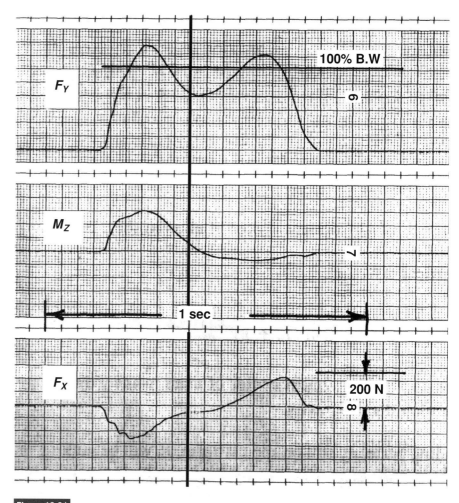

Figure 10.31

For Problem 10.10. Modified from Winter [7]. Reprinted with permission of John Wiley & Sons, Inc.

arm in a horizontal plane with his hand outstretched so as to hit the ball. The muscles in his shoulder exert a constant moment of 10 N m on the arm during this procedure. Assuming his arm starts from rest, that his elbow is locked during the swing, and that the arm swings through 80° before contacting the ball, what is the speed of the hand at the instant of contact with the ball? You may use data in Fig. 10.21 and Table 10.2 and should state assumptions. Hint: in uniformly accelerated rotational motion, the angular velocity, ω, is related to angular position, θ, by:

$$\omega^2 - \omega_0^2 = 2\alpha(\theta - \theta_0) \qquad (10.28)$$

where ω_0 and θ_0 are constants.

10.12   In Section 8.4.2, we computed the weight that could be supported by the forearm of a hypothetical subject. Suppose that the same subject suddenly relaxes all the muscles in his arm while holding a 12.2 kg mass in his hand. As in the text, the line of action of the weight is 31 cm from the elbow axis. (Note that this is not the length of the forearm.)

(a)   Using values given in Section 8.4.2, as well as data from the anthropometric tables (Fig. 10.21 and Table 10.2), calculate the angular acceleration of the forearm at the instant the muscles are relaxed. Assume that the subject's height and body mass are 183 cm and 80 kg, respectively, and make and state other appropriate assumptions as needed.

(b)   How much of a difference does neglecting forearm and hand weight make to the computed value of α?

10.13   The photograph in Fig. 10.32 is of a diver doing a back dive, taken using a stroboscopic flash firing at regular intervals. The images have been numbered to correspond to strobe flashes. The centers of mass of the diver in images 1 and 6 have been shown by a black dot, and the center of mass in image 14 has been shown by a white dot. The vertical scale bar at lower right has a total height of 2 m. The horizontal scale markings are part of the photograph and should not be used for this question.

(a)   One difficulty is that we do not know the flash interval. Assuming that image 6 corresponds to the peak of the diver's trajectory, show that the flash interval must have been about 90 ms.

(b)   When the diver pushes off the board, the board exerts both a vertical and horizontal force on the diver. Assuming that a constant *horizontal* push-off force is exerted for a duration of 400 ms, compute the magnitude of this force. You will need to make some measurements from the picture. You may assume that the diver is a standard American male, with mass 82.2 kg, and that the diver is at rest at the beginning of the push-off phase.

(c)   Assuming that the diver's total rotation from image 1 to 14 was 340°, and that the average moment exerted by the diving board on the diver during push-off was 100 N m, compute the average radius of gyration of the diver during push-off. You may neglect the changes in the diver's posture during the dive.

10.14   A 75 kg soldier is marching with straight, nearly rigid legs ("goose-stepping"). Analysis of video footage shot at 24 frames/s gives the data in Table 10.6, where θ is the angle that the swing leg makes with respect to the vertical (see Fig. 10.33). If the soldier is 1.8 m tall, what is the

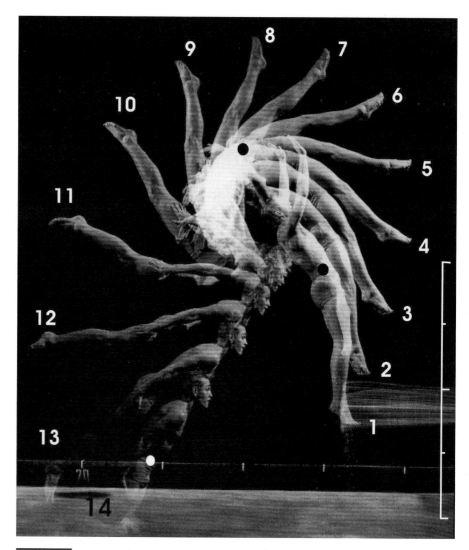

Table 10.6. For Problem 10.14.

Frame number	$\theta(°)$
4	−20
5	−4
6	+7

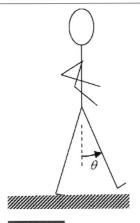

Figure 10.33

For Problem 10.14.

moment exerted on the swing leg at the hip at the instant when frame 5 is taken? Make and state necessary assumptions. Hints: be careful to use radians when appropriate; the greater trochanter can be taken as the center of rotation of the whole leg; the medial malleolus is the "ankle bone;" be careful about how you use Table 10.2 and Fig. 10.21.

References

1. D. I. Miller and R. C. Nelson. *Biomechanics of Sport* (Philadelphia, PA: Lea and Febiger, 1973).
2. H. E. Edgerton. *Stopping Time: The Photographs of Harold Edgerton* (New York: Harry N. Abrams, Inc., 1987).
3. V. T. Inman, H. J. Ralston and F. Todd. *Human Walking* (Baltimore, MD: Williams & Wilkins, 1981).
4. J. B. d. C. M. Saunders, V. T. Inman and H. D. Eberhart. The major determinants in normal and pathological gait. *Journal of Bone and Joint Surgery*, **35A** (1953), 543–558.
5. T. A. McMahon. *Muscles, Reflexes and Locomotion* (Princeton, NJ: Princeton University Press, 1984).
6. G. A. Cavagna, H. Thys and A. Zamboni. The sources of external work in level walking and running. *Journal of Physiology*, **262** (1976), 639–657.
7. D. A. Winter. *Biomechanics of Human Movement* (Toronto: John Wiley, 1979).
8. NASA. *Man–Systems Integration Standards*. Available at http://msis.jsc.nasa.gov/default.htm (NASA Johnson Space Center, 1995).
9. D. A. Winter. *Biomechanics and Motor Control of Human Movement*, 2nd edn (New York: John Wiley, 1990).

10. W. Karwowski (ed.) *International Encyclopedia of Ergonomics and Human Factors* (London: Taylor & Francis, 2001).

11. J. Weimer. *Handbook of Ergonomic and Human Factors Tables* (Englewood Cliffs, NJ: Prentice Hall, 1993).

12. H. E. Edgerton and J. R. Killian Jr. *Moments of Vision* (Cambridge, MA: MIT Press, 1979).

Appendix   The electrocardiogram

The electrocardiogram (ECG) is important for several reasons. First and foremost, it provides information about the activity of the heart muscle, and it is therefore an essential non-invasive clinical tool used to diagnose certain cardiac abnormalities. From the biomechanical viewpoint, it provides a convenient reference signal for time-varying quantities in the vascular tree, specifically pulsatile flow, pressure, and arterial pulsation. Here we briefly describe the main features of the ECG.

The contraction of cardiac muscle is controlled by an intrinsically generated electrical signal. In other words, the heart is responsible for its own stimulation, although the rate of stimulation can be modulated by external factors. In order to understand the behavior of this *pacemaker system*, we must first study several basic facts about the propagation of electric signals in *excitable cells*, a category that includes nerve and muscle cells. The general aspects of the description that follows are true for all excitable cells; however, the details are specific to cardiac muscle cells.

Normally, the cell's interior is held at a negative electrical potential measured with respect to the surrounding extracellular fluid. This small potential difference (the *resting potential*, equal to approximately -90 mV) exists because of a difference in ionic composition across the cell's membrane (Table A.1). This difference in composition is actively maintained by pumps residing in the cell's membrane that transport ions against their concentration gradient. Under the right stimulation, typically the cell's internal potential reaching a *threshold voltage* of approximately -60 mV (measured with respect to the surrounding extracellular fluid), specialized transmembrane channels open and allow ions to flood across the membrane. Because of the timing with which these channels open, the net effect is to force the potential within the cell to become slightly positive (approximately $+30$ mV). The cell then slowly pumps out these ions, eventually returning the intracellular potential to its resting value. Consequently, an electrode inside the cell, measuring the intracellular potential with respect to the surrounding fluid, records a trace similar to that shown in Fig. A.1. This process is known as *depolarization* and *repolarization*. In the context of the electrocardiogram, this process is particularly important since

Table A.1. Ionic concentrations in frog muscle fibers and in plasma. The ionic composition of plasma is representative of extracellular fluid composition. Note the substantial differences in ionic composition across the muscle cell membrane, which is typical of the situation in all cells. From Aidley [1] with kind permission of the author.

Ionic species	Concentration in muscle fibers (mM)	Concentration in plasma (mM)
K^+	124	2.25
Na^+	10.4	109
Cl^-	1.5	77.5
Ca^{2+}	4.9	2.1
Mg^{2+}	14.0	1.25
HCO_3^-	12.4	26.6
Organic anions	~74	~13

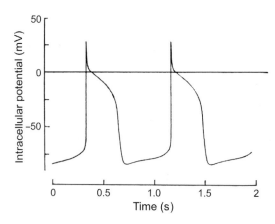

Figure A.1

Typical intracellular potential versus time for a cardiac muscle cell. It can be seen that there is a very rapid depolarization event, during which intracellular potential reaches approximately +30 mV, followed by a slower repolarization to the resting potential of approximately −90 mV. The shape of the depolarization–repolarization signal is different in cardiac muscle cells compared with other excitable cells (e.g., nerve cells). This is from a simulation of the polarization–repolarization process but is very close to that observed experimentally. From Noble [2] with permission from Elsevier.

depolarization of cardiac muscle cells causes them to contract (i.e., causes the heart to beat).

So far we have considered the potential difference across a single small region (or *patch*) of cell membrane. Let us now look at an entire cardiac muscle cell, which is long and spindly (or *fusiform*) in shape. Because of electrical coupling

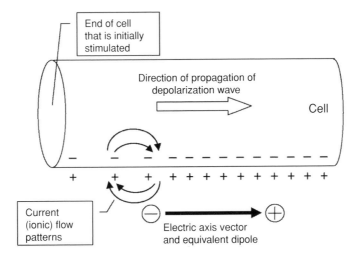

Figure A.2

Schematic diagram of current (ionic) flow and electrical potential in an excitable cell during action potential propagation. The flow of ions is denoted by thin arrows, and + and − signs indicate electrical potential. The depolarization wave is propagating from left to right. The equivalent dipole outside the cell is also shown.

between adjacent regions of membrane, a depolarization event initiated at one end of the cell will cause the adjacent membrane patch to depolarize (Fig. A.2). This, in turn, causes the adjacent patch to depolarize, which results in the passage of a depolarization wave known as the *action potential* along the axis of the cell. Furthermore, the cells in the heart are electrically coupled so that action potentials can jump from cell to cell, and thus propagate throughout an entire tissue mass. The net effect is that excitable cells within the heart act as a "wiring system" to allow electrical information, in the form of action potentials, to be communicated throughout the heart.

The existence of such an electrical signaling system can be exploited for clinical purposes. The fundamental concept is that propagation of electrical signals causes alterations in the electrical potentials in surrounding tissues. Measurement of these surrounding potentials allows information about the state of the heart's electrical signaling system to be determined. The information that results from a measurement of electrical potential at different locations is known as the ECG.

In order to understand the ECG more fully, note from Fig. A.2 that the extracellular fluid in the depolarized area is at a slightly lower potential than the surrounding extracellular fluid. Thus, there is a transient electrical field within the extracellular fluid that causes ions to flow in the extracellular space as shown in Fig. A.2.[1] Far

---

[1] There is also ionic flow within the cell along the cell's axis; however, for present purposes this is not of direct interest.

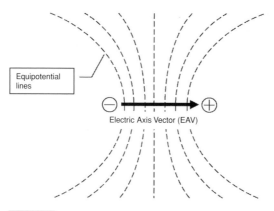

Figure A.3

Equipotential lines resulting from the presence of an electrical dipole in a conducting medium. The dipole is characterized by the EAV. Potential differences exist between any two equipotential lines.

from the cell, this combination of a transient electrical field and ion (current) flow appear to have been generated by an *equivalent electrical dipole* oriented along the axis of the cell, as shown in Fig. A.2. The strength and orientation of this dipole can be characterized by a vector drawn from the negative to the positive terminal of the dipole, known as the *electric axis vector* (EAV).

The presence of an electrical dipole in a conducting medium produces a family of equipotential lines as shown in Fig. A.3. Voltage differences can be measured between two locations within the surrounding conductor. The magnitude of these differences will depend on electrode position, EAV orientation, and EAV magnitude. By measuring differences between several different (known) sites it is possible to deduce the instantaneous state of the EAV.

Clinically, the conductor is the fluid within tissue, which enables voltages to be measured at the surface of the skin. As the mass of cardiac muscle cells depolarize and repolarize, the EAVs from each cell add vectorially to form a time-varying effective EAV for the entire heart, which produces potential variations throughout the body. The measurement of these potential differences at predefined locations is recorded as the ECG. In the normal heart, electrical activity and muscular contraction follow a well-defined sequence. The ECG is therefore clinically useful, since any deviations from the normal sequence are indicative of abnormalities in the heart's operation.

In order for the ECG to be a useful clinical tool, a great deal of standardization in the detailed protocol for acquiring an ECG is essential. This is because all ECGs are interpreted essentially by comparing them with large databases of normal individuals and patients with known conditions. It is clear that, to make this comparison, all ECGs in the database and the patient's ECG must be acquired in a

Table A.2. Standard (Einthoven) electrocardiogram electrode placement locations; the limbs act as conductors so that, for example, an electrode at the left wrist measures a potential at the left shoulder

Electrode	Measurement location	Attachment location	Symbol
A	Left shoulder	Left wrist	LS
B	Right shoulder	Right wrist	RS
C	Left hip	Left ankle	LH

Table A.3. Voltage differences taken to form standard (Einthoven) electrocardiogram leads (see also Fig. A.6 and Table A.2 for abbreviations).

Lead number	Definition
I	$V_{LS} - V_{RS}$
II	$V_{LH} - V_{RS}$
III	$V_{LH} - V_{LS}$

consistent fashion. This is accomplished by placing ECG electrodes in standardized, well-defined locations. In fact, there are a number of different electrode arrangements that are standard, each of which leads to a particular type of ECG. Here we will consider only the simplest of these, known as *Einthoven's triangle*, or *standard limb leads I, II, and III*.

Einthoven's triangle is formed by attaching three electrodes to the patient, as shown in Table A.2. The lead I, II, and III measurements are formed by subtracting electrode voltages as shown in Table A.3. For a given EAV, the signal on a particular lead will be proportional to the projection of the EAV on that lead. In practice, the signal on lead II is most commonly used, since it best demonstrates the large depolarization/repolarization associated with ventricular beating. Einthoven lead II is therefore sometimes called the *standard lead*.

In order to interpret the major features of an ECG, it is essential to know something about the sequence of electrical events in the heart. Depolarization is initiated in a specialized group of cells on the right atrial wall known as the *sinoatrial node*, or the *pacemaker node*. A depolarization wave spreads out over the atria, causing atrial contraction. Because of to the anatomy of the heart, this depolarization wave is then forced to "funnel" into the *atrioventricular node*, located on the junction line between atria and ventricles. It is then conducted by the *bundle of His* down the septum, and out into the ventricular walls by the *Purkinje fibers*, causing

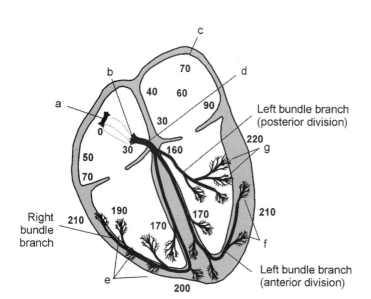

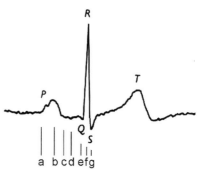

Legend

a   Sinoatrial node
b   Atrioventricular node
c   Remote atrial surface
d   Bundle of His
e   Anterior surface of right ventricle
f   Apical surface
g   Posterior basal area

Left bundle branch
(posterior division)

Right
bundle
branch

Left bundle branch
(anterior division)

Figure A.4

Conduction pathways of the heart. On the left is a heart showing the sinoatrial (pacemaker) node, the atrioventricular node, the bundle of His, and the Purkinje fibers, comprising the right, left anterior, and left posterior bundle branches. Numbers on the heart show typical contraction times after the discharge of the sinoatrial node, in milliseconds. Notice that the impulse is delayed at the atrioventricular node before being conducted down the ventricular septum by the bundle of His. The typical delay, measured from the onset of the P wave (atrial contraction) to the onset of the QRS complex (ventricular contraction), is 120 to 200 ms. The delay is necessary to allow the ventricles to fill with blood during atrial contraction before the ventricles themselves contract. In interpreting the positions of the conducting fibers in this diagram, it should be remembered that the fibers run in the three-dimensional tissue of the heart and therefore can only be shown approximately in a two-dimensional cross-section as shown here. On the right is an Einthoven lead II electrocardiograph (ECG) trace, where lower case letters refer to positions on the diagram of the heart during the conduction and contraction sequence. See the text for explanation of main features of the ECG. Timings in the left panel based on data in Guyton and Hall [3]; timings on the ECG based on data in Greenspan [4]; ECG from Aidley [1] with kind permission of the author.

ventricular contraction. Finally the ventricles repolarize, in preparation for the next heart beat (Figs. A.4 and A.5).

The following main features of the Einthoven lead II ECG can now be identified in Figs. A.4 and A.5:

- The most prominent feature of the ECG is the QRS complex, which corresponds to ventricular depolarization. The prominence of the QRS complex can be explained by the following argument. The heart's net EAV is the vectorial sum of all cardiac cells' EAVs, and since the ventricles contain the bulk of the tissue mass in the heart, the EAV is very large during ventricular depolarization.

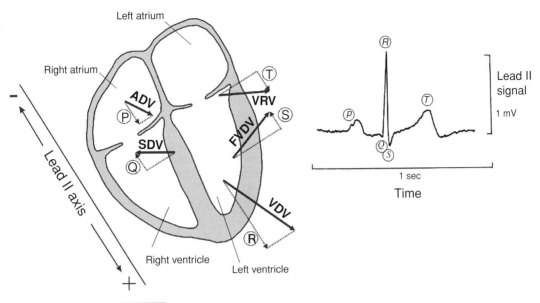

Figure A.5

Electric axis vector orientation and magnitudes at various times during the cardiac cycle (left), and relation of the electric axis vector to specific features of the electrocardiogram (ECG; Einthoven lead II; right). Each electric axis vector (shown as a solid vector) is decomposed into its component along the lead II axis and the orthogonal component (shown as dotted lines). P, projection of ADV (atrial depolarization vector) on lead II; Q, projection of SDV (septal depolarization vector) on lead II; R, projection of VDV (ventricular depolarization vector) on lead II; S, projection of FVDV (final or late ventricular depolarization vector) on lead II; T, projection of VRV (ventricular repolarization vector) on lead II. Vectors are not to scale, but show correct signs for lead II projections. Electrocardiogram from Aidley [1] with kind permission of the author.

- The P wave, which has the same polarity as the QRS complex but is much smaller in magnitude, corresponds to atrial depolarization.
- Atrial repolarization, although present, occurs at the same time as ventricular depolarization, and is therefore largely masked by the QRS complex.
- The T wave, which corresponds to ventricular repolarization, has the same sign as the QRS complex, even though it represents repolarization. This is because the spatial pattern of ventricular repolarization is different than that of ventricular depolarization.

It should also be noted that a *vector ECG* can be created by combining selected ECG leads and plotting their data against one another (instead of as a function of time). For example, the *frontal vector ECG* (or VCG) is obtained by plotting lead I on the x axis, and the average of leads II and III on the y axis (Fig. A.6). Some thought will reveal that the resulting vector is the instantaneous projection of the EAV onto the frontal plane.

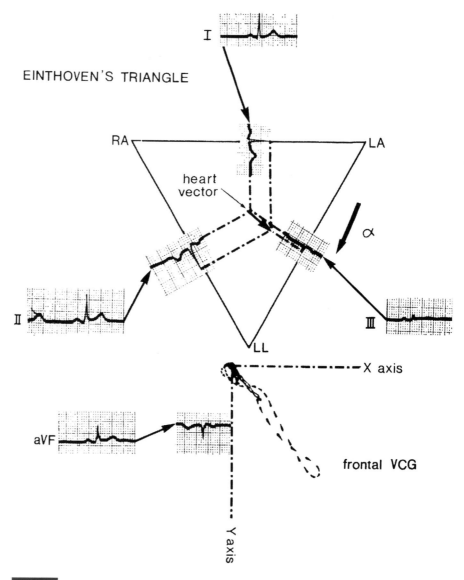

EINTHOVEN'S TRIANGLE

Figure A.6

Projection of the electric axis vector (EAV) onto the Einthoven leads. The electric axis vector is labeled "heart vector." The angle that the EAV makes with the horizontal is denoted by α, measured clockwise, with zero meaning that the EAV points to the patient's left. In the lower portion of the figure, the procedure for creation of the frontal vector ECG (VCG) is summarized. Lead I is plotted on the horizontal axis, while leads II and III are combined (averaged) to produce the aVF lead. This is inverted and plotted on the vertical axis. The result is a point representing the tip of the EAV, which traces out a path during the cardiac cycle as shown. RA, right arm; LA, left arm; LL, left leg. From Dower *et al.* [5] with kind permission of the author.

References

1. D. J. Aidley. *The Physiology of Excitable Cells*, 2nd edn (Cambridge: Cambridge University Press, 1978).
2. D. Noble. Cardiac action potentials and pacemaker activity. In *Recent Advances in Physiology*, 9th edn, ed. R. J. Linden. (Edinburgh: Churchill Livingstone, 1974), pp. 1–50.
3. A. C. Guyton and J. E. Hall. *Textbook of Medical Physiology*, 9th edn (Philadelphia, PA: W.B. Saunders, 1996).
4. K. Greenspan. Cardiac excitation, conduction and the electrocardiogram. In *Physiology*, 4th edn, ed. E. E. Selkurt. (Boston, MA: Little, Brown, 1976), pp. 311–336.
5. G. E. Dower, H. E. Horn and W. G. Ziegler. Clinical application of the polarcardiograph. In *Vectorcardiography – 1965*, ed. I. Hoffmann. (Amsterdam: North-Holland, 1966), pp. 71–91.

Index